Laser Cleaning

Feng Song · Xuechun Lin

Laser Cleaning

Fundamentals and Applications

清華大学出版社
TSINGHUA UNIVERSITY PRESS

Springer

Feng Song
School of Physics
Nankai University
Tianjin, China

Xuechun Lin
Institute of Semiconductors
Beijing, China

ISBN 978-981-97-6630-7 ISBN 978-981-97-6631-4 (eBook)
https://doi.org/10.1007/978-981-97-6631-4

Jointly published with Tsinghua University Press
The print edition is not for sale in China (Mainland). Customers from China (Mainland) please order the print book from: Tsinghua University Press.

The translation was done with the help of an artificial intelligence machine translation tool. A subsequent human revision was done primarily in terms of content.

This Springer imprint is published by the registered company Springer Nature Singapore Pte Ltd.
The registered company address is: 152 Beach Road, #21-01/04 Gateway East, Singapore 189721, Singapore

Foreword

Laser cleaning, as a new surface treatment technology, is receiving more and more widespread attention in today's increasingly environmentally conscious world. In the field of laser cleaning, some research group in the USA and European countries are pioneers and now many countries including China, Korea, and Singapore have paid more attentions on laser cleaning. With the continuous accumulation of technology, scientists have moved from laboratory research to equipment preparation and development. In response to different market demands, the types of lasers used for laser cleaning are different, and the principles and mechanisms are also different. At present, laser cleaning has been used in industries, and certain social and economic benefits have been achieved.

The development of laser cleaning technology has gone through a broad path of theoretical guidance and practical enrichment. Professor Song Feng and his team have been closely following the development of laser cleaning technology in China since 2004. They have been deeply researching, capturing, organizing, condensing, and innovating in the development of laser cleaning technology. They are one of the important practitioners in the research and development of laser cleaning equipment and industry in China. They have applied for patents, published papers, developed laser cleaning prototypes, and carried out laser cleaning research on rust, paint removal, and oil removal. The Prof. Song's team has a certain influence both domestically and internationally.

This monograph, written by Professor Song Feng and Lin Xuechun, discusses comprehensively and systematically, including the basic theory of the interaction between light and matter in laser cleaning process, the mechanism and physical model of laser cleaning to remove particles and film layers, laser cleaning equipment and monitoring in laser cleaning, and various applications of laser cleaning. Readers can delve into the point they are concerned about, find corresponding references, and then delve deeper into solving the problem themselves.

This guiding effect is very important. Sometimes, when encountering a phenomenon in experiments, we really don't know how to go ahead with the vast literature. From here, we can see and get inspiration, roughly which direction to look for, because it discusses very specific and comprehensive.

As a scientific researcher who has been engaged in laser research for decades, I believe that this book not only summarizes the development of laser cleaning technology both domestically and internationally over the past few decades, but also includes the author's foresight and innovative achievements. Therefore, it is a deeply guiding monograph. It has important guiding significance for researchers and equipment developers engaged in laser cleaning research.

Yao Jianquan
Academician of the Chinese Academy of Sciences
Professor of Tianjin University
Tianjin, China

Introduction

Cleaning is a common activity closely related to human life and industry. Laser cleaning has emerged with the development of laser technology. It uses laser as a medium and achieves the purpose of removing pollutants through the interaction between light and substances (including substrate materials and pollutants). Compared with traditional methods such as shot peening and chemical cleaning, it has the advantages of being green and environmentally friendly, accurately locating the parts to be cleaned, and achieving automated operations. This book focuses on the laser cleaning technology and its applications, introducing commonly used lasers for cleaning, the physical foundation of laser cleaning, laser cleaning equipment, and the applications of laser cleaning in electronic components, paint, rust, molds, historical buildings and artworks, as well as other related fields.

Contents

1 Overview of Cleaning Technology 1
- 1.1 Cleaning and Its Significance 1
  - 1.1.1 Concept of Cleaning 1
  - 1.1.2 Industries Using Cleaning 2
  - 1.1.3 The Significance of Cleaning 4
- 1.2 Cleaning Technologies and Classification 4
  - 1.2.1 The Four Elements of Cleaning 4
  - 1.2.2 Common Cleaning Technologies 5
- 1.3 Physical Cleaning Methods 8
  - 1.3.1 Shot Blasting Cleaning Technology 8
  - 1.3.2 High-Pressure Water Jet Cleaning Technology 11
  - 1.3.3 Dry Ice Cleaning Technology 14
  - 1.3.4 Ultrasonic Cleaning Technology 16
  - 1.3.5 Electromagnetic Induction Cleaning Technology 19
- 1.4 Chemical Cleaning Technology 20
  - 1.4.1 Chemical Reagent Cleaning Technology 21
  - 1.4.2 Electrolytic Cleaning Technology 22
  - 1.4.3 Microbial Cleaning Technology 23
  - 1.4.4 Features of Chemical Cleaning Technology 24
- 1.5 Mixed Cleaning Technology 25
  - 1.5.1 Plasma Cleaning Technology 25
  - 1.5.2 Thermal Cleaning Technology 28
- 1.6 Laser Cleaning Technology 29
- References 33

2 Common Lasers in Laser Cleaning 37
- 2.1 Basic Knowledge of Lasers 37
  - 2.1.1 Laser 37
  - 2.1.2 Working Principle of Laser 39

2.1.3 Structure and Types of Laser 42
2.1.4 Laser Parameters 44
2.2 Common Laser Technologies 44
2.2.1 Rate Equation Theory 45
2.2.2 Laser Q-Switching Technology 48
2.2.3 Laser Mode Locking 55
2.2.4 Laser Amplification [10, 13] 61
2.3 Rules for Selecting Lasers in Laser Cleaning 63
2.3.1 Optical Characteristics of the Object to Be Cleaned 63
2.3.2 Impact of Laser Parameters 64
2.3.3 Commonly Used Lasers in Laser Cleaning 65
2.4 Nd:YAG Laser 66
2.4.1 Principle of Light Emission of Nd:YAG Laser 66
2.4.2 Basic Structure of Nd:YAG Laser 69
2.5 Carbon Dioxide Lasers 75
2.5.1 Principle of Light Emission of CO_2 Lasers 75
2.5.2 Basic Structure of CO_2 Laser 77
2.5.3 Output Characteristics of CO_2 Laser 78
2.5.4 Main Types of CO_2 Lasers 79
2.6 Fiber Laser 80
2.6.1 Principle of Light Emission of Fiber Lasers [12] 81
2.6.2 Basic Structure of Fiber Lasers 83
2.6.3 Features of Fiber Lasers 84
2.7 Excimer Lasers 85
2.7.1 Common Types of Excimer Lasers 85
2.7.2 Working Principle of Excimer Lasers 85
2.7.3 Basic Structure of Excimer Lasers 86
2.7.4 Characteristics and Application Fields of Excimer Lasers 87
2.8 Semiconductor Lasers 88
2.8.1 Working Principle of Semiconductor Lasers 89
2.8.2 Classification of Semiconductor Lasers 89
2.8.3 Features and Application Fields of Semiconductor Lasers 90
References 90

3 Physical Basis of Laser Cleaning 93
3.1 Reflection, Scattering, and Absorption of Laser by Matter 93
3.1.1 Reflection and Scattering of Laser by Materials 94
3.1.2 Absorption of Laser by Materials 97
3.2 The Adhesion Force Between Pollutants and the Substrate 101
3.2.1 Adhesion Force 101
3.2.2 Brief Introduction to Principle of Laser Cleaning 103

3.3 Basic Mechanisms of Dry Laser Cleaning 104
3.3.1 Ablation Effect 104
3.3.2 Vibration Effect 105
3.3.3 Thin Film Bending Effect 107
3.3.4 Blast Effect 108
3.4 Mechanism of Wet Laser Cleaning 108
3.4.1 Strong Absorption by the Substrate 108
3.4.2 Strong Absorption by Liquid Film 109
3.4.3 Joint Absorption by Substrate and Liquid Film 110
References 111

4 Laser Cleaning Equipment 113
4.1 Introduction to the Laser Cleaning Equipment System 114
4.2 Example of Nd:YAG Pulsed Laser Cleaning Machine 123
4.2.1 Electro-optic Q-Switched Nd:YAG Laser Cleaning Equipment 123
4.2.2 Quasi-continuous Laser Cleaning Machine 128
4.2.3 Acousto-optic Q-Switched Laser Cleaning Machine Pumped by Semiconductors 132
4.3 International Status of Laser Cleaning Equipment 138
4.3.1 Current Research Status of Laser Cleaning Equipment 138
4.3.2 Laser Cleaning Machine Using Nd:YAG Laser 139
4.3.3 CO_2 Laser Cleaning Machine 145
4.3.4 Laser Cleaning Machine Using Excimer Laser 146
4.3.5 Laser Cleaning Machine Using Fiber Laser 147
4.4 Online Monitoring Methods of Laser Cleaning 148
4.4.1 Laser Cleaning Monitoring Methods 149
4.4.2 Real-Time Monitoring of Acoustic Intensity Measurement 150
4.4.3 Shock Wave Monitoring 152
4.4.4 Real-Time Monitoring of Laser-Induced Fluorescence Spectrum 153
4.4.5 Real-Time Monitoring of Laser Induced Breakdown Spectroscopy (LIBS) 156
4.4.6 Real-Time Monitoring of Photoacoustic Signals 159
4.5 Examples of Online Monitoring of Laser Cleaning 161
4.5.1 Photoacoustic Spectroscopy Detection System 161
4.5.2 Example of Acoustic Monitoring in Laser Cleaning 164
4.5.3 Laser-Induced Breakdown Spectroscopy (LIBS) 166
References 168

5 Application of Laser in Cleaning Electronic Components 171
5.1 Origin of Laser Cleaning of Electronic Components 171
5.2 Main Mechanism for Laser Cleaning of Microelectronic Components 172
5.2.1 Adhesion of Particles 172
5.2.2 Near-Field Optical Focusing Phenomenon 173
5.2.3 Mechanism of Dry Laser Cleaning Particles 174
5.2.4 Mechanism of Wet Laser Cleaning of Particles 175
5.3 Dry Laser Cleaning of Electronic Components 178
5.4 Wet Laser Cleaning of Electronic Components 182
5.4.1 Main Factors Affecting Wet Laser Cleaning of Particles 182
5.5 Application of Laser Cleaning in Electronic Components 186
References 189

6 Laser Cleaning Paint Layer 193
6.1 Basic Introduction to Laser Paint Removal 193
6.2 Theoretical Research on Laser Paint Removal 196
6.2.1 One-Dimensional Thermal Stress Model 198
6.2.2 Cleaning Threshold in Laser Paint Removal 201
6.2.3 Damage Threshold of Laser Paint Removal 202
6.2.4 Three-Dimensional Cleaning Model 203
6.3 Laser Paint Removal Experimental Research Method 204
6.3.1 Laser Paint Removal Equipment 204
6.3.2 Research Parameters in Laser Paint Removal 205
6.3.3 Laser Cleaning of Different Types of Substrates and Paints 206
6.3.4 The Impact of Laser Cleaning on the Performance of the Base Material 210
6.4 Laser Paint Removal Cleaning Threshold and Damage Threshold 211
6.5 The Influence of Laser Cleaning Parameters on Cleaning Effectiveness 220
6.6 Practical Application Research of Laser Paint Removal 230
6.6.1 Laser Paint Removal of Aircraft Skin 230
6.6.2 Laser Paint Removal in Ammunition Repairment 231
6.6.3 Laser Paint Removal in Cultural Relics and Artworks 232
References 234

7 Application of Laser in Rust Removal 237
7.1 Significance of Laser Rust Removal 237
7.2 Rust Formation and Basic Components 239
7.2.1 Corrosion Formation 240
7.2.2 Rust 242
7.2.3 Copper Rust 243

7.2.4 Oxide Skin on Aluminum Plate 246
7.2.5 Silver Rust 246
7.3 Laser Rust Removal Mechanism 248
7.3.1 Absorption of Laser by Metals and Rust Layers 248
7.3.2 Temperature Distribution During Laser Rust Removal 249
7.3.3 Main Mechanism of Laser Rust Removal 250
7.4 Laser Rust Removal Equipment and Characterization Methods 253
7.4.1 Laser Rust Removal Equipment 253
7.4.2 Characterization Methods of Laser Rust Removal Effects 254
7.5 The Impact of Different Laser Parameters on Laser Rust Removal Effect 255
7.6 Research and Practical Applications of Laser Rust Removal 264
7.7 The Impact of Laser Rust Removal on Material Properties 268
References 274

8 Application of Laser in Mold Cleaning 277
8.1 Significance of Mold Cleaning 277
8.2 Commonly Used Mold Cleaning Methods 278
8.3 Laser Cleaning Device for Molds 282
8.3.1 Selection of Laser System 283
8.3.2 The Structure of the Light Guide System and the Design of the Optical Path System 284
8.4 Mechanism of Laser Cleaning Rubber Mold 285
8.5 Impact of Laser Cleaning Parameters on Molds Cleaning Effect 287
8.5.1 Laser Source 287
8.5.2 Laser Energy Density 288
8.5.3 Laser Repetition Frequency 289
8.5.4 Beam Quality 289
8.5.5 Pulse Number 291
References 291

9 Laser Cleaning of Historical Buildings and Artworks 293
9.1 Significance of Cultural Relics Cleaning and Restoration 293
9.2 Laser Cleaning of Stone Artifacts 295
9.2.1 Laser Cleaning Applied to Buildings 296
9.2.2 Stone Artifacts and Cultural Relics 296
9.3 Laser Cleaning of Metal (Bronze) Cultural Relics 303
9.4 Laser Cleaning of Ceramic Artifacts and Artworks 305
9.5 Laser Cleaning of Calligraphy and Painting Artifacts 307
9.6 Partial Monitoring Methods for Laser Cleaning of Cultural Relics and Artworks 309
References 312

10 Applications of Laser Cleaning in Other Areas ... 315
10.1 Laser Cleaning of Oil Stains ... 315
10.1.1 The Significance of Oil Stain Cleaning ... 315
10.1.2 Theory Supplement for Laser Cleaning of Oil Stains ... 316
10.1.3 Research on Laser Cleaning of Oil Contamination ... 317
10.2 Laser Cleaning of Optical Substrate Surfaces ... 321
10.2.1 Significance of Laser Cleaning Optical Devices ... 321
10.2.2 Experimental Research on Laser Cleaning of Optical Devices ... 322
10.3 Laser Cleaning of Buildings ... 327
10.4 Laser Cleaning in Nuclear Power Plants ... 329
10.5 Applications of Laser Cleaning in Other Fields ... 331
References ... 332

Postscript ... 335

About the Authors

Dr. Feng Song is a professor in the School of Physical Sciences at Nankai University. He is mainly engaged in the research on the preparation, characterization, and luminescence of rare earth doped optical materials (glass, powder, nanocrystals, etc.), solid-state laser technology and applications, fiber technology, etc. He has taken charge of 30 scientific research projects and has published over 200 journal papers with about 4000 citations. He is granted more than twenty patents. He has won the Baosteel Excellent Teacher Award, the Second Prize of Tianjin Natural Science Awards, and the Second Prize of the National Teaching Achievement Awards twice. He also served as the deputy editor for the Journal of College Physics, as well as a member of the editorial board of the *Journal of Clean World*, the *Journal of Optoelectronic Technology Applications*, *Journal of Laser* and *Applied Optics* (2010–2016). Now he is also a member of the Teaching Guidance Committee for Basic Physics Courses in Higher Education under the Ministry of Education of China (2006 to present), an executive committee member of the National High School Physics Competition Committee (2011–2020), and the head coach and leader of the Chinese National Team for the Physics Olympiad (2012–2018, 2022), independent referee for the International Youth Physics Competition (IYPT) (2008 to present), and juror chair for the Chinese University Physics Academic Competition (CUPT) (2010 to present). He is also the executive member of the Optoelectronics Professional Committee of the Chinese Optical Society, the vice chairman of the Tianjin Society of Laser Technology (2012–2020), the deputy director and secretary-general of the Tianjin Society of Physics, and the member of the academic committee of the State Key Laboratory of "solid laser technology". He was named by the "Outstanding Young Teacher Funding Program" and the "New Century Excellent Talent Support Program" sponsored by the Ministry of Education of China. He is also the head of the first level innovation team of Tianjin's 131 Talents Program.

Dr. Xuechun Lin received his Ph.D. from the Institute of Physics, Chinese Academy of Sciences (CAS) in 2004, and became the youngest full professor in the Institute of Semiconductors (IOS), CAS in 2007. He is currently the Director of All Solid-State Light Source Laboratory, IOS, CAS. For over two decades, he has devoted himself in the research and development of high-power all solid-state lasers and their applications. His research and inventions are listed as major achievements of the "11th Five-year Plan" by CAS and the Ministry of Science and Technology of China. He won the Second Prize of the National Technological Innovation Awards of China in 2017.

Chapter 1
Overview of Cleaning Technology

1.1 Cleaning and Its Significance

1.1.1 Concept of Cleaning

Cleaning is a labor closely related to human production and life practices. Since ancient times, people have been engaged in cleaning work. With the progress of society and the rapid development of technology, the types and quantities of pollutants in production and life are increasing, and the demand for cleaning is also increasing.

For any type of material (substrate), when exposed to the environment or during use, substances in the environment, or new substances produced by various physical and chemical reactions, will adhere to the substrate. The substances adhering to the substrate are called pollutants, such as dust absorbed on clothes, rust formed by the chemical reaction of iron on an iron plate with oxygen in the air, dust and oil stains stuck on aluminum plates, and so on. In addition, paints and protective layers artificially applied to the substrate, which need to be removed for some reasons (such as damaged), are considered adherent layers on the substrate. For the sake of uniformity, they are also referred to as pollutants. In short, substances that need to be removed from the surface of the substrates are collectively referred to as pollutants or dirt. The process of removing pollutants from the surface of an object and restoring the object to its original surface condition is the cleaning process.

The cleaning process mainly involves removing pollutants by dissolving them in water or chemical solvents, or by physical processes such as mechanical force. In human life, we clean every day, washing clothes, dishes, faces and hands, mopping floors, wiping tables, and so on. In modern industrial production, cleaning is widely used. Sometimes cleaning is a necessary process in production, such as the manufacture of printed circuits, substrate cleaning in the semiconductor and microelectronics industries, cleaning of grease on optical lenses, etc. Some equipment components need to be cleaned and maintained after a period of use, such as cleaning of car tire molds, regular paint removal of airplanes and ships, rust removal operations of

F. Song and X. Lin, *Laser Cleaning*, https://doi.org/10.1007/978-981-97-6631-4_1

train rails, etc. Some fine work requires cleaning, such as the maintenance of ancient artworks. In these cleaning processes, the quality of cleaning often directly affects the performance and preservation of components and products to a large extent.

1.1.2 Industries Using Cleaning

The main purpose of cleaning is to remove dirt from the surface of objects, thereby restoring the original performance of products and the production efficiency of devices, and improving the quality of household and industrial products. In industrial production, cleaning is almost used in all of fields, and many equipment and components in various industries need to be cleaned regularly.

Clothing, food, housing, transportation, energy, manufacturing, all walks of life, all need cleaning. Below are some examples of equipment and components that need to be cleaned in various industries.

1. Textile and Chemical Fiber Industry

Cooling water systems, boilers, refrigeration machine condensers, central air conditioning, aluminum fins, cooling pipes, ion exchangers, printing and dyeing machines, drums, dyeing tanks, water pipes, pulp pools, gas pipes, gas cabinets, coolers, evaporators, humidifier nozzles, humidification water systems, etc.

2. Food and Catering Industry

Fermentation tanks, heat exchange pipes, bottle washers, hot and cold water systems, boilers, condensers, coolers, various storage tanks, crystallizers, cooling towers, evaporation tanks, cold storage water systems, steam pots, water heaters, water bath stoves, tea stoves, ice makers, water systems, biological sludge stripping and cleaning, heat exchangers, range hoods, ducts, stoves, etc.

3. Transportation Industry

Cars: Engine carbon deposits, cooling water systems, fuel injectors, lubrication systems, radiators, vehicle exteriors, etc.

Trains: Car bodies, air conditioning systems, internal combustion engines, cooling systems, lubrication systems, rails, etc.

Ships: Boilers, oil pipelines, seawater desalination systems, oil tanks, flash evaporators, water storage tanks, air coolers, freshwater coolers, seawater coolers, decks, paint stripping and rust removal during maintenance, dry docks, etc.

Aircraft: Aircraft exterior paint stripping, etc.

Train stations, docks, waiting halls: Central air conditioner, ventilation ducts, fan coil units, station cleaning, etc.

4. Construction and Home Industry

Circulating cooling water systems, heat exchangers, condensers, grilles, floors, conveyors, bins, cooling tanks, preheaters, rotary kilns, boilers, extruders, coolers, air conditioners, heat exchangers, stone, rebar, doors and windows, furniture, kitchenware, etc.

5. Energy Industry

Electricity Industry: Power plant ash discharge pipes, ash conveying pipes, condensers, boilers, oil coolers, return water pipes, steam pipe networks, cooling tower sludge, crystallizers (cable factories), dust collector water systems, transformer water cooling systems, refrigeration units, heat exchangers, oil control, deaerators, steam turbines, air preheaters, etc.

Oil Industry: Cooling water systems, heating pipes, heat exchangers, boilers, storage tanks, oil pipelines, casing pipes, guide pipes, water injection pipes, steam pipes, atmospheric system corrosion protection, air compressors, various oil storage tanks, water storage tanks, storage tanks, gas storage tanks, oil pumps, etc.

Coal Industry: Underground water pipes, underground hydraulic systems, rails, coal loading cars (rust removal, corrosion protection), boilers, air preheaters, air compressors, heat exchangers, etc.

Nuclear Industry: Radioactive contamination equipment, air conditioning systems, etc.

6. Various Manufacturing Industries

Electronics Industry: Printed circuit boards, electronic components, etc.

Machinery Industry: Boilers, central air conditioning, air coolers, air compressors, hydraulic machine cooling systems, extruders, mold cooling systems, surface coolers, etc.

Steel Industry: Air compressors, cooling systems, filters, various pipes, coolers, cooling towers, various storage tanks, air compressors, nitrogen compressors, oxygen compressors, oxygen pipelines, nitrogen pipelines, heat exchangers, oil systems, blast furnaces, heat exchangers, process pipeline, smelting furnaces, boilers, melting furnaces, impoverishment furnaces, medium frequency furnaces, high frequency furnaces, sintering furnaces, etc.

Paper and Printing Industry: Black liquor evaporators, black liquor scale, boilers, pulp conveyance pipes, water supply pipes, sewage pipes, heaters, printing machine cooling systems, air conditioning systems, cookers, pulp sterilization and preservation, etc.

Fertilizer and Pharmaceutical Industry: Condensers, pipelines, reaction kettle water jackets, mixers, heat exchangers, evaporators, condensers, reactors, filters, compressors, storage tanks, boilers, distillers, central air conditioning, air ducts, thermal oil furnaces, circulating cooling water systems, mud stripping, air compressors, injection molding machines, extruders, kneaders, crystallizers, absorption towers, reaction towers, regeneration towers, synthesis towers, steam towers, cooling towers, carbonization towers, and so on.

1.1.3 The Significance of Cleaning

Whether in daily life or industrial production, cleaning is of great importance, mainly in the following aspects:

1. Improve the appearance of equipment, purify and beautify the environment. Removing dirt from the interior and exterior surfaces of factories, buildings, and transportation tools can improve their appearance and purify the environment.
2. Maintain normal production and extend equipment life. Removing dirt from the surface of raw materials can restore and maintain the surface properties of the materials, ensuring the implementation of subsequent production processes. Regular or irregular cleaning of production equipment can maintain normal operation of the equipment, control equipment corrosion, and extend its service life and operation cycle.
3. Improve production capacity and product quality. Removing contaminants from the surface of raw materials and regular cleaning of equipment can reduce the impact of dirt on product performance and improve product quality.
4. Reduce energy consumption and production costs. Contaminants on the surface of raw materials and equipment can reduce production efficiency and increase energy consumption. Cleaning contaminants can reduce the consumption of raw materials and energy, improve production efficiency, and reduce production costs.
5. Reduce production accidents and benefit human safety. Cleaning dirt can reduce various accidents caused by production processes and equipment, ensuring the safety of operators.
6. Improve hygiene standards and benefit health. Cleaning can sterilize, disinfect, and remove radioactive contamination, reducing harm to the human body and benefiting health.

From the above concept and significance of cleaning, it can be seen that cleaning is indispensable for production and human life. Continuously developing cleaning technology to adapt to the development of modern science and life, and benefiting mankind, is an important research topic for people.

1.2 Cleaning Technologies and Classification

1.2.1 The Four Elements of Cleaning

To remove contaminants from the surface of the base material that needs to be cleaned (referred to as the cleaning object, or cleaning material), cleaning media such as chemical reagents, iron sand, etc. are needed. The force that promotes the cleaning process using the cleaning medium is called as the cleaning force. Therefore, a cleaning system includes four elements: the cleaning object, contaminants or pollutants, cleaning media, and cleaning force.

Base materials and pollutants are diverse. For example, paint and rust on ships, rubber on molds. The bonding mechanisms between contaminants and the base are also different, some are physical effects, such as pollutants that accumulate on the surface of objects due to gravity (such as very thin dust and grease), particles adhering to the surface of objects due to electrostatic attraction (such as tiny silicon particles on circuit boards); some are the effects of various chemical bonds, such as pollutants that bind to the surface of objects through adsorption between molecules (such as oil pollution), pollutants that are tightly bound due to chemical reactions (such as rust).

1.2.2 Common Cleaning Technologies

For different bases and pollutants, and different bonding mechanisms between the base and pollutants, people have developed different cleaning technologies. There are many commonly used cleaning technologies, such as sandblasting cleaning, pickling, electromagnetic wave cleaning, dry ice cleaning, high-pressure water jet cleaning, and so on. According to whether there is liquid or auxiliary liquid added in the cleaning, it can be divided into dry cleaning and wet cleaning. According to the interaction, it can be divided into physical cleaning technologies, chemical cleaning technologies, biological cleaning technologies, etc., and some cleaning technologies are actually a combination of physical cleaning technologies and chemical cleaning technologies.

I. **Physical Cleaning Technologies**

Based on physical principles, including mechanical, thermodynamical, acoustical, optical, electromagnetic, and thermal principles, the technologies of removing pollutants from the cleaning object surface through mechanical friction, steam, ultrasound, high pressure, ultraviolet radiation, etc., is called the physical cleaning technologies. Physical cleaning technologies do not change the chemical composition and properties of the cleaning object. Common physical cleaning technologies include:

(1) Mechanical cleaning technology: Through scrapers, drilling tubes, shot blasting, etc., the pollutants are peeled off from the surface of the cleaning object by mechanical friction.
(2) Shot blasting cleaning technology: Through a high-pressure vacuum pump, small iron shots or sand shots are sprayed at the cleaning object, and the pollutants are cleaned from the surface of the cleaning object by mechanical force. This is currently the most commonly used technology in industrial cleaning.
(3) Hydraulic cleaning technology: Normal or high-pressure water is sprayed on the cleaning object, and pollutants soluble in water will be removed; if high-pressure water is used, pollutants can also be peeled off from the base surface by pressure drop.
(4) Dry ice cleaning technology: Dry ice particles are used to hit the surface pollutants to be cleaned, and the pollutants are removed from the surface of the

cleaning object through the principle of thermal expansion and contraction, mechanical friction.

(5) Steam cleaning technology: Saturated steam under high temperature and high pressure is introduced into the cleaning object, causing the pollutants on the surface of the cleaning object to dissolve and vaporize, achieving the cleaning purpose. In this technology, the pollutant may undergo a chemical reaction after absorbing heat, so probably it is not a pure physical cleaning process.

(6) Ultrasonic cleaning technology: Ultrasonic waves drive the cleaning liquid, generating tiny vacuum bubbles. When they burst under pressure, they generate powerful shock waves that can peel off pollutants from the surface of the cleaning object. In addition, ultrasound also has the effect of accelerating dissolution and emulsification in liquids. The cleaning liquid can be water or a chemical solvent, so there may also be chemical reactions in this technology, so probably it is not a pure physical cleaning process.

(7) Ultraviolet light cleaning technology: When ultraviolet light is irradiated onto the object to be cleaned, pollutants (such as hydrocarbons) absorb photon energy and peel off from the surface of the cleaning object, thereby achieving the purpose of cleaning. In this technology, the pollutants may also decompose through chemical reactions after absorbing light energy, so probably it is not a purely physical cleaning process.

(8) Electromagnetic cleaning technology: The object to be cleaned is placed in a static electric field or high-frequency electromagnetic field. The molecules of the pollutants polarize under the action of the static electric field, or absorb the energy of the electromagnetic field and their structure changes. The original large molecular structure breaks, and the molecular structure changes, making the pollutants become soft or fragmented and fall off, achieving the effect of descaling. This process may also involve chemical changes.

(9) Laser cleaning technology: Using a laser as a cleaning medium, the interaction between the laser and the pollutants and the cleaning object is utilized. Through physical mechanisms such as ablation and vibration, the pollutants are peeled off from the base surface.

From the above technologies, it can be seen that some technologies involve chemical reactions. If the cleaning process is mainly physical, it does not fundamentally change the chemical composition and physical and chemical properties of the cleaning object, and basically has no corrosive and destructive effect on the cleaning object substrate, all can be classified as physical cleaning technologies. Physical cleaning technologies are difficult to clean some special workpieces.

II. **Chemical Cleaning Technologies**

The chemical cleaning technologies use acids, alkalis, organic chelating agents, dispersants and other chemicals or their solvents, relying on their chemical reactions with the pollutants on the surface of the object, to dissolve and remove the pollutants on the surface of the cleaning object. Strictly speaking, the base object cannot have a chemical reaction with the cleaning medium.

According to the chemical agents, it can be divided into (1) Alkali washing: using sodium carbonate, sodium phosphate, sodium hydroxide, etc.; (2) Acid washing: using sulfuric acid, hydrochloric acid, nitric acid, phosphoric acid, hydrofluoric acid, citric acid, hydroxyacetic acid, ethylenediaminetetraacetic acid, etc.; (3) Neutralization reaction: using sodium nitrite, sodium benzoate, etc.; (4) Dirt stripping: using strippers and quaternary ammonium salts, etc.; (5) Solvent cleaning: using carbon tetrachloride, trichloroethylene, etc. [1].

Electrolytic cleaning technology is also classified as chemical cleaning. It uses the principle of electrolysis, through chemical reactions, to remove contaminants adhered to the cleaning object (metal).

Biological cleaning technology (biology cleaning method) can also be classified as a type of chemical cleaning. The biological cleaning technology uses the catalytic detergent (enzyme) produced by the intracellular of microorganisms with special cleaning functions to decompose the contaminants on the substrate surface, converting them into non-toxic and harmless water-soluble substances. This technology is called microbial cleaning, also known as biological cleaning.

III. **Mixed Cleaning Technologies (Mixed Cleaning Technology)**

Some technologies are difficult to simply categorize as physical cleaning or chemical cleaning. In addition to the steam cleaning technology (chemical reactions occur at high temperatures) and ultrasonic cleaning technology (the cleaning liquid is a chemical solvent) that we mentioned earlier, there are also plasma cleaning technology, etc. These cleaning technologies have both physical effects and chemical reaction processes.

(1) Plasma cleaning technology

Under a certain vacuum condition, the active particles contained in the plasma chemically or physically react with the contaminant molecules to remove the dust and other contaminants on the material surface. Plasma is a kind of electrically neutral, high-energy, fully or partially ionized gaseous substance, which contains ions, electrons, free radicals and other active particles. Plasma cleaning involves plasma physics, plasma chemistry, and gas–solid interface chemical reactions. Commonly used plasmas are divided into three types according to the excitation frequency: ultrasonic plasma, radio frequency plasma, and microwave plasma. The reactions occurring in ultrasonic plasma are mostly physical reactions, the reactions occurring in radio frequency plasma are both physical and chemical reactions, and the reactions occurring in microwave plasma are mostly chemical reactions.

(2) Other mixed cleaning technologies

In fact, many cleaning technologies include both physical technology and chemical technology. Here are two examples: thermal cleaning and ultrasonic cleaning technology.

Thermal cleaning is achieved by heating the cleaning object. Because the thermal expansion coefficients of the substrate and the contaminants are different, thermal stress is generated to peel off the contaminants; or the contaminants become liquid after absorbing heat and leave the substrate. These two situations are essentially physical cleaning. However, in thermal cleaning technology, chemical reagents are often added, because heating can promote chemical reactions, including accelerating the reaction rate and improving the reaction efficiency. At this time, it can be considered as a mixed cleaning that combines physical and chemical cleaning.

In ultrasonic cleaning, the reagent used is often a chemical solvent rather than water. In the cleaning process, in addition to ultrasonic oscillation, chemical reactions also occur, cleaning the oil from the base material. This is also a mixed cleaning. That is to say, in the actual cleaning process, it is possible that multiple action mechanisms work together to achieve the purpose of cleaning. From the above, cleaning technology can be simply divided into physical cleaning and chemical (or biochemical) cleaning technology. There are many technologies in each category, and each cleaning technology has its own characteristics. According to the characteristics of the products to be cleaned and the cleaning requirements, choosing the appropriate cleaning technology can achieve good economic cleaning effects.

1.3 Physical Cleaning Methods

There are many physical cleaning technologies, and here we mainly introduce several physical cleaning technologies commonly used in industry: shot blasting cleaning, high-pressure water jet cleaning, dry ice cleaning, electromagnetic wave cleaning, and ultrasonic cleaning, etc.

1.3.1 Shot Blasting Cleaning Technology

I. **Overview of Shot Blasting Cleaning Technology**

Shot blasting cleaning can be used to remove such contaminants as oxide, rust, molding sand, and old paint film on workpieces or equipment, while the contaminants are generally not less than 2 mm thick or do not require high precision in size and contour shape. The volume of the workpiece is generally large. Shot blasting cleaning is a common cleaning technology before surface coating, widely used in large shipyards, automobile factories, locomotive factories, heavy machinery factories, etc. [2–4]. Shot blasting strengthening is a cold treatment process. Through shot blasting cleaning, the fatigue resistance of metal parts that have been in service under high stress conditions for a long time, such as aircraft engine compressor blades, body structural parts, automobile transmission system parts, etc., can be improved.

The shot in shot blasting cleaning can be steel shot, cast iron shot, glass shot, ceramic shot, quartz sand, etc. Shot blasting cleaning is a mechanical cleaning technology widely used in industrial production at present, with high efficiency and low cost [4].

In industry, shot blasting and shot throwing are sometimes separated. Shot blasting uses high-pressure air or compressed air as power, while shot throwing generally uses a high-speed rotating flywheel to throw steel sand at high speed. Comparatively speaking, shot blasting is more delicate, easy to control precision, but consumes more power and is less efficient than shot throwing. It is suitable for small workpieces with complex shapes; while shot throwing is more economical and practical, easy to control efficiency and cost, can control the granularity of the shot to control the blasting effect, but there will be dead corners, suitable for batch processing of workpieces with a single shape.

In industrial cleaning, shot blasting and sand blasting are sometimes separated. Both shot blasting and sand blasting use high-pressure air or compressed air as power to blow it out at high speed to impact the surface of the workpiece to achieve a cleaning effect, but the selected media are different. Sand blasting uses sandstone, and the surface of the workpiece treated by sand blasting is the original color of the metal, the surface is rough, without metallic luster, and is a dark surface; while after shot blasting, the surface of the workpiece base will be strengthened, and the surface of the workpiece is also the original color of the metal, with a matte effect.

Whether it's shot blasting, sand blasting, or shot throwing, although the blasting power and methods are different, and the media are different, they all achieve the cleaning effect through high-speed impact on the workpiece. Therefore, we write them all here as shot blasting cleaning.

II. The Principle of Shot Blasting Cleaning

The principle of shot blasting cleaning is to use a vacuum pressure pump to spray or throw abrasive materials (such as small iron balls, sand) at a high speed towards the surface of the object to be cleaned. The abrasive materials impact the surface of the object at a high speed, causing friction between the two, which can remove rust, oxidation, paint, oil stains, and other surface contaminants, and achieve a certain roughness on the surface. There are many factors that affect the efficiency of shot blasting cleaning, including air pressure, nozzle diameter and shape, the condition of the surface of the material being cleaned, the quality level of cleaning, shotblasting equipment, the condition of the abrasive, the mixing state of the abrasive and air, the skill level of the operator, the angle of spraying, the distance of spraying, etc. To improve the efficiency of shotblasting cleaning, it is necessary to coordinate the relationship between these factors.

In the shotblasting technology, the shot-spraying method can be divided into dry and wet. During wet spraying, the shot is sprayed out together with a liquid (water or other chemical solvents, if the latter, it is a combination of physical and chemical cleaning). Dry spraying is divided into direct pressure and vacuum suction, the latter is further divided into gravity and siphon, as shown in Fig. 1.1. Depending on the characteristics of different cleaning objects, different forms of shotblasting and types of spraying materials can be chosen.

Fig. 1.1 Classification of shot-spraying methods

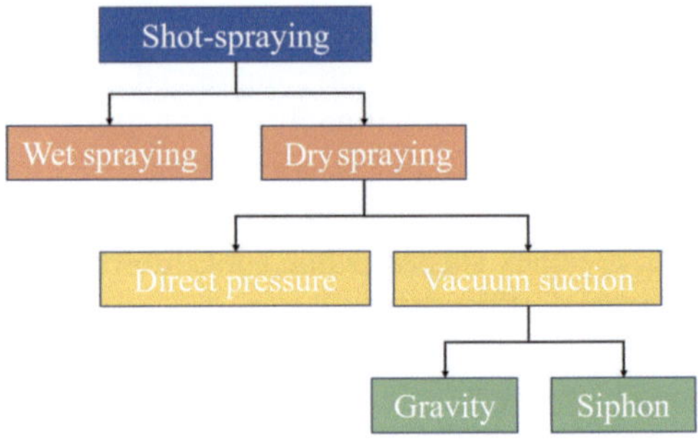

III. **Features of Shot Blasting Cleaning Technology**

Shot blasting cleaning technology is simple, flexible, and low-cost, making it one of the most thorough, universal, rapid, and efficient cleaning technologies currently available. Its main advantages are:

(1) High efficiency and speed. A shot blasting machine can clean more than 10 square meters per hour (the specific amount depends on the thickness and type of the pollution layer);
(2) Due to the use of mechanical friction, as long as the friction force is greater than the adhesion force between the pollutant and the base, all types of pollution layers on the surface of the workpiece, whether rust, oil stains, or paint, can be removed;
(3) Shot blasting cleaning is a physical cleaning method, with no chemical reactions (for wet spraying that adds chemical reagents, its chemical reaction is relatively weak), so it will not change the physical and chemical properties of the object being cleaned;
(4) It can achieve a suitable surface roughness. By choosing different types and sizes of shots, the roughness of the base surface can be controlled, creating a matte surface on the workpiece, which is difficult to achieve with other industrial cleaning methods at the same time. The matte surface can greatly enhance the adhesion of the workpiece to the coating or plating, making the adhesive parts more firmly bonded and of better quality; after shot blasting, mechanical parts can produce uniform and fine uneven surfaces, allowing lubricating oil to remain, thereby improving lubrication conditions, reducing noise and extending the service life of the machinery.

Shot blasting cleaning also has significant drawbacks. For parts with complex shapes and structures, it is difficult to clean thoroughly. It is also difficult to use for some special cleaning requirements (such as materials that are relatively fragile, or situations where uniform cleaning is desired). It is not suitable for thin plate workpieces, as it can easily deform them. For workpieces with oil stains, it cannot completely remove them. Another important point is the issue of environmental pollution, including noise pollution and dust pollution. Even in a relatively enclosed factory, a considerable amount of noise is generated during the cleaning process, and the dust that is raised is harmful to human health. In addition, shotblasting cleaning requires a high level of experience from the operator. When the parameters are not adjusted properly, it can easily cause damage to the base materials. For different cleaning objects, it is necessary to make reasonable choices and adjustments to the type of grinding material, throwing pressure, distance, and other shot blasting processing conditions to minimize damage to the cleaning object.

1.3.2 High-Pressure Water Jet Cleaning Technology

I. **Overview of High-Pressure Water Jet Cleaning Technology**

High-pressure water jet cleaning technology uses water as a cleaning medium, through a high-pressure pump, valves, and a self-control system. A high-pressure generator is used to produce high-pressure water, which gives the water flow a tremendous amount of energy. This water is then transported through pipes to the nozzle, where it is sprayed out as a high-speed jet [5, 6]. As shown in Fig. 1.2. The high-pressure water jet has a great impact kinetic energy. When it is directed at the surface of the cleaning object, it uses the impact force and grinding action to remove the dirt and attachments on the base surface [7, 8]. For parts with complex shapes and structures, as well as narrow spaces, complex environments, and harsh hazardous situations, high-pressure water jet technology can be used. Since the 1970s, many countries have vigorously developed high-pressure water jet cleaning technology, and now the technology is quite mature and widely used [9].

II. **Principle of High-Pressure Water Jet Cleaning Technology**

When the high-pressure water jet reaches the surface of the object, the interaction between the jet and the object changes the speed and direction of water jet and its momentum also changes accordingly. According to the momentum theorem, the change in the momentum of the water jet acts on the surface of the object in the form of impact force. Continuous water jets form a stable impact force, thereby removing

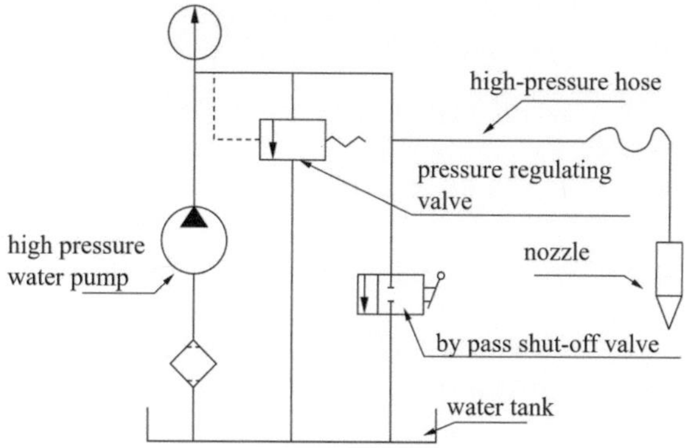

Fig. 1.2 High pressure water jet system

pollutants from the surface of the cleaning object. Assuming the high-pressure water jet is an ideal fluid, it is stably directed at the base surface at a certain speed, v_1 and v_2 are the average flow speeds inside and outside the nozzle outlet cross-section, According to the momentum theorem, we have [10]:

$$F\Delta t = mv_2 - mv_1 \tag{1.3.1}$$

In this equation, F is the impact force acting on the unit volume fluid per unit time, Δt is the time that force F acts on the unit volume fluid, and m is the mass of the unit volume fluid.

When the high-pressure water flow hits the base material at an angle θ, the impact force also satisfies:

$$F = \rho qv - \rho qv \cos\theta$$

the force per unit area of the pollutant is:

$$F = \rho qv - \rho qv \cos\theta$$

In the formula, q is the volume of flow per unit time, DA is the cross section of the flow, ρ represents the density of the water jet, and v represents the velocity of the water jet. The jet stress σ_f calculated by formula (1.3.2) is the theoretical maximum value. Due to the diffusion of the jet and the influence of air resistance, the actual stress produced by the jet is less than the theoretical stress. When the stress σ_f applied on the dirt by water jet is greater than the limited stress σ_p of the dirt itself, the dirt will detach from the base surface.

$$\sigma_f = \frac{F}{\Delta A} = \frac{\rho qv - \rho qv\cos\theta}{\Delta A} = \frac{\rho qv(1-\cos\theta)}{\Delta A} = \rho v^2(1-\cos\theta) \tag{1.3.2}$$

By substituting the limited stress σ_p of the dirt's pressure resistance into formula (1.3.2), we get

$$v = \sqrt{\frac{\sigma_p}{\rho(1-\cos\theta)}} \tag{1.3.3}$$

The effect of high-pressure water jet cleaning is mainly related to water pressure, flow rate, and nozzle diameter. According to the basic theory of small hole outflow, the velocity v of the jet at the nozzle mouth is:

$$v = \varphi\sqrt{2gh} \tag{1.3.4}$$

where v is the exit speed of the nozzle, g is the acceleration of gravity, and φ is the nozzle coefficient. The pressure H of the jet at the nozzle inlet is

$$H = \frac{v^2}{2g\varphi^2} = \frac{\sigma_p}{2g\rho(1-\cos\theta)\varphi^2} \tag{1.3.5}$$

From this, the pressure P is obtained

$$P = \rho g H = \frac{\sigma_p}{2(1-\cos\theta)\varphi^2} \tag{1.3.6}$$

If the pressure p (MPa), flow coefficient φ, and nozzle diameter d (mm) are known, the working flow rate Q (L/min) can be calculated

$$\mathrm{Q} = \varphi d^2\sqrt{P} \tag{1.3.7}$$

III. **High-Pressure Water Jet Equipment**

There are various structures of cleaning equipment for high-pressure water cleaning technology. Generally speaking, high-pressure water jet cleaning equipment mainly includes four systems: high-pressure water generation system, control system, execution system, and auxiliary system. The high-pressure water jet cleaner is the generation device of the high-pressure water jet. Its main technical indicators are pressure and flow rate. High-pressure sealing technology, non-alternating stress pump head technology, and high-pressure rotary sealing technology are the most critical technologies. Figure 1.3 shows the schematic diagram of the high-pressure water jet system [9].

The high-pressure water generation system is the most important system, generally composed of a water tank, level gauge, filter, high-pressure water pump unit, overflow valve, flow meter, pressure sensor, solenoid valve, spray gun, and nozzle. The water tank provides a water source for the system and is equipped with a level alarm to prevent the water level from being too high or too low. The water in the tank passes

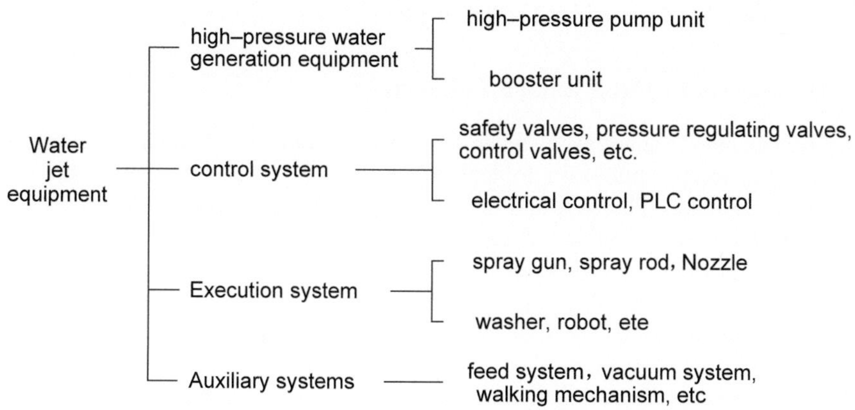

Fig. 1.3 Composition diagram of high pressure water jet cleaning equipment

through the filter and flow meter into the high-pressure pump. The function of the flow meter is to alarm the system when the water flow is too low or there is no water flow. The overflow valve ensures that the entire water circuit works under the set pressure, providing safety protection for the water circuit system. Open the normally closed electric valve as needed, and choose to spray with a water gun or nozzle.

IV. **Characteristics of High-Pressure Water Jet Cleaning Technology**

Water jet cleaning, compared to traditional mechanical and chemical cleaning, has the following advantages:

1. High cleaning efficiency: For the vast majority of dirt scale and blockages, once the appropriate water jet parameters are selected, namely pressure, flow power, and spray gun/nozzle, they can be cleaned quickly and effectively.
2. Low cleaning cost: High-pressure water jets use natural water as a medium, consuming only 3–8 m^3 of water per hour. The high-strength, high-wear-resistant nozzles have very minimal wear. The overall cost is only about 1/3 of the chemical method.
3. Wide application range: There are no special requirements for the size, material, shape, and type of pollutants. It can also easily complete the cleaning of parts with complex shapes and structures that are difficult to complete with conventional cleaning methods.

However, high-pressure water jet technology also has disadvantages, such as causing rust when used to clean metal materials, which is an important reason why it is difficult to apply on a large scale in industry. Its cleaning efficiency is not as good as shotblasting. To improve efficiency, sometimes small pellets or chemical reagents are added to the water. As a result, the water after cleaning is mixed with pollutants and needs secondary treatment, otherwise, it will pollute the environment.

1.3.3 Dry Ice Cleaning Technology

I. **Overview of Dry Ice Cleaning Technology**

Dry ice cleaning technology, also known as dry ice cold jet cleaning technology, is to make low-temperature solid dry ice particles from liquid carbon dioxide through a dry ice granulator [11, 12], and the dry ice particles are put into the dry ice jet cleaning machine, accelerated to a high speed with the high-speed airflow, and hit the surface of the object to be cleaned. On the one hand, the impact generates a force (similar to shot blasting), on the other hand, the surface contaminants suddenly become brittle due to the sudden cooling when they encounter dry ice. At the same time, dry ice will enter the cracked dirt, quickly sublimate into gas, and the volume of the dirt expands about 600–800 times in a very short time. Under the combined action of these factors, the stress generated can remove these shattered pollutants from the surface of the object.

II. Principle of Dry Ice Cleaning Technology

Dry ice cleaning technology mainly relies on high-speed moving dry ice particles hitting the cleaning object surface, and the particles with great kinetic energy overcome the binding force between the dirt layer and the substrate, and combine with the sublimation physical properties of dry ice particles to remove dirt. The mechanism and effects of dry ice cleaning mainly include [13]:

1. Low-Temperature Stripping Action

Figure 1.4a is the pollutants on the substrate material before dry ice cleaning. Figure 1.4b shows the accelerated dry ice particles colliding with the surface of the dirt layer and undergoing heat exchange, the temperature of the dirt layer drops, thereby shrinking, becoming brittle and cracking. Since the expansion coefficients of the dirt layer and the substrate are different, the binding force between the pollutants and the substrate will decrease when cooled, making the dirt layer easier to remove.

2. Impact and Scouring Effect

As shown in Fig. 1.4c, under the action of compressed gas, the high-speed dry ice particles sprayed from the nozzle have a large kinetic energy. Their kinetic energy E and impact force F are respectively [14]

$$E = 1/2mv^2,\ \ Ft = m\Delta v \tag{1.3.8}$$

In the formula, $\Delta v = v_1 - v_2$, where v_1 is the speed of the dry ice particles before they hit the cleaning surface, v_2 is the speed of the dry ice particles after they hit the cleaning surface, m is the mass of a single particle, and t is the time the dry ice particles impact the dirt layer. After the high-speed moving dry ice particles collide with the dirt surface, they have a grinding and shearing force on the pollution layer, overcoming the adhesion between the pollutants and the base surface. The cleaned pollutants are swept away with the high-pressure airflow.

3. Sublimation Effect

After the dry ice particles collide with the dirt layer, they will quickly sublimate into carbon dioxide gas. The fine dry ice particles enter the cracked pollution gaps, and their volume expands instantly, just like an explosion, thereby quickly removing the dirt layer, as shown in Fig. 1.4d [15, 16].

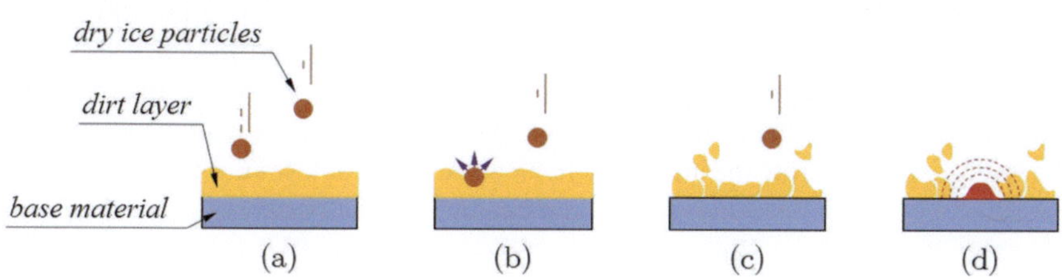

Fig. 1.4 Schematic diagram of dry ice cleaning mechanism

III. **Characteristics of Dry Ice Cleaning**

1. Minimal Damage to the Base: The dry ice cleaning technology is based on the mechanics and thermodynamics mechanism. Dry ice itself is light in weight, causing minimal damage to the base. It can be used to clean parts that require a damage-free surface.
2. High Efficiency: During the dry ice cleaning process, equipment does not need to be shut down or disassembled, allowing for online cleaning of the equipment. This shortens the cleaning time and improves efficiency.
3. No Need for Water: Dry ice cleaning is a waterless cleaning process, therefore, it has unique advantages in waterproof or insulation-required situations, such as power equipment, electronic components, and easily corroded parts.
4. Environmental Protection: After dry ice cleaning, dry ice particles directly sublimate into carbon dioxide gas and enter the atmosphere, leaving only the removed pollutants, which are easy to recycle and do not cause secondary pollution.

Dry ice cleaning also has some disadvantages: firstly, the noise is extremely loud, generally reaching up to 115 dB. Secondly, the cost of cleaning is relatively high, as it requires the preparation of dry ice particles, therefore it is mainly used in areas where the cleaning quality requirements are high such as cleaning of precision parts.

1.3.4 Ultrasonic Cleaning Technology

I. **Overview of Ultrasonic Cleaning Technology**

Ultrasonic cleaning technology radiates ultrasonic waves (sound waves with a frequency higher than 20 kHz) into the cleaning liquid through an ultrasonic transducer, causing cavitation, acceleration, and direct flow in the liquid, so that the dirt layer on the surface of the parts immersed in the liquid is dispersed, emulsified, and peeled off to achieve the cleaning purpose [17].

Ultrasonic cleaning technology first appeared in the 1930s and developed greatly in the 1950s, with the working frequency of ultrasonic waves between 20 and 40 kHz. In the past few decades, medium and high frequency ultrasonic cleaning technology (according to the frequency of ultrasonic waves, those below 40 kHz are called conventional or low frequency ultrasonic waves, and those above 1000 kHz are called high frequency ultrasonic waves, also known as megahertz ultrasonic waves or megasound waves) has developed, which can remove ultrafine dirt particles on the surface of semiconductor materials without damaging the surface of the substrate material [18, 19]. Now, the application range of ultrasonic cleaning technology has gradually expanded from the initial mechanical and electronic traditional industries to multiple fields such as medicine, food and chemical industry, aerospace, and nuclear industry [20, 21].

II. Principle of Ultrasonic Cleaning Technology

The main mechanism of ultrasonic cleaning is the ultrasonic cavitation effect. After the ultrasonic wave enters the cleaning liquid, the tiny bubbles (cavitation nuclei) originally existing in the liquid vibrate under the action of the sound field. When the sound intensity or sound pressure of the ultrasonic wave reaches a certain threshold, the bubbles will rapidly expand and close, and the shock wave emitted can generate thousands of atmospheric pressures and local high temperatures around it, a physical phenomenon known as ultrasonic cavitation. Through ultrasonic cavitation, the pollution layer is directly and repeatedly impacted, on the one hand, destroying the adsorption of dirt on the surface of the cleaning parts, and on the other hand, causing the pollution layer to break and leave the surface of the cleaning parts and disperse them into the cleaning liquid. In addition, the bubbles can also "drill" into the cracks to vibrate, causing the dirt to fall off. Since cavitation phenomena often occur at the interface between solid and liquid, this cleaning method has more advantages when solids and liquids exist. Ultrasonic cavitation can also generate micro sound flow, and the high-speed micro jet flow generated on the surface of solid and liquid, all of which play a role in brushing the workpiece.

Sound pressure p_e of the ultrasound wave can be represented as [22]:

$$p_e = \sqrt{\frac{1}{T}\int_0^T p^2 dt} \tag{1.3.9}$$

In the formula, T is the average time interval, p is the instantaneous sound pressure. The sound pressure at each position in the ultrasonic field will change with time. For plane cosine waves, it can be proven [23, 24]:

$$p = -A\omega\rho_0 c_0 \sin\omega\left(t - \frac{x}{c_0}\right) = \rho_0 c_0 v \tag{1.3.10}$$

In the formula, A is the amplitude of the vibration at that point in the medium, ω is the angular frequency of the vibration at that point in the medium, ρ_0 is the static density of the medium, c_0 is the propagation speed of sound vibration in the medium, x is the distance from the point to the wave source, and ν is the vibration speed of the point in the medium. The ultrasonic wave energy passing through the unit area perpendicular to a certain direction in unit time is called sound intensity. The average sound intensity perpendicular to the direction of sound wave propagation can be proven to be [24]:

$$I = \frac{1}{2}\rho_0 A^2 \omega^2 \tag{1.3.11}$$

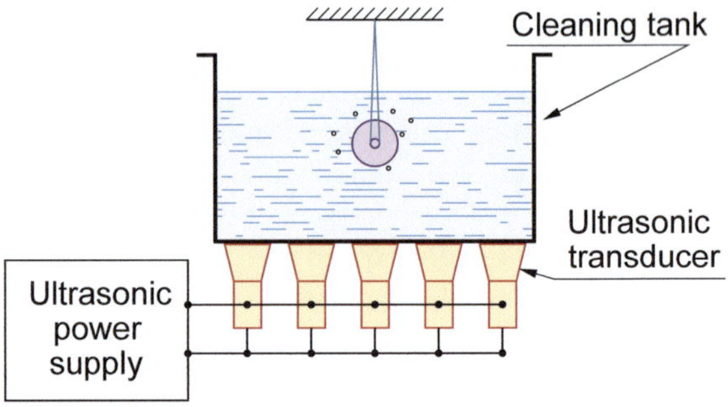

Fig. 1.5 Ultrasonic cleaning schematic diagram

III. **Composition of Ultrasonic Cleaning Equipment**

Ultrasonic cleaning equipment mainly includes ultrasonic generator, ultrasonic transducer, and ultrasonic cleaning tank, as shown in Fig. 1.5. The ultrasonic generator produces electromagnetic oscillation signals, the transducer converts the electromagnetic signal generated by the generator into its own ultrasonic vibration, thereby generating cavitation in the cleaning tank. The cleaning tank is equivalent to a container, used to place the workpieces to be cleaned and the cleaning agent. Coupled with the emulsification and neutralization effects of ultrasonic waves, it can very effectively prevent the cleaned oil from re-adhering to the cleaned workpieces.

IV. **Features of Ultrasonic Cleaning Technology**

Ultrasonic cleaning technology has many advantages. The main ones are:

1. It is particularly suitable for high-precision cleaning of workpieces with irregular surface shapes. Due to the good penetration of ultrasonic waves, it can clean the inner and outer surfaces of the workpiece, every tiny dark hole, slit, micropore, etc., so that the dirt attached to its surface can fall off. Especially for complex parts and assembly components, ultrasonic cleaning can achieve high-precision cleaning effects without causing damage to the device, safe and efficient, easy to automate. This is the most prominent advantage of ultrasonic cleaning;
2. Ultrasonic cleaning is fast and the cleaning effect is stable. Especially for complex parts, there is no need for too many cleaning steps, saving cleaning time, regardless of the shape or size of the cleaned parts, the cleaning effect can remain consistent, which is difficult to achieve with other cleaning methods;
3. Wide range of applications. Cleaning objects, from semiconductor devices, mechanical parts, to tableware, integrated circuits and fiber fabrics, etc., can all use ultrasonic cleaning. Especially for materials with strong sound reflection, such as metal, glass, plastic, etc., the cleaning effect is better. The pollutants that can be cleaned include common pollutants such as dust and oil, as well as polymers, oxides, radioactive pollutants, etc.;
4. Wide range of industry applications. Ultrasonic cleaning is widely applicable in various fields, mainly including: (1) Machinery industry: removal of rust and grease from machines, cleaning of engines, carburetors, and car parts; unclogging

and cleaning of filters and filter nets, etc.; (2) Surface treatment industry: cleaning of glasses lenses, degreasing and rust removal before electroplating or ion plating, phosphating treatment, cleaning of carbon deposits, oxide skin, polishing paste, activation treatment of metal workpiece surfaces, etc.; (3) Instrument industry: cleaning before high cleanliness assembly of precision parts; (4) Electronics industry: removal of rosin and solder spots on printed circuit boards, cleaning of mechanical electronic parts such as high-voltage contacts, high cleanliness cleaning of semiconductor chips; (5) Medical industry: cleaning, disinfection, sterilization of medical equipment, cleaning of experimental utensils, etc.;

5. Environmental protection. In many cases, water can be used as a cleaning agent, reducing environmental pollution and achieving energy conservation and environmental protection.

Ultrasonic cleaning technology also has some disadvantages. Ultrasonic waves have very strong penetration power, and if protection is not in place, it may cause harm to workers; the cleaning process is noisy; sometimes the ultrasonic cleaning use chemical cleaning agents, which may be toxic, and the chemical agents of mixed oil pollutants after cleaning need secondary treatment. In addition, it is difficult for ultrasonic cleaning technology to clean large profiles.

1.3.5 Electromagnetic Induction Cleaning Technology

Electromagnetic induction cleaning can only be used for the cleaning of pollutants (mainly coatings on the surface of metal substrates) on metal substrates. The principle is based on the eddy current skin effect to generate heat on the metal surface. Due to the difference in thermal expansion coefficients between the pollution layer and the substrate material, the pollution layer peels off. If the pollution layer is a polymer coating, the high temperature between the coating and the substrate will also reduce the coordination bond force and van der Waals force, causing the coating to fall off.

During the cleaning process, the initial stage is the induction heating stage. As shown in Fig. 1.6, the induction coil of the induction head transfers electrical energy to the metal substrate, generating eddy currents inside the metal, converting electrical energy into thermal energy. The induction heating power is related to the frequency and magnetic field strength. The larger the current in the induction coil, the larger the magnetic flux. Increasing the current and current frequency in the induction coil can increase the eddy currents in the metal and increase the induction heating power. As the substrate absorbs heat and its temperature rises, the difference in thermal expansion coefficients between the metal substrate and the pollution layer increases with the temperature, the crystal cells in the metal vibrate, and the amplitude increases with the temperature, the pollution layer will be subjected to squeezing force, and finally peel off from the metal substrate [25].

Electromagnetic induction cleaning technology equipment is small in size, flexible and convenient to use, and has a lower cost. In recent years, electromagnetic

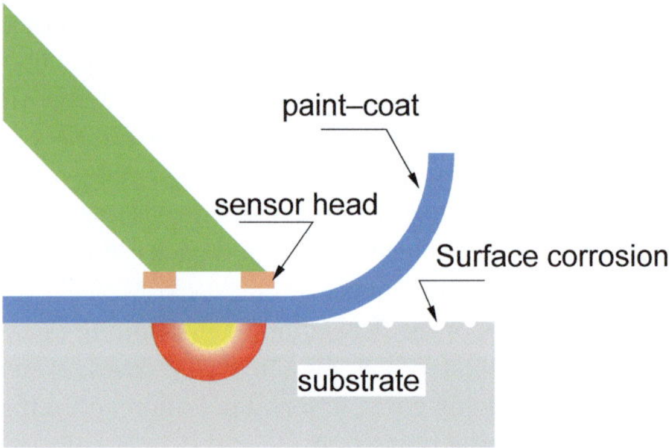

Fig. 1.6 Electromagnetic induction cleaning schematic diagram

paint removal technology has shown a broad application prospect, especially in the field of paint removal. According to relevant information, the US Navy cooperated with a Swedish company to apply electromagnetic induction technology to the rapid removal of coatings on the flight decks of warships and the removal and repair of sound-absorbing tiles on the surface of submarines. In the petrochemical industry, it is used for the cleaning of wax deposits inside oil pipelines, effectively reducing the difficulty of oil pipe cleaning, safe and efficient [26, 27].

Electromagnetic induction cleaning technology, while the efficiency of electric–magnetic conversion is still relatively low, the loss of magnetic field energy during the working process is high. In addition, during the operation process, it is necessary to protect against electromagnetic radiation.

1.4 Chemical Cleaning Technology

Chemical cleaning is a technology of cleaning that uses chemical agents to react with pollutants, causing large molecular pollutants to generate small molecular substances that can be dissolved in the cleaning liquid and detach from the surface of the cleaning object. At the same time, it may also break the chemical bonds between the external pollutants and the surface of the object to be cleaned, restoring the object to its original surface state.

Chemical cleaning is a constantly developing practical science and technology. The professional team engaged in chemical cleaning research and application is constantly expanding, and the field of chemical cleaning is also constantly expanding. It can be used in various industrial fields such as metal processing, food, textiles, petroleum, chemical, power, paper, ships, vehicles, building materials, as well as in the service industry, household life, and other aspects. It has become an indispensable link in production and life, bringing huge economic benefits to society [28].

The most common chemical cleaning uses chemical reagents as the cleaning medium, which react with the pollutants, thereby decomposing the pollutants and peeling them off from the substrate surface. Electrolytic cleaning technology is also a kind of chemical cleaning. Microbial cleaning technology can also be classified as chemical cleaning.

1.4.1 Chemical Reagent Cleaning Technology

I. **Basic Principle of Chemical Reagent Cleaning Technology**

The principle of chemical reagent cleaning is simple. According to the composition of the pollutants, appropriate chemical reagents are added. Through chemical reactions, the chemical agents decompose the pollutants into soluble salts or complexes, which are discharged with the cleaning liquid. At the same time, the reaction products may loosen and peel off the pollutants, accelerating the removal of pollutants.

II. **Selection of Chemical Cleaning Reagents**

The nature of the cleaning object and the pollutants on its surface, especially the chemical reactivity, is the main basis for selecting the type of cleaning reagent. The type, material, and composition of the pollutants of the object to be cleaned should be considered comprehensively. For example, for common scale deposit, the main component of which is calcium carbonate, belonging to alkaline, inorganic acids are usually used for cleaning [29]. Generally, there are three basic standards for selecting cleaning reagents:

(1) The cleaning liquid can quickly and effectively dissolve the pollutants, and it does not damage or corrode the base of the cleaning object, or effective protective measures can be taken (such as adding effective corrosion inhibitors) to avoid damage to the base;
(2) The cleaning reagent selected has good cleaning effect, high efficiency, and relatively cheap price;
(3) It poses no safety hazards to workers and the environment, or protective measures can be taken.

The equipment and components that need to be cleaned have many types of base materials, such as carbon steel, stainless steel, aluminum alloy, etc.; the types of pollutants mainly include rust, paint, oil pollution, scale, etc. Different chemical cleaning reagents are selected according to the base material and pollutants, summarized in Tables 1.1 and 1.2.

III. **Chemical Cleaning Process**

After determining the cleaning reagents, the correct cleaning process should also be chosen according to the actual situation. The cleaning process mainly includes immersion, circulation, and jetting methods. Immersion involves placing the object to be cleaned in a cleaning tank or filling the system to be cleaned with cleaning fluid for soaking, which can be aided by heating to promote chemical reactions. Circulation

Table 1.1 Cleaning object base material and optional cleaning agent

Base material	Optional cleaning reagents
Carbon steel	Hydrochloric acid, nitric acid, hydrofluoric acid, citric acid
Stainless steel	Nitric acid, sulfuric acid, phosphoric acid
Low chromium alloys with little or no nickel	Hydrochloric acid
Copper and copper alloys	Non-oxidizing acids
Aluminum and aluminum alloys	Weak alkali or oxidizing acids
High alloys	Hydrochloric acid, nitric acid
Cast iron	Sulfuric acid

Table 1.2 Types of pollutants and optional cleaning reagents

Types of pollutants	Optional cleaning agents
Carbonate scale	Hydrochloric acid or aminosulfonic acid
Sulfate scale	Alkali boiling followed by hydrochloric acid or aminosulfonic acid
Silicate scale	Hydrofluoric acid or caustic soda
Iron corrosion products	Hydrochloric acid or citric acid
Oil scale	Surfactants
Sludge, mud	Hydrogen peroxide or stripping reagents

involves pumping the cleaning fluid into the cleaning equipment for cyclic cleaning through the flow of the cleaning fluid [30]. Jetting uses a high-pressure pump to spray the cleaning fluid to the area that needs to be cleaned, combining the characteristics of high-pressure jet and chemical reaction. It has the advantages of less cleaning reagent usage and targeted cleaning, and is often used in the cleaning of large equipment, such as tank trucks, ship cabins, etc.

1.4.2 Electrolytic Cleaning Technology

Electrolysis refers to the process of chemical decomposition of substances under the action of electric current. Electrolytic cleaning technology is a technique that uses electrolysis to remove contaminants from the metal surface. Depending on the type of contaminants, it is divided into electrolytic degreasing and electrolytic grinding cleaning.

The electrolyte is placed in the electrolytic tank, and the metal with surface contaminants is placed in it. When the power is connected, the metal base acts as one of the electrodes. When the current passes through the electrolyte, the following chemical reaction occurs:

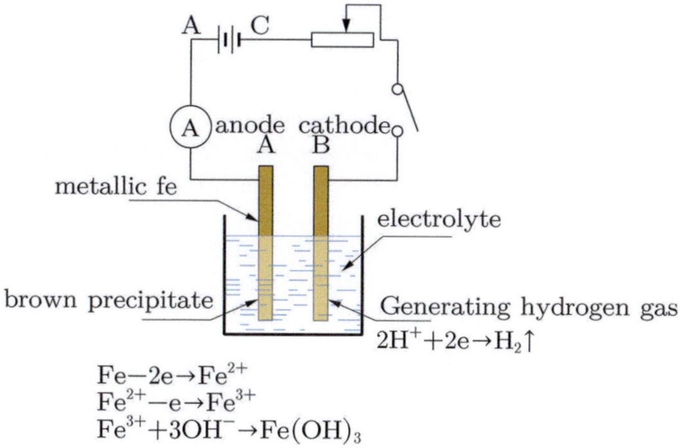

Fig. 1.7 Schematic diagram of electrolysis device

$$2H_2O = 2H_2 \uparrow \text{ (negative electrode)} + O_2 \uparrow \text{ (positive electrode)}$$

As a result, oxygen is generated near the positive electrode of the electrolytic cell, and hydrogen is generated at the negative electrode. These tiny bubbles promote the detachment of dirt from the surface of the object being cleaned. Figure 1.7 shows the schematic diagram of electrolytic device.

When the contaminants on the metal surface are greasy, alkaline aqueous solutions such as sodium hydroxide and sodium carbonate are usually used as electrolytes. These strong alkali electrolyte ions increase the conductivity of water and promote the removal of contaminants. However, if it is a mineral oil type of grease, because sodium hydroxide and other alkaline substances have weak dispersion ability for mineral oil dirt, it is necessary to add sodium metasilicate and a small amount of surfactant to the solution to facilitate the dispersion and removal of mineral oil dirt. When removing rust from the metal surface, the bubbles generated on the electrode are also used to peel off the surface rust, but the electrolyte used may be different.

In electrolytic cleaning, the current can also convert some of the chlorine ions in the water into free chlorine (with residual chlorine in the water up to 0.5 ppm), while generating strong oxidants such as ozone, oxygen radicals, hydroxyl radicals, and hydrogen peroxide, which can achieve sterilization function.

1.4.3 Microbial Cleaning Technology

Microbial cleaning technology is a technique that uses the cells within microbes to decompose contaminants adhering to the surface of the object being cleaned, thereby transforming them into non-toxic, harmless, water-soluble substances. Microbial cleaning technology is also known as biological cleaning technology. Its principle is actually a chemical reaction, so it can be classified as chemical cleaning [31].

Microbial cleaning technology requires the use of enzymes, which are microbial catalysts that can thoroughly decompose pollutants. Enzyme preparations were initially extracted from animal organs and the seeds and fruits of higher plants, typically including trypsin, papain, bromelain, etc. The enzymes used for cleaning can be produced within microbial cells and have special cleaning functions. Generally, there are eight types of microbial catalysts within microbes, with the main four being protease, amylase, lipase, and cellulase. Generally, their catalytic reactions are 10^6–10^{12} times faster than non-enzyme catalysts [32].

Common uses of biological cleaning include washing clothes with laundry detergent, brushing teeth with toothpaste, industrial pipe descaling, and fabric desizing. Sweat, blood stains, and food oil stains on clothes can be greatly improved by using detergent with protease, extending the life of the fabric. Brushing teeth with toothpaste can effectively remove tartar. Drain pipes are often blocked by fruits, vegetables, leftovers, or scale, rust, oil stains, and mud. Adding microbial cleaning reagents (multiple enzymes) can improve the cleaning effect. Clothing and fabrics such as cotton, rayon, viscose, silk, and blended fabrics, especially colored silk and synthetic blend clothes that cannot be desized with alkali, can be cleaned with microbes to achieve efficient and damage-free desizing. The fabric after desizing feels soft and has a strong gloss.

The method of using microbes to degrade organic matter used in microbial cleaning technology is also widely used in the treatment of water pollution and some land pollution. Biodegradation uses natural microbes such as yeast and fungi to decompose or degrade toxic substances into low-toxic or non-toxic substances. The characteristics of microbial cleaning technology are of high efficiency, strong specificity (selectivity), mild reaction conditions, and environmental friendliness. However, its usage is relatively limited.

1.4.4 Features of Chemical Cleaning Technology

Chemical cleaning technology has many advantages and is widely used. Its main features are:

1. Fast cleaning speed and strong removal ability. As long as the selected chemical reagent is suitable, it has a strong effect on pollutants, and the chemical reaction (including electrolytic reaction) is rapid.
2. Easy to use and good cleaning effect. Chemical cleaning uses liquid, which is convenient to use, does not require complex equipment, and can flow through any part that needs to be cleaned, causing a chemical reaction, and there will hardly be any places that cannot be cleaned.
3. Wide range of applications. The liquid chemical reagent has good fluidity and strong penetration, suitable for cleaning objects with complex shapes, and there

will be no “dead corners” that cannot be cleaned. According to the cleaning object and pollutants, suitable chemical cleaning reagents can be found. Therefore, it is widely used in life and various productions.

4. Damage to the base can be controlled. In most cases, as long as the cleaning reagent and corrosion inhibitor are properly selected, the damage to the base material can be controlled to a very small extent, far less than the damage caused by mechanical descaling to the metal.

The disadvantages of chemical cleaning technology are also obvious. The main ones are:

1. Chemical substances are toxic and harmful. Various chemicals used in the cleaning process, some of which are flammable, explosive, poisonous, and corrosive such as acids and alkalis, should be stored and used with caution, otherwise accidents may occur.
2. It is easy to cause secondary pollution. Whether the chemical reagent is an acid cleaning solution, an alkali cleaning solution or a passivation solution, the waste liquid after cleaning cannot be discharged at will, but needs to undergo a strict and complex treatment process, otherwise, it will cause secondary pollution to the environment.
3. It may cause damage to the base. When the chemical cleaning solution is improperly selected, it can corrode and damage the cleaning object base, causing loss. Some chemical reagents penetrate into the base, there may not be any reaction at the time, but perhaps under certain environmental conditions, after a certain period of time, it will cause damage to the base material.

1.5 Mixed Cleaning Technology

1.5.1 Plasma Cleaning Technology

Plasma cleaning is a special cleaning technology. Its cleaning principle has both physical and chemical principles, so we list it as a separate section for introduction.

I. Overview of Plasma Cleaning Technology

Plasma is a kind of electrically neutral, high-energy, fully or partially ionized gaseous substance produced by high-frequency electromagnetic oscillation, radio frequency or microwave, high-energy radiation, corona discharge, laser, high temperature, etc., containing ions, electrons, free radicals and other active particles. Plasma cleaning is a process in which plasma, under a certain vacuum state, uses its active particles to chemically or physically interact with pollutant molecules to clean the surface of the object, effectively removing micro-dust and other pollutants from the material surface without damaging the material surface characteristics. It can achieve molecular-level pollutant removal (generally thickness is 3 ~ 30 nm), and improve the cleaning process of the workpiece surface activity [33]. The structure of the plasma cleaning machine mainly consists of three parts: control unit, vacuum chamber, and vacuum pump.

Plasma cleaning technology is commonly used in the cleaning process of integrated circuits and packaging. There are various kinds of pollutants in the manufacturing and packaging process of integrated circuits, including fluorine, nickel, photoresist, epoxy resin, and oxides, etc. The existence of pollutants will affect the quality of electronic products. As a precise dry cleaning method, microwave plasma cleaning can effectively remove these pollutants, improve material surface performance, and enhance material surface quality. With the rapid development of semiconductor technology and optoelectronic materials, the demand for plasma cleaning will become increasingly large.

II. **Principle of Plasma Cleaning Technology**

Plasma cleaning involves principles such as plasma physics, plasma chemistry, and chemical reactions at the gas–solid interface. Commonly used plasmas are divided into three types according to their excitation frequency: ultrasonic plasma with an excitation frequency of 40 kHz, radio frequency plasma with an excitation frequency of 13.56 MHz, and microwave plasma with an excitation frequency of 2.45 GHz [34]. Generally speaking, the choice of gas type can greatly affect the working mechanism. Ultrasonic plasma mainly involves physical reactions, radio frequency plasma involves both physical and chemical reactions, and microwave plasma mainly involves chemical reactions.

1. Plasma cleaning with chemical reactions as the main surface reaction

A typical plasma chemical cleaning process is oxygen plasma cleaning. Figure 1.8 simply describes the mechanism of plasma cleaning, which mainly relies on the "activation effect" of active particles in the plasma to achieve the purpose of removing surface stains. The oxygen radicals produced by the plasma are very active and easily react with hydrocarbons to produce volatile substances such as carbon dioxide, carbon monoxide, and water, thereby removing surface contaminants.

When the pressure in the vacuum chamber reaches a certain range, the process gas such as oxygen, argon, methane, and hydrogen is filled in. When the pressure in the chamber is dynamically balanced, the high-frequency alternating electromagnetic field generated by the microwave source oscillation ionizes the process gases to generate plasma. The equation is $O_2 + e^- \rightarrow 2O + e^-$. The free radicals in the plasma further react chemically with the material surface. $O^* + organic(pollutant) \rightarrow$

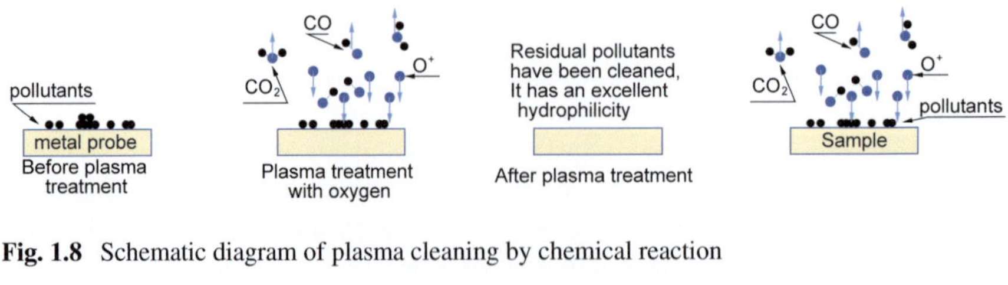

Fig. 1.8 Schematic diagram of plasma cleaning by chemical reaction

$CO_2 + H_2O$. The greater the pressure, the more favorable the generation of free radicals. So it is necessary to control a higher pressure to promote the activity of the chemical reaction. The material on the surface of the cleaned object is turned into particles and gaseous substances through chemical reactions, and is discharged after vacuum extraction, thereby achieving the purpose of cleaning.

As can be seen from the reaction formula, oxygen plasma can turn non-volatile organic matter into easily volatile water and carbon dioxide through chemical reactions, making it easy to leave the substrate surface [35].

2. Plasma cleaning with physical reactions as the main surface reaction

The physical reaction in plasma cleaning mainly uses the ions in the plasma to make pure physical impacts, knocking off the atoms on the material surface or adhering to the material surface. A typical plasma physical cleaning process is to add inert gas argon to the reaction chamber as an auxiliary. Argon plasma does not react with the surface material, and Ar^+ is accelerated under self-bias voltage or externally applied bias voltage to generate kinetic energy, and then bombards the surface of the cleaned workpiece placed on the negative electrode. It is generally used to remove oxides, epoxy resin overflow or micro-particle contaminants, and at the same time activate the surface energy. As shown in Fig. 1.9.

$$\mathrm{Ar} + e^- \rightarrow \mathrm{Ar}^+ + 2e^-,\ \mathrm{Ar} + pollutants \rightarrow volatilized\ pollutants$$

3. Both physical and chemical reactions play important roles in surface reactions

In plasma cleaning, both physical and chemical reactions sometimes play important roles. The above-mentioned physical and chemical processes coexist, that is, the plasma physically bombards the object being cleaned, and at the same time, the free radicals in the plasma react chemically with the surface of the material. Under the dual action of physics and chemistry, the purpose of cleaning is achieved. The effect of plasma cleaning is closely related to its process parameters. The mechanism of plasma cleaning should be selected according to the material of the cleaning material, and appropriate process parameters should be set.

III. **Characteristics of Plasma Cleaning Technology**

Plasma cleaning is a precise and effective cleaning process with characteristics not found in other cleaning methods.

1. Plasma cleaning technology is applicable to a variety of substrate types, including metals, semiconductors, oxides, and most high polymer materials, such as polypropylene, polyester, polyimide, polyvinyl chloride, epoxy, and even polytetrafluoroethylene can be well treated [36, 37]. It can achieve overall, local,

Fig. 1.9 Schematic diagram of plasma cleaning by physical reaction

and complex structure cleaning, and can also clean the inside of fine pores and depressions;

2. Plasma cleaning is carried out in a vacuum, does not pollute the environment, and ensures that the cleaning surface is not secondarily polluted. Plasma avoids the use of acids, alkalis, or organic solvents in wet chemical cleaning, and there is no need to transport, store, and discharge cleaning fluids. It is safe and reliable, produces no waste liquid, and is beneficial to environmental protection;
3. Plasma cleaning is easy to implement automation with numerical control technology, has high-precision control devices, and correct plasma cleaning will not damage the material surface;
4. Plasma cleaning is efficient. The entire cleaning process can be completed within a few minutes, and the cleaned object is dry after plasma cleaning, and does not need to be dried before being sent to the next process, which can improve the efficiency of the entire process line;
5. Plasma cleaning can achieve high cleaning quality, good uniformity and repeatability, and can also improve the surface properties of the material itself.

However, plasma cleaning technology has high operational requirements, and its cleaning scope is mainly limited to precision cleaning.

In summary, the unique superiority of plasma cleaning is gradually being recognized by various industries at home and abroad, and it has gradually begun to be applied in semiconductor manufacturing, microelectronics packaging, surface activation of plastics and ceramics, precision machinery, rubber and plastics, medical instruments, automobiles, textile fibers and other industries [38, 39].

1.5.2 Thermal Cleaning Technology

The so-called thermal cleaning technology is to use thermal energy to remove contaminants from the surface of the object being cleaned, mainly based on physical principles. Thermal cleaning technology is often used in conjunction with other technologies such as chemical cleaning technology, so chemical principles also play a role [40, 41].

I. **Principle of Thermal Cleaning**

1. Using the Mechanism of Physical Action

For different substances, the melting point, thermal conductivity, and thermal expansion coefficient vary. The cleaning object and pollutants, after absorbing heat, have different temperature rise ranges, different melting temperatures, and generate different thermal stresses. If during the temperature rise, the change in stress causes the force between the pollutant and the cleaning object to decrease and fall off, or the pollutant melts and flows away from the cleaning object, a cleaning effect can be achieved.

2. Promoting Chemical Reactions

For ordinary chemical reactions, generally speaking, for every 10 °C increase in temperature, the reaction rate can approximately double. Therefore, increasing the temperature can significantly speed up chemical reactions. In chemical cleaning, thermal cleaning technology is often used. Some pollutants are easier to decompose when heated, and some cleaning reagents' dissolution speed and amount of dirt increase proportionally with temperature. Therefore, increasing the temperature is beneficial for the solvent to exert its dissolving effect and remove the pollutants and cleaning medium adsorbed on the surface of the cleaning object.

II. **Features of Thermal Cleaning Technology**

Thermal cleaning technology is simple, convenient, and fast. It can be used alone or in combination with other cleaning technologies. It should be noted that when the cleaning object is a metal material, oxides will form on the surface during the heating process, especially when water or aqueous solution is used as the cleaning medium. Therefore, when using heating to remove organic dirt, the effect of the newly formed oxide film on the metal should be considered.

Steam cleaning technology can also be classified as thermal cleaning. In addition to the heat brought by high-temperature steam itself, the heat released by the steam during cooling will be absorbed by the cleaning object, pollutants, and cleaning medium, thereby generating physical and chemical effects [42].

1.6 Laser Cleaning Technology

Laser cleaning was born with the development of laser technology. As the requirements for environmental protection become higher and higher, the research and application of laser cleaning are getting more and more attention. Laser cleaning can be classified as a physical cleaning technology, but due to its unique characteristics and future development prospects, we will introduce it separately here.

In 1965, Sholokhov [38] first proposed the concept of "laser eraser". By quickly heating the target material, laser pulses can quickly vaporize the ink (carbon ink pigment) on the surface of the paper without damaging the paper itself, as shown in Fig. 1.10.

In 1969, Bedair and others [43] used a Q-switched laser to remove oxygen and sulfur pollution on the surface of nickel, studied the damage of the laser to the sample surface, and the experiment showed that high-power Q-switched lasers can be used to clean surfaces. However, due to inappropriate laser parameters, the substrate under the action of a laser with a power density of 100 MW/cm^2, the surface suffered irreversible damage. Bedair and other researchers first used the term "laser cleaning", and summarized the advantages and disadvantages of laser cleaning in the article.

Subsequently, people have conducted related research on laser cleaning technology for various pollutants, such as laser cleaning of artworks, laser removal of oil

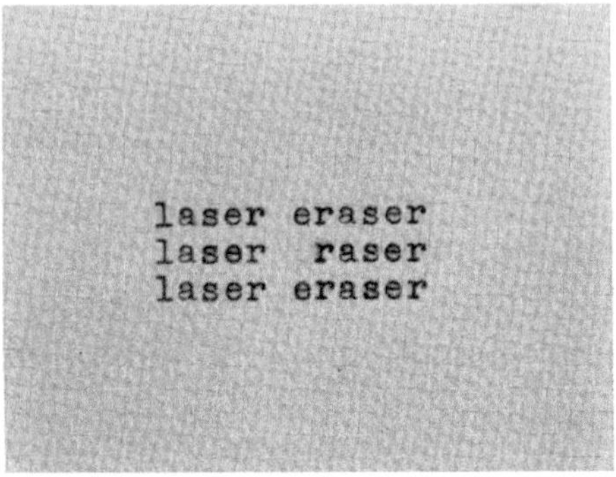

Fig. 1.10 Laser erase—laser irradiation to remove ink

floating on water, laser cleaning of tiny particles on electronic circuit boards, laser paint removal from aircraft skin and other materials, and laser rust removal from metal materials, developing related principles [44].

I. **Concept of Laser Cleaning Technology**

Laser cleaning uses laser as a medium, through the interaction of light and matter (including base materials and pollutants), to achieve the purpose of removing pollutants. In conventional physical cleaning technologies, mechanical methods (such as shot blasting) cannot meet the requirements for high cleanliness, and although ultrasonic cleaning has good effects, it is powerless for sub-micron particles, and the size of the cleaning tank limits the range and complexity of the parts to be processed; chemical cleaning methods can easily cause environmental pollution. Laser cleaning is a new surface cleaning technology that is more in line with the current needs for environmental protection and sustainable social development. Since its birth in the last century, it has made rapid progress in theory and technology in just a few decades. Various types of laser cleaning equipments have been developed and successfully applied in industrial cleaning, achieving good benefits.

II. **Features of Laser Cleaning Technology**

Compared with traditional cleaning technology, laser cleaning technology has the following advantages:

(1) In most cases, it can adopt "dry" cleaning, that is, it does not need cleaning fluid or other liquids, and the cleaning medium is the laser itself;
(2) Using the characteristics of lasers, it can accurately locate the parts to be cleaned, achieving precise cleaning;
(3) By controlling laser process parameters, pollutants can be effectively removed without damaging the surface of the base material, restoring the surface to its original state;
(4) It can effectively remove extremely small pollutant particles. Some pollutants may be extremely fine, reaching the micron or even sub-micron level, with strong adsorption, and it is difficult for methods other than laser cleaning to remove these pollutants;

(5) It can easily achieve automated operation, real-time monitoring, and timely feedback;
(6) The range of pollutants removed and the range of applicable base materials are very wide;
(7) The equipment can be used for a long time, with low operating costs;
(8) It is a "green" cleaning process, the eliminated waste is mainly solid powder, small in volume, easy to recycle and process, and basically does not pollute the environment.

Laser cleaning also has some shortcomings: the cost of laser cleaning equipment is relatively high; lasers are a high-tech technology, and a certain amount of specialized knowledge is required to operate; lasers have a certain harmful effect on the human body, and protection should be taken into account when using.

III. **Classification of Laser Cleaning Technology**

There is no fixed unified standard for the classification of laser cleaning technology internationally. Generally speaking, from whether auxiliary materials are used, it can be divided into several types:

1. Dry Laser Cleaning (DLC) Method: The laser directly irradiates the surface of the object. After the pollutant particles or film layer absorbs the laser energy, the particles leave the surface through mechanisms such as thermal diffusion, photodecomposition, vaporization, vibration, etc. Figure 1.11 is a schematic diagram.
2. Wet Laser Cleaning (WLC) Method, or Steam Laser Cleaning (SLC) Method: A non-polluting liquid (mainly water, sometimes alcohol) is sprayed onto the surface of the material to be cleaned, and then irradiated with a laser. After the liquid medium or substrate material absorbs the laser energy, it explosively vaporizes, pushing the surrounding contaminant particles away from the material surface. Figure 1.12 is a schematic diagram.
3. Laser Compound Cleaning Method: The laser energy is absorbed by the substrate material, and then through thermal convection, the adsorbed intermediate medium is heated. The intermediate medium explosively vaporizes, and

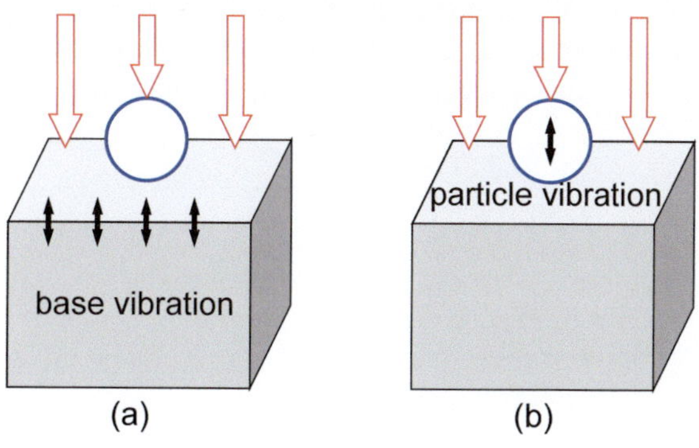

Fig. 1.11 Schematic diagram of laser dry cleaning

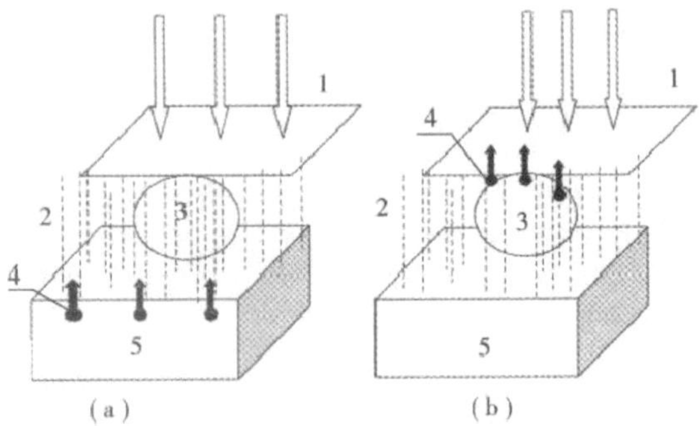

Fig. 1.12 Laser wet cleaning schematic diagram. *1* Laser beam, *2* liquid film, *3* interaction zone, *4* vaporization, *5* base material

under the push of high airflow, the pollutants are detached from the substrate surface along with the intermediate medium. For example, the method of laser plus inert gas, that is, inert gas is blown onto the workpiece surface while the laser is radiating. When the dirt is peeled off from the surface, it is blown far away from the surface by the inert gas, avoiding re-contamination and oxidation of the clean surface.

4. Laser Hybrid Cleaning Method: After the dirt is loosened by the laser, non-corrosive chemical methods are used to remove the contaminants.

Among the above methods, the first two are the most basic laser cleaning methods, and the last two are compound or hybrid methods combined with laser cleaning. At present, the first three cleaning methods are mainly used in industrial production, among which dry laser cleaning and wet laser cleaning are used the most.

IV. Brief Introduction to Laser Cleaning Equipment

Laser cleaning equipment mainly consists of the following parts: laser, power supply, water cooling, beam shaping and transmission, control unit, workbench, detection unit, recovery unit, etc., as shown in Fig. 1.13. The laser is the main body, the control unit controls the operation of the entire equipment through a computer or a single-chip microcomputer, and the detection system is used to check whether it is cleaned. If it is offline detection, this unit can be omitted.

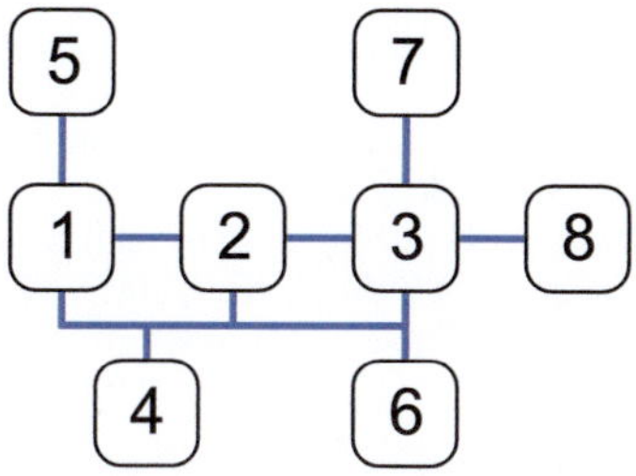

1--Laser, 2--Beam Shaping and Transmission, 3--Workbench, 4--Power Supply, 5--Water Cooling, 6--Control Unit, 7--Detection Unit, 8--Recycle Unit

Fig. 1.13 Laser cleaning equipment diagram. *1* Laser, *2* beam shaping and transmission, *3* workbench, *4* power supply, *5* water cooling, *6* control unit, *7* detection unit, *8* recycle unit

References

1. Zhao Y H. Chemistry washes application in machinery washed of technology [J]. Coal Mine Machinery, 2006, 27(8): 136–138.
2. Wei M, Ji X C, Zhou X Y, Song Z Y, Zhao K K, Wang Y. Research on quick cleaning and pretreatment of equipment surface by micro-blasting technology [J]. Chinese Materials Science Technology & Equipment, 2012, 1: 24–27.
3. Chen Y Z, Hu Y H. Research and application of ship hull fouling cleaning technologies [J]. Surface Technology, 2017, 46(10): 60–71.
4. Xue W J, Liu L S, Wang K Y, Wang L, Yang Y P. Application of shot peening technology and its development [J]. Material Protection, 2014, 47(05): 46–49.
5. Zhou Y W, Xu B C, Han F. China's industrial cleaning technology application status and development trend [J]. China Cleaning Industry, 2010, 1: 33–36.
6. Wu X Y. Analysis of high-pressure water jet cleaning in ship building and repairing application [J]. Cleaning World, 2008, 24(9): 28–30.
7. Liu P, Huang Y Z. Application and Development of Water Jet Technology [J]. Coal Mine Machinery, 2009, 30(9): 10–12.
8. Jin Y B, Dong K Y. Introduction of applications and market prospect analysis of high-pressure water cleaning equipment [J]. Cleaning World, 2015, 31(1): 30–34.
9. Xue S T, Wang Y Q, Yu L, Zhu H Q. Engineering application of ultra-high pressure and high-power water jet technology in China [J]. Cleaning World, 2005, 21(2): 4–9.
10. Zhang J F, Gao S B, Chang T, Wu J L. Study of high-pressure water-jets cleaning technology [J]. Development & Innovation of Machinery & Electrical Products, 2011, 24(5): 24–26.
11. Zheng L, Li M J. Cleaning principle of dry ice cleaning machine and using cautions [J]. Cleaning World, 2015, 31(2): 15–18.
12. Dai W K. Using dry ice to clean ceramic cultural relics [J]. Sciences of Conservation and Archaeology, 2015, 27(1): 116–120.
13. Liu S N, Ma C R, Su W, Wei Z F. Flow field simulation of quadrate dry-ice blasting nozzle for substation insulators [J]. Electric Power, 2014: 47(6): 75–79.
14. Liu M, Wang J L, Ma X L, Yu J H, Huang Y J, Zhou W J. Cleaning experiment of substation insulators by dry ice [J]. High Voltage Engineering, 2011, 37(7): 1649–1655.
15. Guo X H, Wang L, Jing Y P. Carbon dioxide snow jet cleaning technology [J]. Micronanoelectronic Technology, 2012, 49(4): 258–262.
16. Zheng J J, Zou J M, Jiang F, Yu J H, Han J J, He X X, Xu H Z. Vehicle-based dry ice cleaning system for live-line insulators in substations of 220 kV [J]. Electric Power, 2014, 47(4): 118–122.

17. Li W L. Principle and process of ultrasonic cleaning [J]. Journal of Shandong Industrial Technology, 2018, 12: 48.
18. Wang W H, Du H L. Development of Automatic Ultrasonic Industrial Cleaning Machine [J] Machine Tool & Hydraulics, 2019, 47(2): 49–53.
19. Li Q, Lu L H, Ding A J. Design of cleaning machine and cleaning effect research based on ultrasonic [J]. Physical Experiment of College, 2017, 30(5): 31–35.
20. Ni Z H, Wang G, Ma J, Bai J Y, Qian J. Application of ultrasonic cleaning technology in electrode cleaning of Zn-AgO batteries [J]. Chinese Journal of Power Sources, 2018, 42(5): 674–677.
21. Hua Y K. Ultrasonic cleaning technology in the application of precision instruments [J]. China Science & Technology Overview, 2018, 22: 80–82.
22. Guo H M. Research on the distribution law of ultrasonic cleaning field intensity [D]. Qinhuangdao: Yanshan University, 2018.
23. Zhai F L. The study on complex ultrasound field measurement and modeling methods [D]. Lanzhou: Lanzhou University, 2015.
24. Jiang W P, Fang J. Ultrasonography [M]. Wuhan: Wuhan University of Surveying and Mapping Science and Technology Press, 1991: 1200.
25. Bai Q L, Ma E T, Bai L J, Zhang H K. Application of induction heating technology for cleaning used oil pipe [J]. Oil Field Equipment, 2004, 33: 54–56.
26. Zhang J W, Zhao J X, Fang Q C, Lu Y F. The application of electromagnetic induction heating and atmospheric pressure cleaning technology for oil pipeline [J]. China Petroleum Machinery, 2003, 31(9): 60–61.
27. Zhao X F, Li S B, Zhang H L, Luo X L, Tang C H. Research on cleaning process and device for atmospheric pressure electromagnetic induction heating oil pipe [J]. Drilling & Production Technology, 2007, 33(3): 85–87.
28. Liu L, Han Z Y. Chemical cleaning of a 300 MW coal-fired power plant [J]. Cleaning World, 2019, 35(1): 1–3.
29. Liu L J. Application of a sulfate detergent in chemical cleaning of calcium sulfate scale in reverse osmosis membrane system [J]. Cleaning World, 2019, 35(2): 12–17.
30. Huang H, Cang M J, Wang G. Chemical cleaning of a 600 MW subcritical controlled cycle operating unit [J]. Cleaning World, 2018, 34(12): 12–17.
31. Zhang D X. Microbiological cleaning technology [J]. Cleaning World, 2004, 20(3): 32–34.
32. Mcnally T W, Guo Z B. Exploration of biological cleaning technology- microbial technology opens up a new field of parts cleaning [A]: Proceedings of the 2003 International Forum on Cleaning Technology [C]. Beijing: 2003 International Forum on Cleaning Technology, 2003: 113–122.
33. Jia C X, Wang Q, Pu Y W. Plasma Cleaning Technology and Its Application in Composites Filed [J]. Aeronautical Manufacturing Technology, 2016, 18: 95–108.
34. Yang L, Wu A D, Lu L, Li Y M, Xu X B, Chen L, et al. Research on plasma discharge for plasma processing of a 1.3 GHz single-cell SRF cavity [J]. Nuclear Physics Review, 2018, 35(3): 278–286.
35. Cheng P J, Li H. Plasma Cleaning Process in the Application of Electroacoustic Device Package [J]. Equipment for Electronic Products Manufacturing, 2016, 45(9): 28–33.
36. Sha C P, Lu S W, Zhao X Y, Jia C X. Applications and prospects of plasma cleaning in the aerospace manufacturing industry [A]: Proceedings of the 19th National Conference on Paint Coating and Surface Protection [C]. Chongqing: The 19th National Conference on Paint Coating and Surface Protection, 2017: 13–16.
37. Kumar A, Bhatt R B, Behere P G, Afzal M, Kumar A, Nilaya J P, et al. Laser-assisted surface cleaning of metallic components [J]. Pramana-Journal of Physics, 2014, 82(2): 237–242.
38. Schawlow A L. Lasers [J]. Science, 1965, 149(3679): 13–22.
39. Raza M S, Das S S, Tudu P, Saha P. Excimer laser cleaning of black sulphur encrustation from silver surface [J]. Optics and Laser Technology, 2019, 113: 95–103.
40. Bandara P C, Ferrara F I, Nguyen H, Santos G, Shih W C, Rodrigues D F. Investigation of thermal properties of graphene-coated membranes by laser irradiation to remove biofoulants [J]. Environmental Science & Technology, 2019, 53(2): 903–911.

41. Li H S, Yang H W, Wang Z B, Li D M, Song Q. Research and application of dry medium frequency heating cleaning technology for oil pipes [J]. Industry, 2015, 19: 195.
42. Wang Z L, Zhang J, Li R L, Ji S C. Technique of cleaning by water vapor for live power equipment [J]. Insulating Materials, 2009, 42(3): 69–72.
43. Bedair S M, Smith H P. Atomically clean surfaces by pulsed laser bombardment [J]. Journal of Applied Physics, 1969, 40(12): 4776–4781.
44. Tian Z, Lei Z L, Chen X, Chen Y B. Evaluation of laser cleaning for defouling of marine biofilm contamination on aluminum alloys [J]. Applied Surface Science, 2020, 499: 144060.

Chapter 2
Common Lasers in Laser Cleaning

Laser cleaning uses laser to irradiate the substrate material with attached contaminants. Through the interaction of light and matter, the contaminants are removed. So, what is the laser? What lasers are used in laser cleaning? Why choose these lasers? What laser technology needs to be used for the lasers used? In this chapter, we first introduce the working principle of lasers and the basic characteristics of lasers, then introduce the main characteristics of lasers and the widely used laser technology, and finally introduce the main lasers used in laser cleaning.

2.1 Basic Knowledge of Lasers

2.1.1 Laser

Laser is widely considered as one of the most important achievements in the history of human technology. Laser stands for Light Amplification by Stimulated Emission of Radiation.

The world's first laser was born in 1960. Theodore Maiman of the United States successfully invented a ruby laser, and since then, research and development of lasers have quickly followed. Today, not only is research on lasers very active in laboratories and scientific research, but it is also widely used in many fields such as industry, agriculture, military, and biomedical, such as laser cutting, laser welding, laser drilling, laser ranging, laser breeding, laser guidance. In our daily life, laser is everywhere, such as laser cosmetic, laser printers, laser copiers, laser communication. It can be said that laser has penetrated into many areas of society and have become indispensable in human production and life. Laser is one of the greatest advancements in technological history and is hailed as the "light of the century" of the twentieth century [1–4].

F. Song and X. Lin, *Laser Cleaning*, https://doi.org/10.1007/978-981-97-6631-4_2

Laser is also a type of light, but it is very different from the common light sources we see daily, such as sunlight, LED light, candlelight, etc. Compared with ordinary light sources, lasers have unique characteristics. Specifically, they have the following outstanding features.

1. Good Monochromaticity

The color of the laser is very pure, with an extremely narrow frequency range, which in professional terms, means good monochromaticity. This allows energy to be concentrated in a very narrow wavelength or frequency range. The sunlight we commonly see has a wide spectral wavelength or frequency range, covering areas from ultraviolet to infrared.

2. Good Directionality

The light emitted by a laser source has a very small divergence angle, generally only a few milliradians, and it appears almost as a straight line. The flashlight we commonly use usually diverges at a few meters away, and the searchlight, after propagating tens or at most hundreds of meters, also diverges. When a laser is emitted to the moon, after a journey of 384,000 km, the diameter of its diffused light spot is only about 2 km.

3. Extremely High Brightness

The laser beam is a highly parallel beam of light, with the energy of the laser concentrated in a very small light spot. The laser can produce extremely narrow light pulses. Using mode-locking technology, it can produce light pulses of 10^{-14} s. Because the energy is emitted in a very short time, the light power is extremely high. It can be said that the laser is currently the brightest light source. The brightness of a giant pulsed ruby laser can reach 1 $\mathrm{kW}/\left(\mathrm{cm}^2\ \mathrm{sr}\right)$, which is many times higher than the brightness of the solar surface.

4. High Coherence

Coherence includes temporal coherence and spatial coherence. The radiation mechanism of a laser light source is different from that of a common light source. It is stimulated radiation, and the photons it emits are consistent in phase. In addition, the mode selection effect of the resonant cavity makes the laser beam have a fixed phase relationship between each point on the cross-section, so the spatial and temporal coherence of the laser is very good. The ordinary light produced by spontaneous radiation is incoherent light.

2.1.2 *Working Principle of Laser*

The reason why the laser has these characteristics is related to its generation mechanism.

I. **Energy Levels**

According to quantum mechanics, microscopic particles (atoms, molecules, or ions. For simplicity, they are uniformly written as particles or atoms) have a discrete energy level structure. The lowest energy level in an atom is called the ground state, while the rest are called excited states. Atoms can only exist on these discrete energy levels. Usually, in a thermally balanced atomic system, the distribution of the number of atoms according to energy levels follows the Boltzmann distribution rate. A large number of atoms are in the ground state, and the number of atoms at higher energy levels is always less than the number of atoms at lower energy levels. When it absorbs external energy, it can transition to the excited state. For example, when an atom absorbs radiation with a frequency of ν, it will transition from a lower energy state (its energy is E_1) to a higher energy state (its energy is E_2). When transitioning from a higher energy state to a lower energy state, it will emit a photon of a certain wavelength (or frequency) (called light radiation).

II. **Stimulated Absorption, Spontaneous Radiation, and Stimulated Radiation**

In 1916, Einstein published "The Quantum Theory of Radiation", which proposed the core foundation of laser theory—laser radiation theory, so Einstein is considered the father of laser theory.

Stimulated absorption: It refers to the transition of particles at a lower energy level to a higher energy level after absorbing external photons of a certain frequency (as shown in Fig. 2.1). The energies of the lower and higher energy levels are E_1 and E_2 respectively. The frequency of the absorbed photons must satisfy: $h\nu = E_2 - E_1$, where h is the Planck constant [5].

Spontaneous emission: Particles in a higher-energy state are unstable and cannot stay for too long. They will spontaneously return to a lower-energy state and emit photons. When transitioning from a higher-energy state to a lower-energy state, they will emit photons at a frequency of ν, and the frequency ν satisfies $h\nu = E_2 - E_1$, as shown in Fig. 2.1b. The mechanism of light emission from ordinary light sources is spontaneous emission, which is random and spontaneous. Each particle's transition is spontaneous and independent, and the process is not affected by the outside world. There is no correlation between them, and in professional terms, there is no coherence between photons.

Stimulated emission: Particles at a higher energy level (energy E_2), under the action of external photons with a certain frequency ν (frequency ν satisfies $h\nu = E_2 - E_1$), transition to a lower energy level (energy E_1), while emitting a photon with the same frequency as the external photon, as shown in Fig. 2.1c. Unlike spontaneous emission, particles at higher energy levels are "stimulated" or "induced"

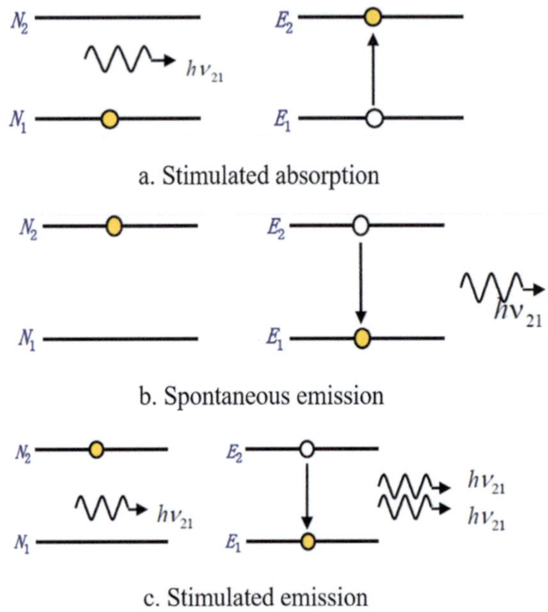

Fig. 2.1 Process of stimulated absorption, spontaneous radiation and stimulated radiation of light

by special photons to transition to lower energy levels, and emit a photon with the same frequency as these "stimulated" photons. Through one stimulated emission, the photon released by the particle is exactly the same as the original "stimulated" photon in frequency, direction, phase, etc. Stimulated emission is the light emission mechanism of laser.

III. **Population Inversion, Pumping, and Active Medium**

In a certain material, spontaneous emission, stimulated emission, and stimulated absorption always exist simultaneously in the interaction between light and the atomic system. In a thermal equilibrium state, the number of particles at lower energy levels is always more than the number of particles at higher energy levels. Therefore, when a beam of light is incident on this material, stimulated absorption will generally occur, and the particles at the higher energy level often randomly spontaneously emit back to the lower energy states. To get laser, stimulated emission must be formed, and various losses must be overcome. That is to say, in order to achieve stimulated emission, there is a threshold. Only when the inverted particle number (the number of particles at the upper energy level minus the number of particles at the lower energy level) exceeds the threshold, it is possible to generate the laser. This requires making the number of particles at higher energy level more than the number of particles at lower energy level, that is, the so-called realization of particle number inversion (population inversion).

So, how can we achieve a state of population inversion? We must use certain methods to excite the atomic system, increasing the number of particles at the upper energy level and maintaining it at a state higher than the lower energy state. There are various methods of excitation, such as using gas discharge to excite with kinetic electrons, known as electrical excitation; using light source irradiation, known as light excitation; and also thermal excitation, chemical excitation, etc. Various excitation methods are vividly referred to as pumping, because exciting particles from lower energy levels to higher energy levels is like using a pump to pump water from a low place to a high place. In order to continuously obtain laser output, we must continuously "pump" to ensure that the number of particles at the upper energy level is greater than the number of particles at the lower energy level.

We call materials that can provide suitable energy level structures to produce lasers as active media or gain media. It can be a gas, a solid or liquid. The active medium generally dopes active particles (such as rare earth ions Nd^{3+}, Er^{3+}, transition metal ions Ti^{3+}, Cr^{3+}, gas molecules N_2, etc.) into the matrix host (such as yttrium aluminum garnet, quartz), and through pumping, particles in the lower energy state can transition to the higher energy state. The laser medium also needs to have good optical, thermal, electrical, and mechanical properties.

IV. **Laser Resonator and Light Amplification**

In the definition of laser, the term "light amplification" is mentioned. That is to say, the photons produced by stimulated radiation need to be amplified to reach a sufficiently high energy to overcome various losses and form a laser output. Light amplification is achieved through a structure called a "resonator". The simplest resonator consists of two mirrors (a resonator composed of multiple mirrors and components can also be equivalent to two mirrors), one of which is a total reflection mirror, that is, an optical mirror coated with a film that completely reflects the wavelength of stimulated radiation light. When this wavelength of light hits the film, it immediately reflects back. The other mirror is the output mirror, which is coated on the mirror so that the laser wavelength light can partially pass through, and the other part is reflected back into the resonator. The light returned from the total reflection mirror and the output mirror, after passing through the gain medium, will undergo the process of stimulated radiation amplification again.

Let's take a look at the process of laser formation. After the particles in the gain medium absorb photons, they transition to a higher energy level. The higher-energy level particles will randomly generate some spontaneously radiated photons. If one of the photons passes through the active medium, it will "stimulate" a coherent photon of the same frequency. These two photons hit the mirror and are blocked back. After passing through the active medium again, they "stimulate" two new coherent photons. In this way, two become four and four become eight. So repeatedly, in an instant (the speed of light is extremely fast), it will become many. This is the so-called light amplification.

We refer to a cavity composed of two mirrors that strictly parallel each other and reflect and transmit lasers as a laser resonant cavity. It can amplify light of a certain wavelength, while suppressing light in other directions and frequencies, and only allows stimulated radiation photons to advance in a fixed direction, reflect back and forth to amplify and eventually output. In this way, the monochromaticity and directionality of the laser can be ensured, as shown in Fig. 2.2.

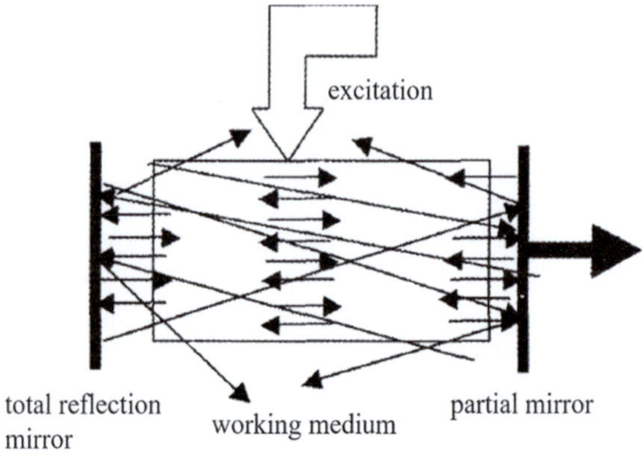

Fig. 2.2 Schematic diagram of light amplification in laser resonator

The process of generating laser in the resonant cavity can be summarized as: excitation (pumping)—particle number inversion of the gain medium—occasional spontaneous emission—stimulated emission of other particles—photon amplification—photon oscillation and photon amplification—laser generation.

2.1.3 Structure and Types of Laser

I. **Basic Structure of Laser**

Laser consists of three basic components: working material (active medium), resonant cavity (for amplifying light feedback), and excitation energy (pumping particles from the lower energy level to the upper energy level of the active medium, forming particle number inversion). Figure 2.3 is a schematic diagram of the basic structure of the laser.

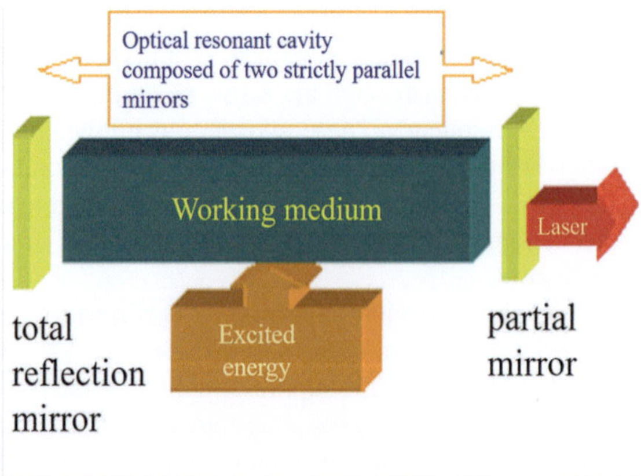

Fig. 2.3 Basic structure diagram of laser

II. Classification of Lasers

There are different classification methods for lasers: according to whether the laser output is continuous, it can be divided into continuous wave (CW) lasers and pulsed lasers. Pulsed lasers include free-running pulsed lasers, Q-switched lasers, and mode-locked lasers. According to the physical state of the active medium, it can be divided into solid, gas, and liquid types (sometimes semiconductor lasers are singled out from solid-state lasers as a separate type, because semiconductor lasers have some unique properties, and their research and application are very extensive). They can also be classified according to the type of particles in the active medium, such as atomic, ionic, molecular, and free electron lasers. They can also be classified according to the frequency and power of light emission. Below we will give a brief introduction to some commonly used lasers [6, 7].

1. Solid-state lasers: Generally small and sturdy, with high pulsed radiation power and a wide range of applications. Such as ruby lasers, Nd:YAG lasers, neodymium glass lasers, erbium glass lasers, etc. These types of lasers are the most widely used, among which Nd:YAG lasers are the most common in laser welding, laser cutting and other applications, and are also the main force in laser cleaning.
2. Semiconductor lasers: Small in size, light in weight, simple in structure, and cheap in price, but relatively poor in the beam quality and monochromaticity. Such as gallium arsenide, aluminum gallium arsenide and zinc telluride lasers. Semiconductor lasers actually belong to solid-state lasers, but due to their unique characteristics and large usage, they are often classified as a separate category.
3. Fiber lasers: Using fiber as the gain medium, under the action of pump light, high power density is easily formed in the fiber, resulting in "population inversion", forming laser oscillation output. Its characteristics are simple configuration, high reliability, and good beam quality. It can also be classified as a solid-state laser, but due to its unique configuration and increasing application, it is often listed as a separate category.
4. Gas lasers: They have strong monochromaticity, for example, the monochromaticity of a He–Ne laser is a hundred million times higher than that of a common light source. Moreover, gas lasers have a wide variety of working materials, such as nitrogen molecule lasers, argon ion lasers, copper vapor lasers, etc. Therefore, they can produce lasers of many different frequencies. However, due to the low density of gas, the output power of the laser is correspondingly small. (However CO_2 laser output power is much higher.)
5. Liquid lasers: Mainly dye lasers, about 500 types of dyes that can produce lasers have been discovered so far. These dyes can be dissolved in alcohol, benzene, acetone, water or other solutions. They can also be included in organic plastics to appear in solid form, or sublimated into vapor to appear in gaseous form. The outstanding feature of dye lasers is that their wavelength is continuously adjustable. They are diverse, relatively cheap, efficient, and their output power can compete with gas and solid lasers. They are used in fields such as spectroscopy, photochemistry, medicine, and agriculture.

6. X-ray lasers: They have important value in scientific research and military. They have advantages in laser anti-missile weapons. Biologists can use X-ray lasers to study the molecular structure in living tissues or understand cell functions in detail. Photos of molecular structures taken with X-ray lasers have high contrast in the images of biomolecules obtained.
7. Chemical lasers: Some chemical reactions produce enough high-energy atoms to release strong energy. Chemical lasers have important applications in the military.
8. Free electron lasers: These lasers are more suitable than other types for producing high-power light radiation. Its working mechanism is different. It obtains high-energy electron beams of tens of millions of volts from accelerators, forms energy levels of different energy states through a periodic magnetic field, and produces stimulated radiation.

Considering the mechanism of laser and material interaction, as well as the price, durability, and convenience of lasers, the lasers mainly used in laser cleaning are Nd:YAG lasers, CO_2 lasers, fiber lasers, and excimer lasers etc.

2.1.4 Laser Parameters

When a laser is applied, it is necessary to understand basic parameters such as wavelength, power or energy of the laser. The main physical quantities describing the laser are summarized in Table 2.1.

For example, in laser cleaning, a certain type of Nd:YAG laser is used. Its main technical parameters are: wavelength of 1.06 μm, repetition rate of 20 kHz, average power of 300 W, pulse width of 100 ns, peak power of 150 kW, and single pulse energy of 15 mJ. Knowing these parameters, you can know what objects can be cleaned. Or conversely, if you know what technical parameters are needed for cleaning, then you can choose the appropriate laser.

2.2 Common Laser Technologies

In laser cleaning, different parameters of lasers need to be selected according to the characteristics of pollutants and substrate materials. Q-switching technology, mode-locking technology, and amplification technology can help to achieve narrow pulse width and high peak power laser output.

Table 2.1 Main physical quantities describing the laser

Physical quantity	Common letter	Description	International unit	Other units
Wavelength	λ	The product of frequency equals the speed of light	m	nm, μm
Speed of light	c	Speed of light in vacuum	m/s	
Frequency	ν	Number of light vibrations per unit of time	Hz	
Power	P	Used for continuous lasers	W	
Single pulse energy	E	Used for pulsed lasers	J	
Pulse width	τ	Width of a single pulse	s	ns, ps, fs
Repetition rate	f	Number of output pulses per unit of time, used for pulsed lasers	Hz	kHz, MHz
Peak power	P_p	Maximum power of a single pulse, used for pulsed lasers	W	
Average power	$\overline{P}$	Average value over a period of time, used for pulsed lasers	W	
Input power	P_{in}	Power injected by the power source, the electrical or optical power required for the laser to operate	W	

2.2.1 Rate Equation Theory

To generate stimulated radiation, population inversion is required. The excitation conditions required for the population inversion depend on the characteristics of the energy levels of the laser active gaining medium. The rate equations of the laser are the equations satisfied by the change in particle number density (number of particles per unit volume) between the relevant energy levels of the laser active medium. It is related to the structure of the energy levels and the transition characteristics of the particles between these energy levels. The energy level structure and particle transition characteristics of different laser media are different, so the situation is complex. But we can still summarize some common and main physical processes, and write out the rules of particle number density and photon number density changes over time, which is called rate equations.

In Einstein's theory of stimulated radiation, two energy levels are involved, referred to as the upper and lower energy levels of the laser. However, in reality, the activated ions in the laser gain medium often have many energy levels. In the absorption and radiation process, in addition to the upper and lower energy levels of the laser, other energy levels are also needed for assistance. According to the operation rules of the laser, people have summarized two situations: three-level systems and four-level systems.

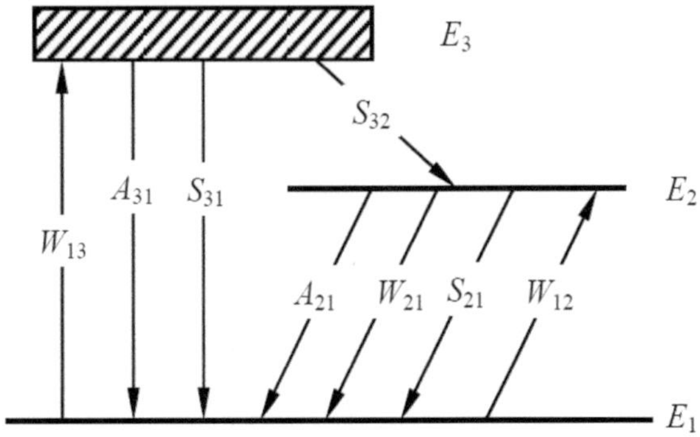

Fig. 2.4 Energy level diagram of three-level laser system

I. **Three-Level Laser System**

As shown in Fig. 2.4, it is a schematic diagram of the energy levels of the three-level system laser active media where absorption and radiation is shown. The three energy levels involved in the generation of the laser are: the ground state, which also serves as the lower energy level of the laser, with energy E_1, the metastable state energy level, with energy E_2, and the upper energy level of the laser, with energy E_3.

In thermal equilibrium, the particle distribution satisfies the Boltzmann distribution, with most particles in the ground state. When the external environment provides excitation energy, the particles in the ground state (energy state 1) absorb the energy and transition to the metastable state (energy state 3). The lifetime of the metastable state is generally short, and the particles will quickly non-radiatively relax to the upper energy level of the laser (energy state 2). The lifetime of the particles at this energy level is relatively long, so particles can be gathered at the upper energy level of the laser, achieving population inversion, and then these particles return to the ground state, radiating the laser. In addition, there are other processes, including spontaneous radiation and non-radiative relaxation from energy state 3 to energy state 1, and spontaneous radiation and non-radiative relaxation from energy state 2 to energy state 1. In this process, W_{13} represents the number of particles pumped from the ground state (energy state 1) to the metastable state (energy state 3) per unit time, called the pumping rate; A_{ij} represents the number of particles spontaneously radiating from energy state i back to energy state j per unit time, called the spontaneous radiation probability, which characterizes the number of photons in spontaneous radiation; S_{ij} represents the number of particles non-radiatively relaxing from energy state i to energy state j per unit time, called the non-radiative relaxation rate, which characterizes the energy consumed in heating; W_{12} represents the number of particles stimulated from energy state 1 to energy level 2 per unit time, called the stimulated absorption rate; W_{21} represents the number of particles that are stimulated from energy state 2 to energy level 1 per unit time, called the stimulated radiation rate. These two physical quantities characterize stimulated absorption and stimulated radiation.

N_i represents the particle density of energy level i. We can write out the rate equations of the three-level system [6, 8]:

$$\begin{aligned} \frac{dN_3}{dt} &= W_{13}N_1 - (A_{31} + S_{31} + S_{32})N_3 \\ \frac{dN_2}{dt} &= S_{32}N_3 - (W_{21} + A_{21})N_2 + W_{12}N_1 \\ N &= N_1 + N_2 + N_3 \end{aligned} \tag{2.2.1}$$

In a steady state, the particle population density at each energy level does not change with time, so we have:

$$\begin{cases} \frac{dN_3}{dt} = 0 \\ \frac{dN_2}{dt} = 0 \end{cases} \tag{2.2.2}$$

The solution is:

$$\frac{N_2}{N_1} = \frac{\frac{W_{13}S_{32}}{A_{31}+S_{31}+S_{32}} + W_{12}}{W_{21} + A_{21}} \tag{2.2.3}$$

Considering the simplified approximation, and assuming that the statistical weights of energy levels E_2 and E_1 are equal, according to Einstein's relation, to experimentally reverse the number of particles between the upper and lower energy levels of the laser, it needs to satisfy:

$$W_{13} > A_{21} \tag{2.2.4}$$

II. **Four-Level Laser System**

The characteristic of the four-level system compared to the three-level system is that the lower energy level of the laser is not the ground state, so the number of particles at the lower energy level of the laser is very small in the thermal equilibrium state, which is conducive to the formation of population inversion between the upper and lower energy levels of the laser. Figure 2.5 shows the energy level and transition process of the four-level system, where the ground state energy level is E_0, the lower and upper energy levels of the laser are E_1 and E_2, and the metastable state is E_3. Similar to the three-level system, we can write out the rate equations of the four-level laser system, and the meanings of each letter are the same as those of the three-level laser system [7, 9, 10].

Four-level system rate equations are:

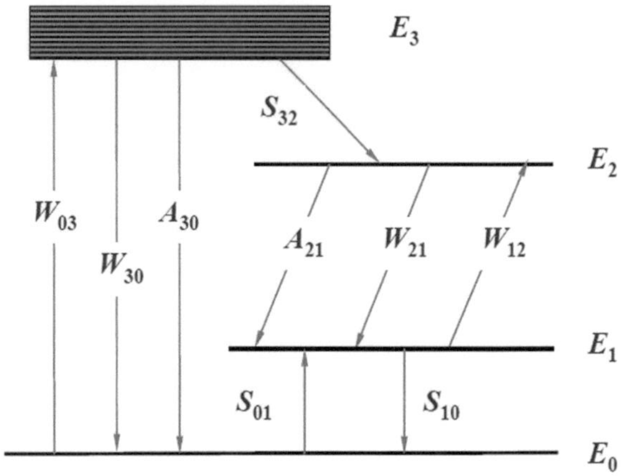

Fig. 2.5 Energy level diagram of four-level laser system

$$\begin{cases} \dfrac{dN_3}{dt} = W_{03}N_0 - (W_{30} + A_{30} + S_{32})N_3 = 0 \\ \dfrac{dN_2}{dt} = S_{32}N_3 - (A_{21} + W_{21})N_2 + W_{12}N_1 = 0 \\ \dfrac{dN_1}{dt} = S_{01}N_0 + (A_{21} + W_{21})N_2 - (S_{10} + W_{12})N_1 = 0 \\ N = N_0 + N_1 + N_2 + N_3 \end{cases} \tag{2.2.5}$$

2.2.2 *Laser Q-Switching Technology*

Through Q-switching technology, the continuous or pulsed laser energy output can be compressed into extremely short pulses for emission, thereby increasing the peak power of the laser by several orders of magnitude to gigawatts, and the pulse width is shortened to the level of 10^{-7}–10^{-10} s. This kind of output laser pulse is often referred to as a giant pulse.

I. **Principle of Q-Switched Laser**

Q-switching means adjusting the Q value. The Q value is an indicator of the quality of the optical resonant cavity in the laser, also known as the quality factor. It is defined as: the ratio of the total energy W stored in the laser resonant cavity to the energy loss dW/dt per unit time in the cavity [8, 11].

$$Q = 2\pi \nu_0 \frac{W}{\frac{dW}{dt}} = \frac{2\pi}{\lambda \alpha_{total}} \tag{2.2.6}$$

In the formula, ν_0 is the center frequency of the laser, α_{total} is the total one-way loss of the resonant cavity. It can be seen that the Q value is inversely proportional to the total loss α_{total} in the cavity.

Q-switching technology is a technique that changes the loss of the resonant cavity over time in a certain program. By changing the loss in the cavity, the Q value in the cavity can be adjusted.

Normally, the loss of the resonant cavity of a stable laser is fixed. Once the number of inverted particles reaches or slightly exceeds the threshold through pumping, the laser begins to oscillate. Subsequently, the number of particles in the upper energy level of the laser decreases due to stimulated radiation, and continuous pumping and laser output prevent the accumulation of a large number of inverted particles in the upper level, which can only be limited to around the threshold population inversion. This is why the peak power of ordinary lasers (generally on the order of tens of kilowatts) is not very high.

When the laser starts pumping initially, if we adjust the loss in the laser resonant cavity to be very high, that is, the Q value is very low. At this time, the oscillation threshold is correspondingly very high, and the laser oscillation cannot be generated. In this way, the number of inverted particles in the upper energy level of the laser can continue to accumulate and increase. When the number of inverted particles accumulates to a sufficiently large extent, suddenly adjust the loss to a very low value, that is, adjust the Q value to be very high, and the threshold also becomes very low accordingly. At this time, a large number of particles accumulated in the upper energy level will avalanche transition to the lower energy level, releasing energy in a very short time, and obtaining a giant pulse laser output with extremely high peak power. The time required to adjust from low Q value to high Q value is called Q-switching time.

II. **Characteristics of Q-Switched Lasers**

1. Peak Power

Ignoring the influence of the switching time and the spontaneous radiation and pump excitation during the pulse duration, starting from the rate equations, the maximum photon number density of the three-level system can be obtained [9, 11]

$$N_m \approx \frac{1}{2}\Delta n_t\left(\frac{\Delta n_i}{\Delta n_t} - \ln\frac{\Delta n_i}{\Delta n_t} - 1\right) \tag{2.2.7}$$

In the formula, Δn_i and Δn_t are the initial inverted particle number density and the threshold inverted particle number density, respectively. Assuming the cross-sectional area of the active material is S, one of the two resonant cavity mirrors is a total reflection mirror, and the other has a transmission rate of T, then the peak power P_m of the laser is

$$P_m \approx \frac{1}{2}h\nu_{21}N_m vST \tag{2.2.8}$$

In the formula, h is Planck's constant, ν_{21} is the laser frequency, v is the speed of light in the working medium, N_m is the peak number of inverted particles. Both N_m and P_m increase with the increase of $\Delta n_i/\Delta n_t$. $\Delta n_i/\Delta n_t$ depends on the following factors:

(1) The larger the maximum loss in the cavity α_{max}, the larger the allowable Δn_i. The smaller the minimum loss in the cavity α_{min}, the smaller the Δn_t. Therefore, efforts should be made to increase the value of $\alpha_{max}/\alpha_{min}$;
(2) The higher the power of the pump source, the larger the $\Delta n_i/\Delta n_t$;
(3) The longer the lifetime of the upper energy level of the laser, the larger the $\Delta n_i/\Delta n_t$.

Generally, the lifetime of the upper energy level of gas lasers is short and not suitable for Q-switching devices (such as He–Ne with an upper energy level lifetime of only 20 ns). Only CO_2 lasers with a longer upper energy level lifetime (1 ms) can use Q-switching technology.

2. Giant Pulse Energy

At the beginning of the giant pulse, the initial energy that can be converted into a laser in the active material with a volume of V is:

$$E_i = \frac{1}{2} h\nu_{21} V \Delta n_i \tag{2.2.9}$$

At the end of the giant pulse output, the remaining energy and the energy of the giant pulse in the cavity are

$$E_f = \frac{1}{2} h\nu_{21} V \Delta n_f \tag{2.2.10}$$

$$E_w = E_i - E_f \tag{2.2.11}$$

The output giant pulse energy and energy utilization rate are

$$E = \frac{T}{T+\alpha}\left(E_i - E_f\right) = \frac{T}{T+\alpha}\mu E_i \tag{2.2.12}$$

$$\mu = \frac{E_w}{E_i} = 1 - \frac{\Delta n_f}{\Delta n_i} \tag{2.2.13}$$

In the above formula, it can be seen that the larger the stored energy, the larger the giant pulse energy.

3. Time Characteristics

The time characteristics of the Q-switched pulse are shown in Fig. 2.6. The number of inverted particles N in the resonant cavity rises from $N_m/2$ to N_m in time t_r, and N falls from N_m to $N_m/2$ in time t_e, the total time is $\Delta t = \Delta t_r + \Delta t_e$.

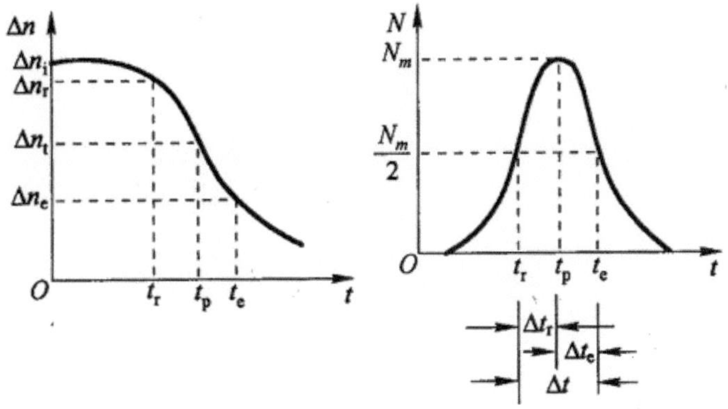

Fig. 2.6 Time characteristics of Q-switched pulse

(1) When $\Delta n_i/\Delta n_t$ increases, both the leading and trailing edges of the pulse narrow simultaneously, with the leading edge narrowing more significantly. This is because the larger $\Delta n_i/\Delta n_t$, the higher the net gain coefficient in the cavity, and the faster the growth of the number of photons in the cavity and the decay of the number of inverted particles. So the pulse establishment and extinction processes are shorter.
(2) The pulse width is directly proportional to the length of the resonant cavity and inversely proportional to the loss in the resonant cavity. Therefore, it is appropriate to use a short cavity and low-loss resonant cavity structure.

III. **Q-Switching Methods**

The Q value of the resonant cavity is inversely proportional to the total loss α_{total} in the cavity. The losses of the resonant cavity generally include: reflection loss, diffraction loss, absorption loss, etc. Therefore, any element that can cause a sudden change in the loss of the resonant cavity can be used as a Q-switching element, also known as a Q switcher. Common Q-switching methods are divided into active Q-switching and passive Q-switching. In active Q-switching, the resonant cavity loss can be actively controlled by an external driving source; in passive Q-switching, the resonant cavity loss depends on the laser intensity in the cavity and the Q switcher, and cannot be actively controlled. Common active Q-switching includes rotating mirror Q-switching, electro-optic Q-switching, and acousto-optic Q-switching. Passive Q-switching includes saturable absorption Q-switching.

1. Active Q-Switching

Active Q-switching can actively control the loss or Q value in the resonant cavity. The electro-optic Q-switching and acousto-optic Q-switching technologies are most commonly used. In addition, there is rotating mirror Q-switching, which produces

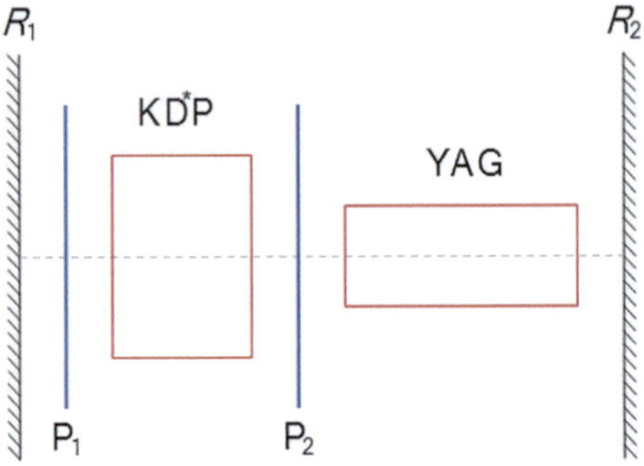

Fig. 2.7 Schematic diagram of high-power Nd:YAG concave cavity structure

relatively wide pulse width and is prone to multiple pulses, so it is rarely used except for some special lasers. In laser cleaning, rotating mirror Q-switching lasers are generally not used. Therefore, we only introduce electro-optic Q-switching, and acousto-optic Q-switching here.

(1) Electro-optic Q-Switching

Electro-optic Q-switching uses the electro-optic properties of specific crystals to change the polarization direction of polarized light (additional polarizers are required in the cavity to form linearly polarized light), making it unable to pass through the polarizer again, at which point the loss in the cavity is large, as shown in Fig. 2.7. The voltage applied to the electro-optic crystal is generally very high, requiring thousands of volts. It is difficult for the electro-optic crystal to achieve high repetition rate Q-switching, so it is difficult to obtain high repetition rate laser output with electro-optic Q-switching.

Taking the commonly used Nd:YAG laser as an example, explain the device and principle of electro-optic Q-switching. In Fig. 2.7, Nd:YAG is the laser medium, KD^*P is the Q-switching crystal, P_1 and P_2 are polarizers with perpendicular polarization directions, and R_1 and R_2 are the two resonator mirrors of the resonant cavity. Due to the presence of P_2, the light in the laser resonator is linearly polarized. The light emitted after passing through the KD^*P crystal is still linearly polarized, and the polarization direction remains unchanged. However, because the polarization direction of the P_1 polarizer is perpendicular to that of P_2, the linearly polarized light emitted from the KD^*P crystal cannot pass through the P_1 polarizer, so the light cannot reach the reflector R_1 and propagate back and forth in the cavity to form stimulated radiation light amplification (i.e., laser). The inability of light to pass through P_1 is equivalent to a large loss of photons in the cavity, with a low Q value, and this state is called the "closed door" state.

Due to continuous pumping, the number of particles is still being continuously pumped to a high energy state. When the number of inverted particles reaches a certain level, a voltage is suddenly applied to the KD^*P crystal. The crystal with voltage has a characteristic: the vibration direction of the linearly polarized light passing through it changes by 90°, and becomes consistent with the polarization direction of P_1, so the linearly polarized light emitted from the crystal can pass through P_1. The light can pass through the components in the cavity, which is equivalent to an immediate decrease in cavity loss or an immediate increase in Q value. The light immediately forms oscillations in the cavity, forms stimulated radiation light amplification, and outputs a large energy giant pulse through the output mirror.

In the above process, there was no voltage on the crystal at the beginning, and then voltage was added, which is called pressure-increasing electro-optic Q-switching. There is also a type of electro-optic Q-switching, which is first applying voltage and then withdrawing voltage, called voltage-withdrawal electro-optic Q-switching. The principles of the two are the same.

Electro-optic Q-switching has the advantages of high accuracy, good repeatability, and narrow pulse width. The width of the output giant pulse is about 10 ns, and the narrowest can reach 2–3 ns. The disadvantage is that the applied voltage often needs to reach several thousand or even close to ten thousand volts.

(2) Acousto-Optic Q-Switching

Acousto-optic Q-switching utilizes the acousto-optic effect. When ultrasonic waves propagate in a medium, the refractive index of the medium will change according to the sine or cosine rule. At this time, when the laser passes through this medium, the diffraction of light will occur, that is, the acousto-optic diffraction effect. The medium used is the acousto-optic Q switcher.

Let's take the commonly used Nd:YAG laser as an example to illustrate the device and principle of acousto-optic Q-switching. In Fig. 2.8, YAG is the laser medium. An acousto-optic Q-switcher is inserted into the laser cavity. This switcher consists of an acousto-optic medium, absorbing material, and a high-frequency oscillator. The oscillator is controlled by a pulse modulator. When powered, it generates ultrasonic waves. Thus, when the light in the cavity passes through the acousto-optic medium, diffraction occurs. The diffracted light cannot propagate along the optical axis in the resonant cavity, and therefore cannot reach the mirror, making it impossible to form effective laser oscillation. At this time, the cavity has a large loss, and the Q value is very low, that is, the Q switcher is in the "off" state. When the number of population in the upper energy level of the laser accumulates to a certain extent, the ultrasonic wave is removed and the diffraction effect disappears immediately. So the loss decreases, the Q switcher opens ("on" state), and the light immediately propagates along the axis, forming a giant pulse of Q-switched laser.

The accuracy of acousto-optic Q-switching is high, with good repeatability and narrow pulse width. It can be applied to a continuous laser to obtain a Q-switched pulse laser output. The width of the output giant pulse is on the order of 100 ns.

2. Passive Q-Switching

Some media have the characteristic of sudden changes in the absorption of light waves of specific wavelengths. When the wavelength of the incident light signal is near its absorption peak, for weaker incident signal light, the medium will strongly absorb, so the incident light cannot pass through the medium (equivalent to being in the "off" state); when the incident light signal is enhanced to a certain extent, suddenly the medium almost no longer absorbs the signal light, showing a clear

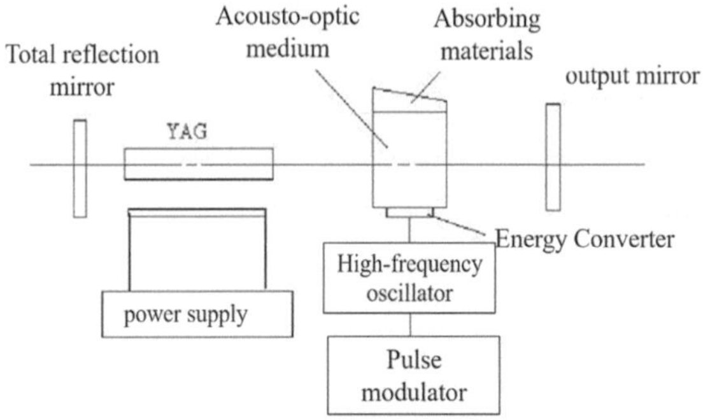

Fig. 2.8 Acousto-optic Q-switch laser schematic diagram

absorption saturation trend. At this time, the signal light can directly pass through the medium, as if the medium is "transparent" (equivalent to the "on" state). Using this feature, it can be placed in the laser resonant cavity for Q-switcher, called passive Q-switcher. Initially, the stimulated radiation signal of the laser gain medium is weak, and the Q switcher almost completely absorbs the signal light, the cavity loss is large, and the Q value is low. When the population inversion in the gain medium reaches its maximum, the stimulated emission light intensity increases to a sufficient level, making the Q switcher in the absorption saturation state. Then the signal light can pass through, and can propagate and oscillate in the cavity, forming a Q-switched laser output.

Common passive Q-switcher include dye-saturated absorbers, Cr^{4+}: YAG saturable absorbers, semiconductor saturable absorbers, etc. The advantages of passive Q-switching are low cost with simple structure. Its pulse width is from tens to hundreds of ns. However, its pulse repeatability is not good, making it easy to produce multiple pulses, and each pulse's energy and peak power fluctuate, and the accuracy is not high enough.

People are also constantly developing new laser Q-switching technologies, such as technology combined with active and passive Q-switching, dual passive Q-switching technology, Q-switched mode-locking technology.

IV. **Key Points of Q-Switching Technology**

Regardless of the types of Q-switching technology, the following key points should be noted:

1. Dynamic loss: When the Q switcher is in the "off" state, the resonant cavity should have the maximum loss to ensure that no laser is generated before the Q switcher is opened;
2. Insertion loss: When the Q switcher is in the "on" state, the Q switcher itself will introduce reflection and scattering losses;
3. Switching time: Q-switcher should have excellent on–off switching performance. A fast switching time will produce narrow and high peak power pulses, while a slow switching time will cause the stored energy to rapidly deplete before the switcher is fully opened;

4. Synchronization performance: The Q-switcher should be able to precisely control and maintain synchronization with external signals.

Q-switching technology is one of the main foundational technologies for high-power pulsed lasers. For commonly used pulsed solid-state lasers, the application of Q-switching technology can achieve laser pulses with peak powers ranging from tens of kilowatts to hundreds of megawatts and pulse widths of only a few nanoseconds to a few hundred nanoseconds, making the laser a very strong coherent light source. At the same time, it has also promoted the development of technologies such as laser radar, laser ranging, high-speed photography, nuclear fusion, and has achieved important applications in laser cleaning.

2.2.3 Laser Mode Locking

The width of the laser pulses output by the Q-switched laser is theoretically limited and cannot be further compressed, with the narrowest pulse width of only a few nanoseconds. With the development of science and technology, in many fields such as telemetry technology, biomedical research, high time-resolution spectroscopy, nonlinear optics, optoelectronics, chemical dynamics, and controlled nuclear fusion, there is a demand for laser pulses with narrower pulse widths and higher peak powers. At this time, laser mode-locking technology is necessary. Mode-locking technology can produce ultra-short light pulses on the order of (10^{-12}–10^{-15}) s, with peak powers reaching the terawatt level [12].

I. **Working Principle of Mode-Locked Lasers** [10]

The output of a free-running laser generally contains several longitudinal modes that exceed the threshold, as shown in Fig. 2.9.

The amplitudes and phases of these modes are not fixed, and the laser output changes over time as a result of their irregular superposition, which is a time-averaged statistical value. The frequency interval between adjacent modes of the laser is $\Delta\upsilon_q = c/2L'$, where L′ is the optical cavity length of the resonator, and c is the speed of

Fig. 2.9 Laser modes

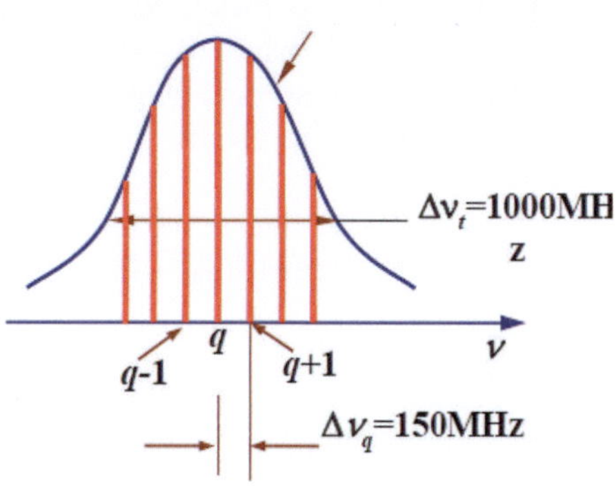

light in a vacuum. Then the angular frequency interval is: $\Delta\omega = 2\pi\Delta\upsilon_q = \pi\upsilon/L'$, where q is the longitudinal mode order. Assuming that there are (2N + 1) longitudinal modes within the net gain linewidth of the laser gain medium, then the optical wave field output by the laser is the sum of the electric fields of the (2N + 1) longitudinal modes, that is

$$E(t) = \sum_{q=-N}^{N} E_q \cos(\omega_q + \varphi_q) \tag{2.2.14}$$

In the formula, q = 0, ± 1, ± 2, …, ± N, is the order of the qth longitudinal mode in the (2N + 1) oscillating modes inside the laser; E_q is the field strength of the longitudinal mode with the order of q; ω_q and φ_q are the angular frequency and phase of the longitudinal mode with the ordinal number q, respectively. The initial phases of all longitudinal modes have no definite relationship with each other, and are completely independent and random. The frequency intervals between adjacent longitudinal modes are not strictly equal. The longitudinal modes are incoherent. The output light intensity presents random and irregular fluctuations due to the incoherent superposition between the longitudinal modes. Assuming that the electric field amplitudes of all longitudinal modes are the same, then

$$\bar{I}(t) = \bar{E}^2(t) = \frac{1}{t}\sum_{q=-N}^{N} E^2(t)dt = (2N+1)\frac{E_0^2}{2} \tag{2.2.15}$$

If some measures are taken to synchronize these independent longitudinal modes in time, that is, to lock the relationship between their phases to satisfy a certain relationship ($\varphi_{q+1} - \varphi_q$ = constant), in other words, the phases of the modes of the laser are locked according to the relationship of $\varphi_{q+1} - \varphi_q$ = constant, then the output of the laser will be laser pulses with extremely narrow pulse width and high peak power. This type of laser is called a mode-locked laser, and the corresponding technology is called "mode-locking technology".

Next, we will theoretically derive the output light intensity and pulse width of the mode-locked laser. For simplicity, assume that the initial positions of the (2N + 1) longitudinal modes are always equal, and have $\varphi_q = \varphi_{q-1} = 0$, the amplitudes of each mode are equal, that is, $E_q = E_0$, and the light intensities are equal $I_q = I_{q-1} = I_0$, then the total light field output of the laser is the coherent result of (2N + 1) longitudinal modes. The electric field intensity of the (2N + 1) modes is

$$E(t) = \sum_{q=-N}^{N} E_q \exp[i(\omega_0 + q\Omega)t + \varphi_q] = E_0\left(\sum_{q=-N}^{N} e^{iq\Omega t}\right)e^{i\omega_0 t} \tag{2.2.16}$$

In the formula, $\Omega = \omega_q - \omega_{q-1}$. Expand (2.2.16) in exponential form, and then use trigonometric functions to get:

$$E(t) = E_0 \left(\sum_{q=-N}^{N} e^{iq\Omega t} \right) e^{i\omega_0 t}$$

$$= E_0 \frac{\sin\left[\frac{1}{2}(2N+1)\Omega t\right]}{\sin \frac{1}{2}\Omega t} e^{i\omega_0 t} = A(t)e^{i\omega_0 t} \quad (2.2.17)$$

In the formula,

$$A(t) = E_0 \frac{\sin\left[\frac{1}{2}(2N+1)\Omega t\right]}{\sin \frac{1}{2}\Omega t} \quad (2.2.18)$$

The output laser intensity obtained is

$$I(t) \propto E_0^2 \frac{\sin^2\left[\frac{1}{2}(2N+1)\Omega t\right]}{\sin^2 \frac{1}{2}\Omega t} \quad (2.2.19)$$

From the above formula (2.2.18) and (2.2.19), it can be seen that the laser field becomes a frequency modulated wave with a frequency of ω_0. The amplitude A(t) is a periodic function that changes with time, and the laser intensity I(t), which is proportional to $A^2(t)$, is also a function of time, and the laser intensity is modulated. According to Fourier analysis, the total laser field consists of (2N + 1) longitudinal mode frequencies, so the laser output pulse is a light wave including (2N + 1) longitudinal modes. As long as the change rule of amplitude A(t) is known, the persistence of the output laser can be understood. Both the numerator and denominator of Eq. (2.2.18) are periodic functions, so A(t) is also a periodic function. As long as its period and zero point are obtained, the change rule of A(t) can be obtained.

1. When the numerator is 0 and the denominator is not 0, it is the zero point of A(t): $t = 0, 1/(\frac{1}{(2N+1)}\frac{2L}{c}, \frac{2}{(2N+1)}\frac{2L}{c} \cdots \frac{2L}{c})$.
2. The points with the denominator to be 0: $t = 0, \frac{2L}{c}, \frac{4L}{c} \cdots \frac{2Ln}{c}$.
3. Both numerator and denominator of A(t) equal to zero, and the maximum value of A(t) at this time can be obtained using the Robita method

$$A_{max} = (2N+1)E_0$$

There are 2N zero points and 2N + 1 extreme points in one period. Since the laser intensity is proportional to $A^2(t)$, the maximum values at t = 0 and t = 2L/c are called the main pulses. Between two adjacent main pulses, there are 2N zero points and 2N − 1 secondary maximum values, called secondary pulses.

II. **Characteristics of Mode-Locked Lasers**

The following properties can be known through analysis:

1. Peak Value of Output Pulse (Maximum Light Intensity)

$$I_m = (2N + 1)^2 E_0^2 \tag{2.2.20}$$

If the phases of each mode are not locked, the modes are incoherent, and the output power is simply the sum of the powers of each mode, that is, $I \propto (2N + 1)E_0^2$. After mode-locking, the peak power of the pulse is (2N + 1) times higher than that before mode-locking. The longer the cavity, the larger the fluorescence linewidth, the more longitudinal modes oscillate in the cavity, and the greater the peak power of the mode-locked pulse.

2. Period (T)

The time interval between adjacent pulse peaks is T = 2L/c, therefore, the output of the laser is a regular pulse sequence with an interval of 2L/c. 2L/c is the time for light to travel back and forth in the cavity once, which is equal to having only one pulse oscillating in the cavity, and one pulse contains the number of locked longitudinal modes. The maximum value at t = 0 and t = 2L/c is called the main pulse. Between two adjacent main pulses, there are 2N zero points. Therefore, mode-locked oscillation can also be understood as having only one light pulse propagating back and forth in the cavity, as shown in Fig. 2.10.

3. Pulse Width (τ)

The width of the pulse, also known as the pulse width, is the width between the two half-maximum values of the laser pulse, which can generally be approximated by the time between the maximum value and the zero point. t = 0 is the maximum value, and the first zero point is: $\frac{1}{(2N+1)}\frac{2L}{c}$, so $\tau = t_1 \frac{1}{(2N+1)}\frac{2L}{c} = \frac{T}{2N+1} = \frac{1}{\Delta\upsilon}$. In Q-switching lasers, the narrowest output pulse width is from the transmissive Q-switched laser, with the minimum pulse width being 2L/c, while the pulse width τ of the mode-locked laser is narrower than the pulse width of the Q-switched laser by a factor of (2N + 1). The more longitudinal modes that are locked, the narrower the pulse

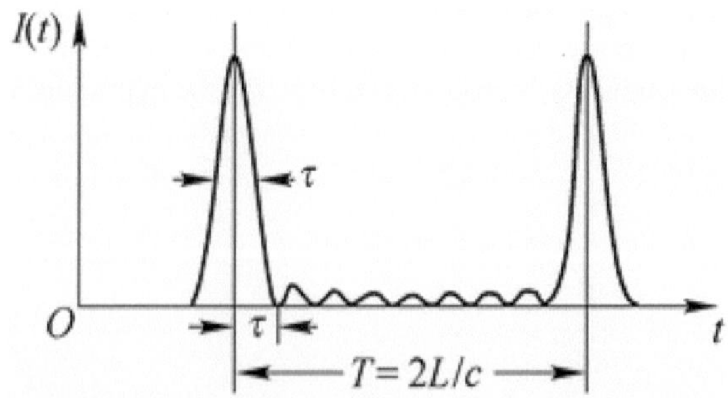

Fig. 2.10 Mode-locked light intensity pulse (2N + 1 = 9)

width τ. In the above formula, Δv is the bandwidth of mode-locked lasers, and it is obviously impossible to exceed the gain bandwidth of the laser gain medium. The spectral line width of gas lasers is small, and the width of their mode-locked pulses is about the nanosecond level; Solid state lasers have a larger spectral line width and can generate picosecond pulses with a pulse width narrower than 10^{-12} s. For example, the oscillation spectrum width of a neodymium glass laser reaches 25–35 nm, and its mode-locked pulse width can reach 10^{-13} s; the fluorescence spectral line width of a titanium sapphire laser exceeds 300 nm, and its mode-locked pulse width can reach a few femtoseconds, which is the narrowest pulse width laser obtained by mode-locking technology so far.

III. **Mode-Locking Methods**

In order to obtain ultra-short mode-locked pulses, measures must be taken to force the initial phase of each longitudinal mode to be locked, and the frequency intervals between adjacent modes are equal. The currently used mode-locking methods can be divided into active mode-locking, passive mode-locking, synchronous mode-locking, injection mode-locking, and collision mode-locking.

1. Active Mode-Locking

Active mode-locking uses the method of periodically modulating the parameters of the resonant cavity. An acousto-optic modulator is inserted into the laser resonant cavity to generate periodic alternating changes in the loss γ in the resonant cavity. If the change period of γ is $T = 2L/c$, which is the time required for light to go back and forth in the cavity, then the alternating frequency $f = c/2L$ of the loss γ, which is the longitudinal mode interval Δv_q, then mode-locked pulses can be formed in the resonant cavity. Narrow pulses periodically appear at the minimum value of the loss γ. This structure looks similar to acousto-optic Q-switching at first glance. The main difference is that the driving signal applied to the acousto-optic modulator during mode-locking must ensure that the changing frequency of the loss it generates must strictly equals Δv_q. It can be divided into phase modulation (PM), frequency modulation (FM) mode-locking, and amplitude modulation (AM, also known as loss modulation) mode-locking.

2. Passive Mode-Locking

The passive mode-locking device is very simple. Just insert a saturated absorption dye into the cavity. The dye must meet the following conditions: First, the absorption line of the dye should be very close to the laser wavelength; second, the linewidth of the absorption line should be greater than or equal to the laser linewidth; third, the relaxation time should be shorter than the time for the pulse to go back and forth in the cavity, otherwise it becomes a passive Q-switched laser.

3. Injection Mode-Locking

Using a high-quality mode-locked pulse as a seed, it regenerates and is amplified many times in the resonant cavity, and finally a mode-locked light pulse with narrow pulse width and high output power is obtained. If this seed is injected into the regenerative amplifier laser from another mode-locked laser, it is called external injection mode-locking. If the same laser first forms a seed pulse and then amplifies, it is called self-injection mode-locking. Self-injection is much simpler in structure than external injection.

4. Synchronous Pump Mode-Locking

A mode-locked laser pulse sequence is used to pump another laser, and mode-locking is achieved by modulating the gain within the cavity. The key point to achieve synchronous pumping mode-locking is to make the resonant cavity length of the pumped laser equal to or an integer multiple of the resonant cavity length of the pumping laser. Under certain conditions, the gain is modulated, and the modulation period equals the light's cycle period in the resonant cavity. Similar to loss modulation, a short pulse is formed in the cavity when the gain is at its maximum, and its pulse width is much narrower than the pumping pulse width used.

5. Collision Mode-Locking

Collision mode-locking is currently the best method for generating ultra-short pulses with a pulse width shorter than 100 fs. In the ring cavity, two light pulses propagating in the clockwise and counterclockwise directions collide exactly at the saturable absorption dye, forming a standing wave there, thus causing a spatial distribution of particle numbers between the upper and lower energy levels of the saturable absorption medium, known as the "optical grating" of particle number distribution. Since the refractive index of the medium is related to the particle number distribution between its energy levels, a spatial periodic distribution of refractive index is formed in the medium. During the formation of the "optical grating", the rising edge energy of the two light pulses is absorbed. Since the relaxation time of the saturable absorption medium is much longer than the relaxation time of the pulse itself, when the trailing edge of the pulse passes through the saturable absorption medium, the modulation degree of the particle number distribution grating is still very high, and the trailing edge of the light pulse undergoes strong backward scattering. In this way, both the leading and trailing edges of the light pulse are weakened, and after multiple cycles, the light pulse can be compressed very narrowly.

Compared to Q-switched lasers, mode-locked lasers have a more complex configuration and higher price. Although their pulse width is shorter and peak power is higher, making them more suitable for some cleaning tasks, they are currently mainly used in laboratory research in laser cleaning. In the future, as research on ultra-short pulse laser cleaning deepens and the price of mode-locked lasers decreases, it is expected to gradually increase its application in the field of laser cleaning, or play a unique role in some special cleaning tasks.

2.2.4 *Laser Amplification [10, 13]*

Using Q-switching and mode-locking techniques, extremely high peak power (10^9–10^{12} W) can be achieved, but the energy output of such high peak power lasers is not necessarily large. The output power or single pulse energy of the laser sometimes cannot meet the needs. For example, in laser cleaning, lower average power or energy will result in low cleaning efficiency, or even inability to clean because the cleaning threshold is not reached. For a single laser, it is difficult to significantly increase the energy or power of the laser output just by increasing the size of the gain medium and increasing the pump power. Moreover, even if it is increased, the divergence angle of the output laser will increase, and the mode of the laser will deteriorate, which is not conducive to practical applications. Not to mention, in actual situations, some laser gain media cannot be made in large sizes, and the pump power density cannot be too large otherwise it will damage the laser medium.

Laser amplification technology can achieve high-performance, high-energy or high-power lasers without changing the polarization state, divergence angle, monochromaticity, and other characteristics of the laser.

I. **Working Principle of Laser Amplifiers**

A laser amplifier is a device that amplifies the energy (power) of laser through stimulated radiation. Its principle is also based on the amplification of light by stimulated radiation. Like lasers, they have gain media and pump sources. However, unlike lasers, the initial light signal of a general laser is spontaneous radiation, while the initial signal of an amplifier is the laser output from the laser. Amplifiers generally do not have a resonant cavity.

II. **Structure of Amplified Lasers**

Amplified lasers generally include oscillators and amplifiers, as shown in Fig. 2.11.

1. Oscillator: The laser output from the oscillator serves as the input light signal for the amplifier. Different lasers are selected according to the actual application requirements, so that the laser characteristics output by the laser (oscillation stage) meet the application requirements. Depending on the pulse width of the laser output from the oscillation stage, amplifiers can be divided into three categories. (1) Continuous or long-pulse laser amplifiers: The output of the general oscillator laser is continuous or long-pulse laser. The number of particles in the upper energy level can be replenished in time after being consumed by the pump. At this time, the photon density in the cavity and the population inversion in the gain medium can be considered as not changing with time, which is called a steady-state process. (2) Short-pulse amplifiers: When the oscillator is a Q-switched laser, the pulse width τ is about 10^{-8} s. Due to the small pulse width, during the pulse signal amplification period, the population inversion and photon density of the gain medium change with time, which is a non-steady-state process. (3) Ultra-short pulse amplifiers: That is, the pulse width is only on the order of 10^{-11}–10^{-12} s for laser amplification.

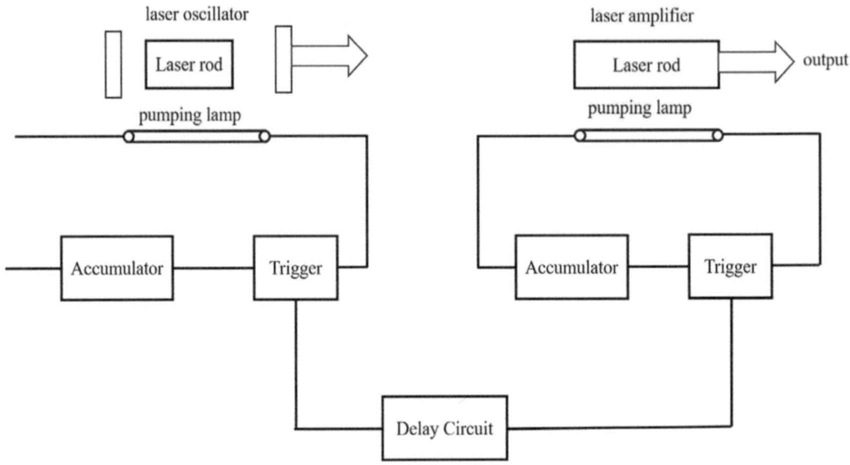

Fig. 2.11 Working schematic diagram of laser amplifier

2. Amplifier: The role of the amplifier is to amplify the energy (or power) of the light signal output from the oscillator. To achieve effective amplification, the amplifier must meet the following two conditions:

(1) The energy levels of the oscillator and the amplifier must match with each other

Because the light signal of the amplifier comes from the oscillator, their energy levels must match (the energy level of the working material) to achieve light amplification. For example, if the gain medium of the oscillator is ruby, which produces light at 694.3 nm, the gain medium of the amplifier must also be ruby, not YAG.

(2) The amplifying medium is in a state of population inversion

When the light signal output from the oscillator enters the amplifier, the amplifying medium is in a state of population inversion. This can achieve a large gain when the light signal passes through, thereby obtaining a lot of energy amplification. This is ensured by the delay part, which is the time difference between the pump of the amplifier and the oscillator. For different gain media, the delay time between the oscillator and the amplifier is different.

III. **Types and Characteristics of Laser Amplifiers**

1. Types of Laser Amplifiers

   (1) Traveling wave amplifier: The light signal only passes through the gain medium once, and the general amplifier does not add a resonant cavity, only the gain medium exists.
   (2) Multi-pass amplifier: The optical signal travels back and forth multiple times within the gain medium.

(3) Regenerative amplifier: A weak signal of good quality is injected into the laser, where it acts as a "seed" to control the generation of laser oscillation and is then amplified.

2. Characteristics of Laser Amplifiers

(1) Compared to the oscillator, the energy and power of the amplifier are amplified.
(2) The amplified laser retains the characteristics of the incident light.

With the development of laser cleaning, high-power laser cleaning machines have become a necessity. Many high-power laser cleaning machines have amplifiers, so laser amplifiers will be increasingly used in the laser cleaning industry.

2.3 Rules for Selecting Lasers in Laser Cleaning

When choosing the appropriate laser for laser cleaning, one must consider the optical characteristics of the object to be cleaned and the output characteristics of the laser, such as power, wavelength, pulse width, repetition frequency, single pulse energy, beam mode, and other factors.

2.3.1 Optical Characteristics of the Object to Be Cleaned

When choosing a laser, the optical characteristics of the pollutants and the base material are very important. Detailed content will be introduced in the next chapter. Here we present the main conclusions.

1. Based on the absorption characteristics of the pollutants: If the composition of the surface pollutants is known in advance, the type of laser can be chosen based on the optimal absorption wavelength of the pollutants. Since the absorption rate of different materials to the laser is closely related to the wavelength, the wavelength has a great impact on the laser cleaning process. For metals, generally, the absorption of light by the material decreases as the wavelength increases.
2. Characteristics of the object to be cleaned: Generally speaking, when the requirement for precision is higher than the requirement for cost, or when cleaning within a small area, an excimer laser is a more suitable tool; in other cases, CO_2 and solid-state lasers are preferable.

For the microelectronics industry, excimer lasers, such as KrF excimer lasers, are generally chosen; for surface cleaning, the use of CO_2 pulsed lasers is quite effective, especially for removing rubber residues. For mold cleaning, CO_2 or YAG lasers are better; for the removal of some coatings, CO_2 lasers are more effective, especially TEA CO_2 lasers, which have great potential in this field; however, since CO_2 lasers cannot be transmitted by optical fibers, their application in remote cleaning is greatly limited, which provides a broad space for the application of YAG lasers.

2.3.2 *Impact of Laser Parameters*

Different types of lasers, with their different wavelengths, repetition frequencies, pulse widths, and peak powers, have some differences in the mechanisms of interaction between light and matter.

I. **The Impact of Laser Wavelength**

The laser cleaning process is actually a process of interaction between the laser and the material, which largely depends on the material's absorption coefficient, the pulse energy of the laser, and the duration of the action. The absorption coefficient of the material and the penetration ability of the laser are closely related to the laser wavelength. Generally speaking, the shorter the wavelength, the greater the energy of the photon ($h\nu$, where h is Planck's constant, and ν is the frequency), and the chemical action under atomic excitation will increase accordingly. Conversely, if the wavelength is longer, the photon energy gradually decreases, but the thermal effect produced will increase, and the radiation penetration ability of the laser will also increase. Figure 2.12 shows the change in the absorption rate (coefficient) of the material with the wavelength.

II. **Power Selection**

During laser cleaning, the average power of the laser (related to the repetition rate and single pulse energy) and peak power (related to pulse width and single pulse energy) are very important. There are two thresholds in laser cleaning, one is the cleaning threshold, and the other is the damage threshold. The former is the minimum power (energy) that can make the pollutants detach from the substrate material, and the latter is the power (energy) that causes damage to the substrate. Different powers need to be selected for different substrate materials and different pollutants. If the power is too high, it may damage the substrate. If the power is too low, it will not achieve the cleaning effect.

III. **Selection of Other Parameters**

In addition, according to the substrate material and pollutants, it is also necessary to consider parameters such as pulse width, repetition rate, and mode. If the repetition rate is too low, the work efficiency is low. If the pulse width is very narrow, the peak power will increase accordingly, which may damage the substrate material; on the contrary, if the pulse width is large, the peak power is small, the cleaning effect is

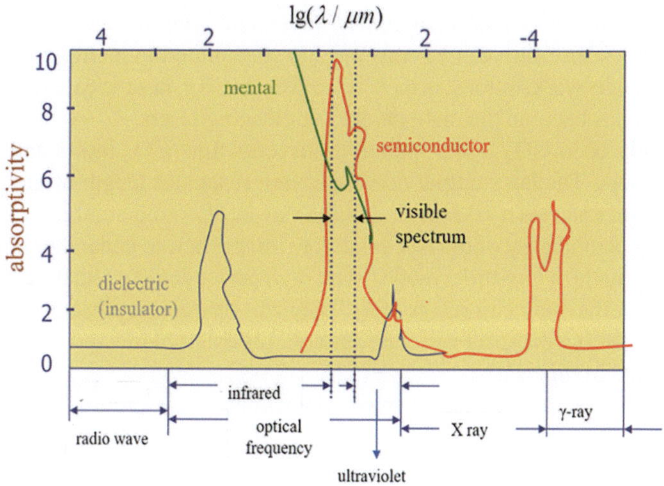

Fig. 2.12 Material absorption rate (coefficient) varies with wavelength

affected. If the mode of the laser is not good enough, the divergence angle is large, and it is difficult to focus; if the mode is very good, the intensity at the center of the laser spot is very high, which can easily cause damage to the substrate at the center and the surrounding area is not cleaned. These all need to be considered comprehensively according to the actual situation.

2.3.3 Commonly Used Lasers in Laser Cleaning

The laser is the core of the laser cleaning equipment. Lasers that have been used in cleaning include: 193 nm ArF excimer laser, 248 nm KrF excimer laser, 308 nm XeCl excimer laser, 351 nm XeF excimer laser, 337 nm N_2 laser, 255.3 nm frequency-doubled copper vapor laser, 266 nm fourth harmonic Nd:YAG laser, 292 nm frequency-doubled dye laser, 349 nm Nd:YLF third harmonic laser, 355 nm Nd:YAG third harmonic laser, 400–800 nm wavelength-tunable OPO pulsed laser, 400–900 nm tunable Ti:Sapphire laser, 523 nm Nd:YLF frequency-doubled laser, 583 nm dye laser, 1064 nm Nd:YAG laser, 1047 nm Nd:YLF laser, 9.22 μm, 9.317 μm, 9.6 μm and 10.6 μm TEA CO_2 laser, 2.94 μm Er:YAG laser, etc.

Although there are many types of lasers, the lasers commonly used in laser cleaning are mainly neodymium-doped yttrium aluminum garnet lasers (Nd:YAG), high-power CO_2 lasers, and excimer lasers. Now, the use of fiber lasers is also gradually increasing. Among them, Nd:YAG laser is very mature, which has such advantages as high fluorescence quantum efficiency, low threshold, and high thermal conductivity. It can operate continuously and in pulse mode. The laser wavelength is suitable for cleaning, maintenance is relatively simple, and the cost is relatively low.

Especially, it can be coupled with optical fibers, and with the help of the transmission system, a laser beam can be conveniently transmitted to multiple workstations or long-distance workstations, which is convenient for laser cleaning operations. Therefore, it has become the main choice for cleaning lasers.

Commonly used CO_2 lasers include transverse flow CO_2 lasers and fast axial flow CO_2 lasers. The laser output power per unit resonator length of the transverse flow CO_2 laser can reach 10 kW/m, and the power of commercial lasers can reach 25 kW. The beam quality of the fast axial flow laser is better, generally outputting in fundamental mode or low-order modes, with high power density, and the electro-optic transforming efficiency can reach 26%. They can operate in pulse and continuous wave modes, with a compact structure suitable for external use and a wide range of application environments.

Excimer lasers refer to lasers whose gain media are excimers. The excimer medium is an unstable compound, which combines into molecules in the excited state and dissociates into atoms in the ground state. Commonly used excimer gain media include gaseous substances such as XeCl, KrF, ArF, and XeF. The laser wavelengths emitted by these excimer media are in the ultraviolet band, with a wavelength range of 193 ~ 351 nm, among which the laser wavelength of XeCl is 308 nm; the laser wavelength of KrF is 248 nm. The main working mode of the excimer laser is the pulse mode, with an average output power of 100–200 W and a maximum power of 750 W. The use of ultraviolet band excimer lasers for laser cleaning mainly utilizes the characteristic of larger excimer laser energy, which is usually larger than the energy of the material molecule or atomic bond, and can be used to remove particles on the solid surface, and can also be used to peel off organic coatings.

New types of laser systems, such as high-power semiconductor lasers (LD), fiber lasers, are beginning to be used in the laser cleaning industry. The comparison with traditional types of lasers is as shown in Table 2.2.

2.4 Nd:YAG Laser

2.4.1 Principle of Light Emission of Nd:YAG Laser

Nd:YAG, short abbreviation for neodymium-doped yttrium aluminum garnet, is a crystal in which trivalent neodymium ions Nd^{3+} are doped into the yttrium aluminum garnet crystal, replacing Y^{3+} ions. Nd:YAG not only has very superior spectral and laser characteristics, but the lattice of its substrate material also has very attractive physical, chemical, and mechanical characteristics, which are particularly conducive to the generation of laser operation. Lasers with Nd:YAG as the laser gain medium are currently the most mature, most applied, and most widely used solid-state laser [14].

The simplified energy level structure of Nd^{3+} in Nd:YAG is shown in Fig. 2.13, with three spectral lines at 1050–1120 nm ($^4F_{3/2} \rightarrow {}^4I_{11/2}$), 870–950 nm ($^4F_{3/2} \rightarrow$

Table 2.2 Comparison of mainstream lasers in cleaning applications

Technical parameters	CO_2 laser	Flash pumped Nd:YAG	LD pumped Nd:YAG	High power LD	Fiber laser
Wavelength (μm)	10.6	1.06	1.06	Broadband	1.00 ~ 1.10
Output power (kW)	1 ~ 20	0.5 ~ 5	0.5 ~ 5	Broadband	0.5 ~ 50
Beam quality (mm mrad)	> 10	50 ~ 80	25 ~ 50	Poor	1 ~ 20
Fiber transmission	No	Yes	Yes	Yes	Yes
Volume	Huge	Large	Medium	Micro	Small
Energy conversion efficiency (%)	5 ~ 15	1 ~ 3	5 ~ 10	> 40	> 20
Maintenance cycle (kh)	1 ~ 2	< 1	3 ~ 5	Long	40 ~ 50
Maintenance cost	High	Relatively high	Medium	Low	Almost none
Energy consumption	Medium	High	Medium	Extremely low	Low

$^4I_{9/2}$), and 1310–1350 nm ($^4F_{3/2} \rightarrow {}^4I_{13/2}$). Among them, the energy levels of $^4F_{3/2}$ and above levels can serve as the absorption energy level E_3 of the pump light; $^4I_{15/2}$ is the metastable energy level E_2; $^4I_{13/2}$, $^4I_{11/2}$, and $^4I_{9/2}$ can all serve as the lower energy level E_1 of the laser. Because the $^4I_{13/2}$ energy level is far from the $^4I_{9/2}$ energy level, according to the Boltzmann distribution, at room temperature equilibrium, most of the neodymium ions are at the ground state energy level $^4I_{9/2}$, and the number of neodymium ions at the $^4I_{13/2}$ and $^4I_{11/2}$ energy levels is close to zero. When the $^4I_{13/2}$ energy level and the $^4I_{11/2}$ level are used as the lower laser level, Nd:YAG is a four-level system; when the $^4I_{9/2}$ level is used as the lower laser level, then Nd:YAG is a three-level system. The laser with a wavelength of 1064.1 nm is a typical four-level system, the upper level of the laser transition is the R_2 component of the $^4F_{3/2}$ level, and the lower laser level is the Y_3 component of the $^4I_{11/2}$ level. The fluorescence line corresponding to 1.06 μm is shown in Fig. 2.14.

Nd:YAG crystal has several strong absorption bands in the visible and near-infrared spectrum regions. The 0.81 μm band and 0.75 μm band are the two main absorption bands of Nd:YAG. The absorption coefficient near the strongest absorption band of 0.81 μm is about 740 m^{-1}, and the bandwidth of each absorption band is about 30 nm. Therefore, a flash lamp or laser diode with a wavelength of about 800 nm can be selected as the pump source.

Different neodymium ion doping concentrations can be chosen according to different needs to improve certain properties of the Nd:YAG laser. The general principle is that for Q-switching operation, in order to obtain high energy storage, the

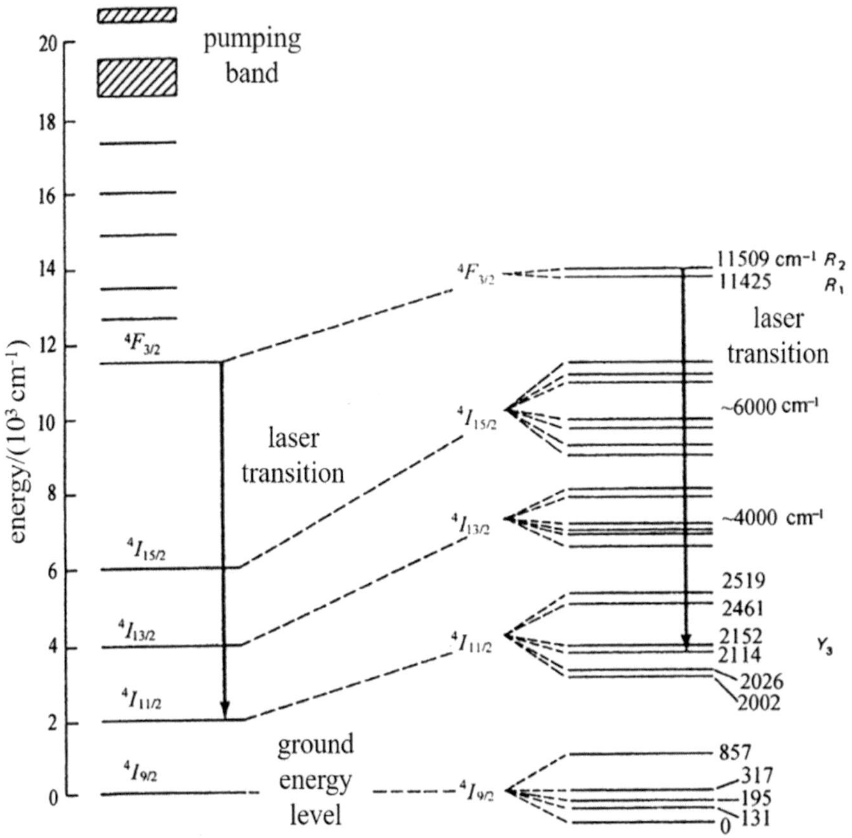

Fig. 2.13 Simplified energy level diagram of Nd:YAG

Fig. 2.14 Fluorescence spectrum in the 1.06 μm region produced by Nd^{3+} in YAG at 300 K temperature

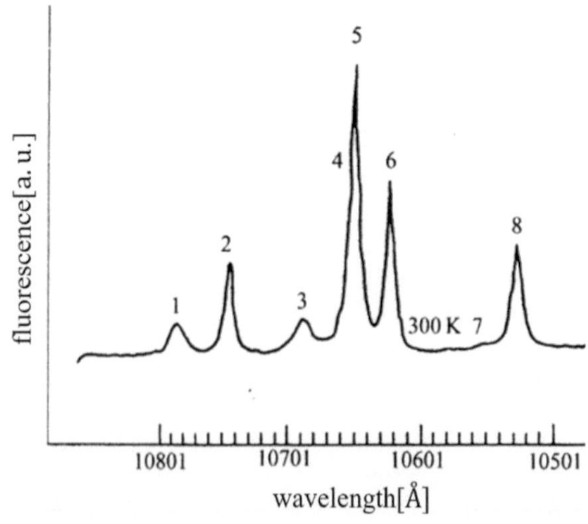

neodymium doping concentration is generally higher, close to 1.2%. For continuous operation, the neodymium doping concentration is generally chosen to be lower, about 0.6–0.8%, to obtain excellent beam quality. Since the radius of Y^{3+} ions and Nd^{3+} ions differ by about 3%, when a large amount of Nd^{3+} ions are doped, it will cause fluorescence concentration quenching phenomenon ($^4F_{3/2} - {}^4I_J$). Therefore, the concentration of neodymium atoms in YAG is usually limited to 1 ~ 1.5%. When the concentration of doped Nd^{3+} is $\leq 1\%$ (atomic ratio), the fluorescence lifetime of the fluorescence line near 1.06 μm is about 230–250 μs, the fluorescence linewidth is about 0.7–1 nm, and the fluorescence quantum efficiency is about 0.6.

2.4.2 *Basic Structure of Nd:YAG Laser*

The Nd:YAG laser consists of gain medium, pumping system, resonant cavity, cooling system, power supply system, etc. If it is a Q-switched or mode-locked laser, there is also a Q-switching or mode-locking system.

Figure 2.15 is a schematic diagram of the basic structure of a traditional flash lamp pumped (LD pumped, which is replaced by LD and divided into end-face pumped and side pumped) Nd:YAG laser.

I. **Structure of the Laser Resonant Cavity**

The simplest Nd:YAG laser resonator consists of an output coupler and a total reflector, with the gain medium Nd:YAG crystal inside the cavity. According to

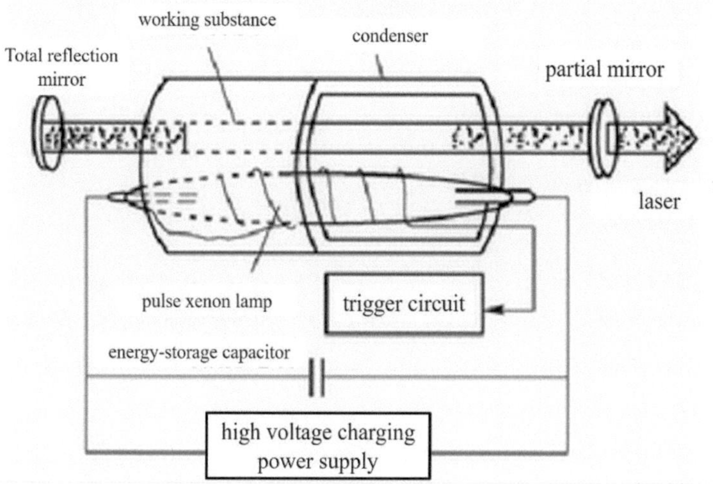

Fig. 2.15 Basic structure diagram of Nd:YAG laser

the structure of the cavity mirror, there are confocal cavity, concentric cavity, plane-concave cavity, concave-convex cavity, and plane-plane cavity, among which the plane-plane cavity can make good use of the volume of the gain medium, and the output laser has good directionality and higher power. In addition, the light-to-light conversion efficiency is also good, so the plane-plane cavity is often chosen.

Sometimes, for special needs, such as mode requirements, other retroreflective elements are placed in the cavity. In this way, according to the cavity type, there are linear cavity, V-shaped cavity, and Z-shaped cavity. The V-shaped and Z-shaped cavities are used more in end-pumping, which can effectively separate the pump light and the output light, reducing the damage to some light-sensitive elements.

II. **Q-Switching System**

The laser used for cleaning needs to have a higher peak power, but the peak power of the pulsed laser pumped by the flash lamp and the semiconductor is not high. At this time, Q-switching technology is needed to compress the pulse and increase the peak power of the laser pulse. The Nd:YAG laser used for laser cleaning mainly uses electro-optic Q-switching and acousto-optic Q-switching technologies. The parameters of the devices used in the two technologies are shown in Table 2.3.

Table 2.3 Comparison of electro-optic Q-switching devices and acousto-optic Q-switching devices

Characteristics	Acousto-opto Q-switching devices	Eletro-opto Q-switching devices
Shut-off ability (diffracted light intensity/incident light intensity)	Low (> 1:10)	High (< 1:200)
Insertion loss	Low (< 0.02)	High (> 0.05)
Switching time (ns)	> 50	2–5
Maximum repetition frequency	MHz	< 200 kHz
Minimum repetition frequency	> 1/τ	Several Hz
Beam quality requirements	Moderate	High
Polarization requirements	None	High
Temperature stability requirement	Low	High
Drive supply method	Radio frequency	High voltage
Destruction threshold	High (1 > GW/cm^2)	Medium (< 1 GW/cm^2)

The electro-optic Q-switching has large losses and high requirements for beam parameters, limited repetition frequency, and pulse width generally in the order of 10 ns. The repetition frequency of acousto-optic Q-switching is high, generally reaching even kHz, and the pulse width is about 100 ns. In laser cleaning, considering the cleaning efficiency, the Nd:YAG laser with acousto-optic Q-switching is most widely used.

Figure 2.16 is a schematic diagram of the concave cavity structure of the acousto-optic Q-switched Nd:YAG. Figure 2.17 is a photo of an acousto-optic Q-switcher device, the main body of which is a quartz crystal and a transducer. The main function of the Q driving source is to generate high-frequency radio frequency signals, which are transmitted to the acousto-optic Q-switcher through shielded wires. Since most of the radio frequency signals transmitted by the Q driver are absorbed by the crystal and the temperature rises, a cooling system is used to control the crystal temperature, and the common method is water cooling. Table 2.4 gives some parameters of the selected acousto-optic Q-switcher. Table 2.5 gives some parameters of the Q driving source.

III. **Cooling System**

As shown in Fig. 2.18, the actual photo of the cooling system of the laser is a two-way water cooling system. Path A is used for cooling the gain medium of the laser resonator, which requires a higher cooling capacity. Path B is used for cooling the acousto-optic Q switcher, which requires a lower cooling capacity. Path A consists of a main water tank, a magnetic pump, and a heat dissipation water tank. The tank stores a sufficient amount of water. After the gain medium absorbs the pump energy, a part of it will be deposited as heat. Through the circulation of cooling water, the heat in the gain medium is converted into the internal energy of the water, causing the water temperature to rise. Then, through the forced convection of the cooling fan,

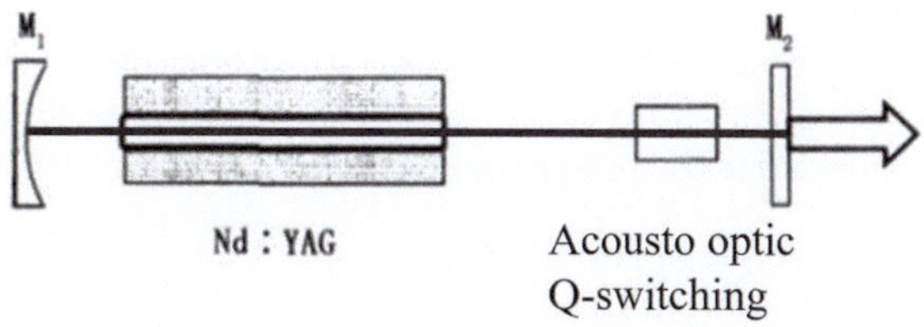

Fig. 2.16 Schematic diagram of high-power acousto-optic Q-switching Nd:YAG concave cavity structure

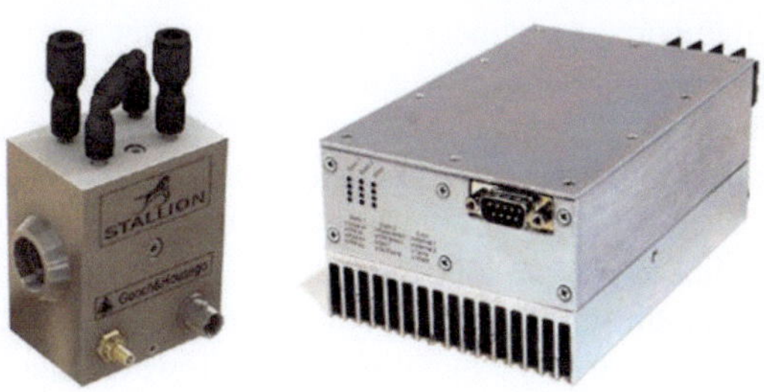

Fig. 2.17 Schematic diagram of acousto-optic Q-switcher and Q driving source

Table 2.4 Acousto-optic Q-switch parameters

	Fused quartz	Laser wavelength	1064 nm
Anti-reflection coating	Multilayer dielectric hard film	Transmittance	> 99.6% (typical >> 99.9%)
Film damage threshold	> 1 GW/cm^2	Static insertion loss	≤ 6% (typical < 5%)
VSWR	≤ 1.2:1	Maximum drive RF power	100 W
Overheat protection point	55°	Waterway material	Stainless steel 316
Cooling water flow rate	> 190 ml/min	Working water temperature	22–32°

heat exchange with the environment is carried out to dissipate the heat stored in the water. The cooling water used is pure water. Path B consists of a secondary water tank, a magnetic pump, a cooling fan, etc. It adopts a two-stage cooling method. First, the acousto-optic Q switcher is cooled by circulating cooling water (pure water) at a flow rate exceeding 190 cc/min, taking out the heat. Then, a refrigerant is used to cool the primary circulating water for the second time, and convection cooling is carried out through the cooling fan to ensure the cooling demands of the acousto-optic Q switcher.

IV. **Circuit System**

The entire laser requires a circuit system for power supply and control. The power supply part includes the laser power supply, the acousto-optic driver, and the cooling part. Figure 2.19 shows the laser power supply and the acousto-optic Q driving power supply, which can change parameters such as laser pump current and repetition frequency. They can also be modulated by external control input of square wave pulses with a duration of about 5 μs or other forms of modulation signals for laser Q-switching. Other electrical appliances such as water pumps and fans can be directly connected to the power supply circuit.

V. **Example of Nd:YAG Laser**

In laser cleaning, the Nd:YAG laser is the main force. Different lasers are selected according to different cleaning objects and types of pollutants. The average power is an important parameter, and many products are also distinguished by the average power of the laser cleaning machine.

For low-power (< 20W, referring to average power, the same below) laser cleaning systems, pulse-pumped electro-optic Q-switched or acousto-optic Q-switched Nd:YAG lasers can be selected as light sources. Taking the electro-optic Q-switched laser as an example, this type of laser has a narrow pulse width (about 10 ns), high single pulse energy (> 500 mJ), which can easily remove the paint layer on the surface of the metal substrate. However, due to the small average power and

Table 2.5 Q-switcher driving source parameters

Electrical parameters	Minimum	Maximum	Unit
Output power (with 50 Ω load impedance)	10	120	W
Supply voltage	23.5	24.5	V
Input current	4.5	9.5@100 W	A
Power loss	80	120	W
Maximum allowed VSWR value		1.25	
Electrical parameters	Minimum	Maximum	Unit
Maximum repetition rate when first pulse is suppressed	DC	500	kHz
Modulation frequency difference	DC	2@50 Ω	MHz
Linearity—offset		7	%
Operating frequency	24	46	MHz
Dynamic response	Minimum	Maximum	Unit
Fall time	80@40.68 MHz	120@27 MHz	ns
Rise time		100	ns
Harmonic suppression	< 40		dB
Dynamic response	< 40		dB
Thermal characteristics	Minimum	Maximum	Unit
Temperature drift	–	0.1	W/K
Time to reach thermal equilibrium		120	s
Environmental, installation, and transportation requirements	Minimum	Maximum	Unit
Airflow under standard heat sink at 40°	3.5 (2)	na	l/s
Storage temperature	− 5	+ 80	Celsius
Storage relative humidity		90	%
Operating ambient temperature	+ 5	+ 45	Celsius
Operating relative humidity		75	%
Ambient atmosphere		Atmospheric	
Main body dimensions (length, width, height)		152 × 100 × 40	mm
Weight	1100	1800	g

low repetition frequency (tens of Hz) if it is an electric tube Q switcher, the cleaning efficiency is relatively low. So the finished laser cleaning equipment generally does not use this low-power laser.

Power (20–200 W) laser cleaning systems have multiple options: (a), high repetition frequency (> 100 Hz) pulse-pumped electro-optic Q-switched Nd:YAG laser; (b), continuous krypton lamp pumped acousto-optic Q-switched Nd:YAG laser; (c), continuous LD (laser diode) module pumped acousto-optic Q-switched Nd:YAG laser. For example, the French Company QuantelLaserBlast adopted plan (a), using

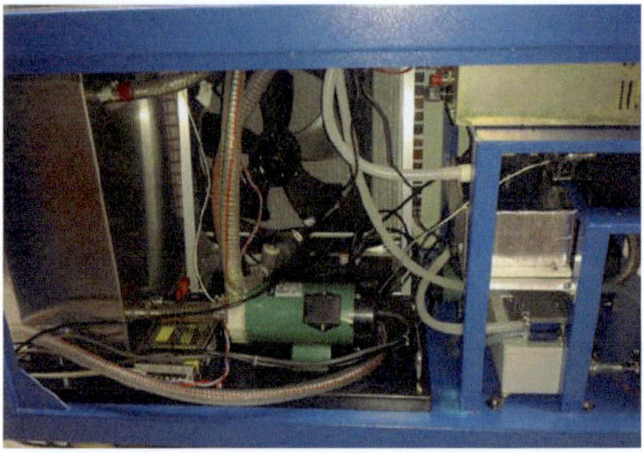

Fig. 2.18 Photo of the cooling system

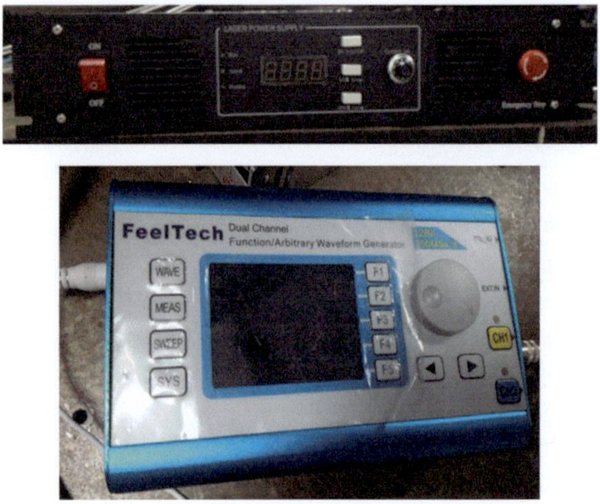

Fig. 2.19 Laser power supply and acousto-optic driver

a lamp-pumped laser with a repetition frequency of 120 Hz and a single pulse energy of 350 mJ, which is one of the best choices in moderate-power lasers with similar cleaning efficiency but the best cleaning effect. As for most professional laser cleaning companies, represented by AdaptLaser Company in the United States, either plan (b) or plan (c) is adopted. The cost of continuous krypton lamp pumping in this power range is slightly lower, but due to the low light conversion efficiency (about 1–2%), the cooling device in the laser system is heavily burdened. At the same time, the lifespan of the krypton lamp is short, requiring more frequent replacements than LD modules. The cost of continuous LD module pumping is higher, but in recent years, due to improvements in module production technology, the price of continuous LD modules has significantly decreased, especially for low-power modules, making the cost of manufacturing lasers using LD modules close to the cost of manufacturing traditional krypton lamp pumped lasers. Moreover, LD pumped light conversion efficiency is higher (> 10%), allowing for the use of cooling devices with lower refrigeration capacity. Although the lifespan of LD modules is longer than that of krypton lamps, both the LD module and the focusing cavity need to be replaced, unlike lamp pumping which only requires the replacement of the krypton lamp. With technological advancements and price reductions, continuous LD (laser diode) module pumped acousto-optic Q-switched Nd:YAG lasers have become mainstream.

High-power (> 200W) laser cleaning systems are basically continuous LD module pumped high repetition rate laser systems. Under higher power, it is necessary to increase the amplifier or adopt in-cavity multi-pump module series technology. Traditional acousto-optic Q-switching can be used to obtain laser pulses with narrower pulse widths. The lasers used by Germany's Rofin Company for cleaning, under acousto-optic Q-switching, achieve a narrow pulse output of 38 ns at a repetition rate of 6 kHz. With the maturity of new electro-optic crystal preparation technology, high repetition rate electro-optic Q-switching is also used in lasers for cleaning. The American Company AdaptLaser uses electro-optic Q-switching to achieve high power output with narrower pulse widths.

2.5 Carbon Dioxide Lasers

The carbon dioxide (CO_2) laser can emit invisible infrared lasers with a wavelength of 10.6 μm. Its research and development are relatively earlier, and commercial products are mature, widely used in material processing, medicine, military weapons, environmental measurement, and other fields [15].

2.5.1 Principle of Light Emission of CO_2 Lasers

The main gain media consist of CO_2, nitrogen, and helium. Among them, CO_2 is the gas that produces laser radiation, and nitrogen and helium are auxiliary gases. Figure 2.20 shows the molecular energy level diagram of the CO_2 laser, where laser transitions occur between two vibrational–rotational energy levels of the electronic ground state of CO_2 molecules. The purpose of adding nitrogen is to transfer energy in the CO_2 laser, allowing more particles to accumulate at the laser's upper energy level, improving the excitation efficiency of the laser's upper energy level, and helping to deplete the laser's lower energy level. The addition of helium has two functions: one is to accelerate the thermal relaxation process of the 010 energy level, which is beneficial to the depletion of the laser energy levels 100 and 020; the second is to achieve effective heat transfer. The excitation pathway of the CO_2 laser is as follows:

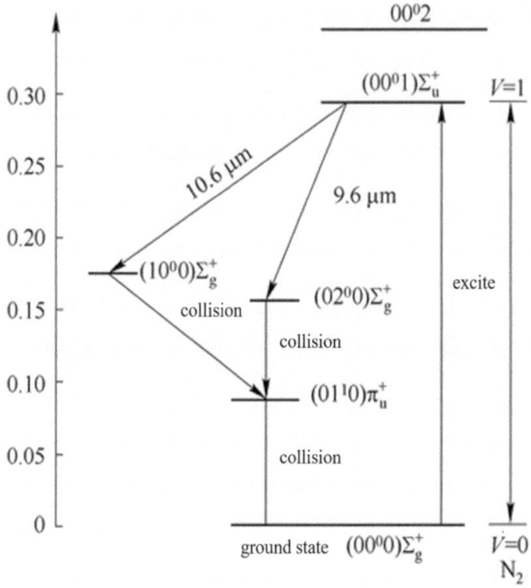

Fig. 2.20 Energy level diagram of CO_2 molecules related to laser generation

1. Direct Electron Collision

Electrons collide with CO_2 molecules in the ground state (00^00), exciting them to the laser energy level, which can be represented as: CO_2 (00^00) + e → CO_2 (00^01) + e.

2. Cascade Transition

Electrons collide with CO_2 molecules in the ground state, causing them to transition to the 00^0n energy level. Ground state CO_2 molecules then collide with high-energy level CO_2 molecules, transitioning to the laser upper energy level. This process can be represented as: CO_2 (00^00) + CO_2 (00^0n) → CO_2 (00^01) + CO_2 (00^0n − 1).

3. Resonance Transfer

N_2 molecules (v = 0) collide with electrons and transition to the v = 1 vibrational energy level. This energy level has a longer lifetime and is known as a metastable energy level, which can accumulate a large number of N_2 molecules. Ground state CO_2 molecules collide with metastable N_2 molecules in an inelastic collision and transition to the laser upper energy level. This process can be represented as: CO_2 (00^00) + N_2 (v = 1) → CO_2 (00^01) + N_2 (v = 0). The CO_2 molecule's 00^01 energy level is very close to the N_2 molecule's v = 1 energy level, so energy transfer is very rapid. In addition, the v = 2 ~ 4 energy levels of N_2 molecules are also very close to the 00^02 ~ 00^04 energy levels of CO_2 molecules, and resonance transfer can occur between them. CO_2 molecules at the $00^02 \sim 00^04$ energy levels can be excited to the 00^01 energy level by colliding with ground state CO_2 molecules.

Among the three excitation pathways mentioned above, resonance transfer has the highest probability and is the most significant. The number of CO_2 molecules at the energy level under molecular lasers mainly relies on the collision between gas molecules for pumping. Once the population inversion between (00^01) and (10^00), (02^00) is achieved, a laser with a wavelength of 10.6 μm can be generated through stimulated radiation transition.

2.5.2 *Basic Structure of CO_2 Laser*

As shown in Fig. 2.21, it is a schematic diagram of a typical CO_2 laser structure. It consists of a laser tube, an optical resonator, and a light source.

I. The Laser Tube

The laser tube usually consists of three parts: a gas storage tube, a discharge tube, and a water cooling tube. One end of the gas storage tube has a small hole connected to the discharge tube, and the other end is connected to the discharge tube through a spiral return tube, which can make the gas circulate in the discharge tube. The discharge tube is generally made of hard glass and often uses a layered sleeve structure. The gas entering the discharge tube is excited by the voltage between the cathode and anode, which can affect the output of the laser and the power of the laser output. The output power is proportional to the length of the discharge tube. Within a certain length range, the power output per meter of discharge tube length increases with the total length. Generally speaking, the thickness of the discharge tube has almost no effect on the output power. The water-cooled jacket tube is also made of hard glass, and its function is to cool the working gas, making the output power stable.

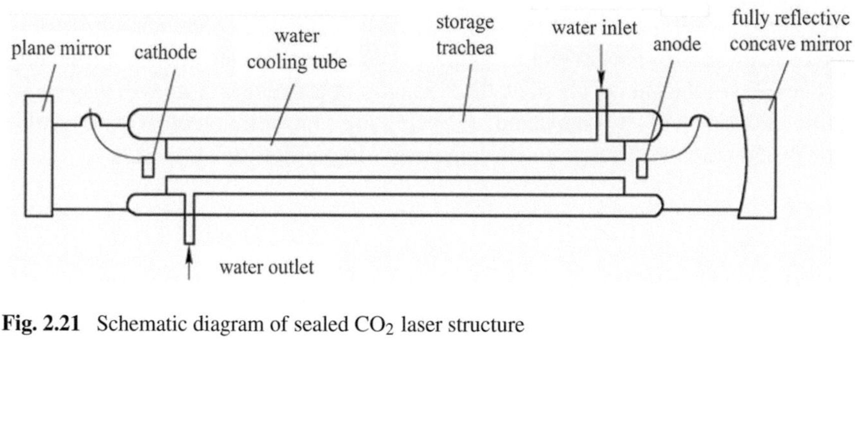

Fig. 2.21 Schematic diagram of sealed CO_2 laser structure

II. **The Optical Resonator**

The optical resonator is composed of a total reflection mirror and a partial reflection mirror. It is an important part of the CO_2 laser, and the two reflection mirrors that make up the CO_2 laser resonator can be directly pasted on both ends of the discharge tube. A flat-concave cavity is often used.

III. **The Pump Power Supply**

The pump power supply can provide energy to achieve population inversion between the upper and lower energy levels in the gain medium. The AC voltage is boosted by a transformer, and a high-voltage DC is obtained through high-voltage rectification and high-voltage filtering, which is applied to the laser tube, so that the CO_2 gas in the tube is excited.

2.5.3 *Output Characteristics of CO_2 Laser*

I. **Discharge Characteristics**

The CO_2 laser has an optimal discharge current, which is related to the diameter of the discharge tube, the total pressure in the tube, and the gas mixture ratio. Experiments show that as the tube diameter increases, the optimal discharge current also increases. For example, when the tube diameter is 20 ~ 30 mm, the optimal discharge current is 30 ~ 50 mA; when the tube diameter is 50 ~ 90 mm, the optimal discharge current is 120 ~ 150 mA.

II. **Temperature Effects**

The conversion efficiency of a CO_2 laser is very high, reaching up to nearly 40%. This means that over 60% of the energy is converted into thermal energy of the gas, causing the temperature to rise. The increase in gas temperature will cause the de-excitation of the upper energy level of the laser and the thermal excitation of the lower energy level, both of which will reduce the number of population inversion. Moreover, the increase in gas temperature will cause spectral line broadening, leading to a decrease in the gain coefficient. The increase in gas temperature will also cause the decomposition of CO_2 molecules, reducing the concentration of CO_2 molecules in the discharge tube. Therefore, cooling is very important for CO_2 lasers.

2.5.4 Main Types of CO_2 Lasers

III. Transverse Flow CO_2 Laser

The transverse flow CO_2 laser currently mainly adopts a three-axis orthogonal structure, that is, the directions of gas flow, discharge, and optical axis are mutually perpendicular. The special feature of the transverse flow CO_2 laser is that it can obtain a volumetrically uniform glow discharge, thus obtaining a larger laser output power, which is widely used in industry. Figure 2.22 is a schematic diagram of the external shape and internal structure of the transverse flow CO_2 laser.

IV. Axial Flow CO_2 Laser

In the axial fast flow CO_2 laser, the gain gas flows at high speed along the axial direction in the laser tube, excited by direct current or radio frequency, characterized by high power and high beam quality. The current laser output has reached 20 kW.

The output power of the fast axial flow carbon dioxide laser mainly depends on the gas flow in the discharge tube. Whether the gas in the discharge tube is laminar or turbulent affects the stability of the discharge and the maximum stable discharge current, thereby determining the size of the injectable power. The loss of the optical resonant cavity directly affects the photoelectric conversion efficiency of the laser and the transverse mode of the beam, but these two cannot be balanced. The larger the aperture of the discharge tube, the smaller the diffraction loss, the higher the efficiency, and the higher the order of the transverse mode.

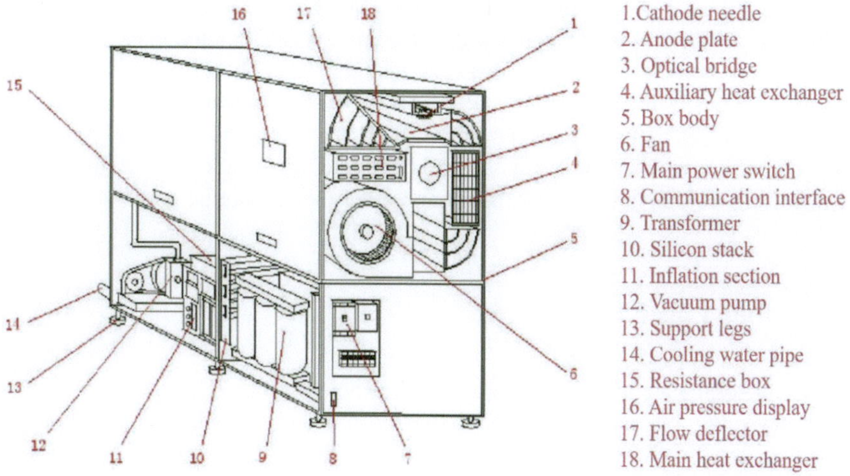

Fig. 2.22 Schematic diagram of the external shape and internal structure of the transverse flow CO_2 laser

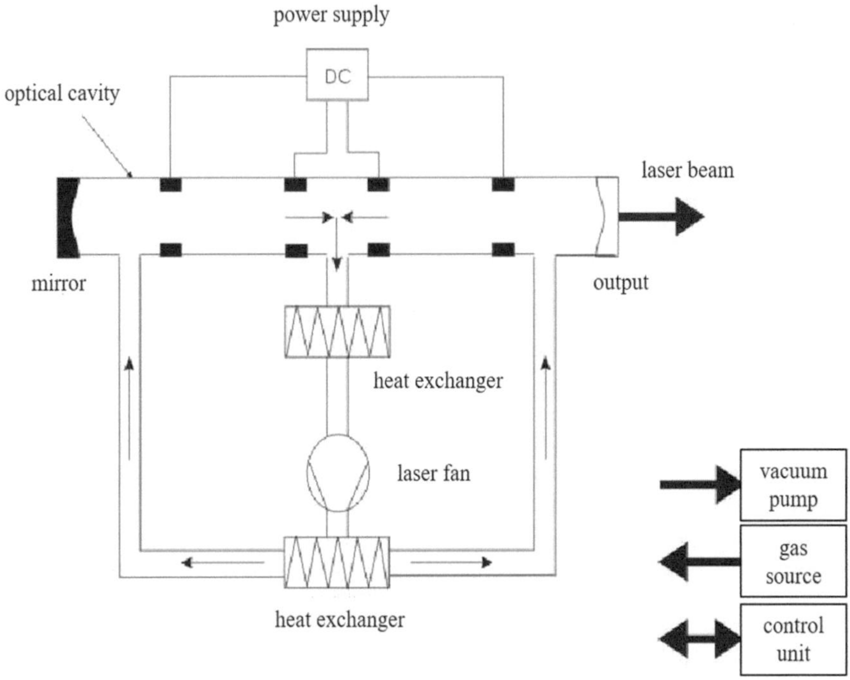

Fig. 2.23 Direct current excited fast axial flow CO_2 laser

Figure 2.23 is a schematic diagram of the axial flow CO_2 laser. The axial flow carbon dioxide laser is developed based on the early sealed circular glass tube longitudinal excitation CO_2 laser. The fast flow technology is introduced into the laser, and the gain gas is cooled by high-speed flow to obtain a higher power laser output. Its characteristics include: high power and high beam quality. The axial flow CO_2 laser output has reached 20 kW with good beam quality, which is more suitable for laser cutting welding and heat treatment and other metal material processing.

2.6 Fiber Laser

In general solid-state lasers, the coupling efficiency of pump light and laser is relatively low, and mode control is also relatively difficult. Due to dust adhesion on the reflectors, lenses, etc. of the optical system, and the deviation of the optical axis caused by the heat and mechanical effects of the surrounding environment, it is easy to cause phenomena such as reduced output power and deteriorated beam quality. In order to pursue high laser output power, it is necessary to increase the pump power, and the resulting thermal lens effect, thermally induced birefringence effect and other thermal effects become significant, which deteriorates the beam quality. Fiber lasers use rare-earth-doped fibers as the gain medium, overcoming the shortcomings of ordinary solid-state lasers. Although it is a type of solid-state laser, because the shape of the medium is very different, lasers that use fibers as the gain medium are generally considered separately from solid-state lasers that use bulk crystals or glass as the gain medium [16].

2.6.1 *Principle of Light Emission of Fiber Lasers [12]*

I. **Structure of the Fiber**

Fiber is the abbreviation for optical fiber, which is a very small diameter fiber made of transparent dielectric materials such as quartz glass (made from melted SiO_2) and plastic. The typical structure of a fiber, as shown in Fig. 2.24, is divided into the core (core layer), cladding, and coating (from inside to outside). The core is made of light-guiding material with a high refractive index (denoted as n_1). The core material of the SiO_2 fiber is often quartz glass, which is doped with a certain concentration of GeO_2 and P_2O_5. Its function is to basically transmit the light signal along the axial direction within the core. Rare earth elements are often doped in the core, not only as activating ions, but also to increase the refractive index, making it higher than the cladding, so that the light propagating in the core can undergo total reflection. The cladding of the fiber is made of light-guiding material with a slightly lower refractive index (denoted as n_2) than n_1, and its function is to confine light. Due to the refractive index of the core and cladding, satisfying $n_1 > n_2$, total reflection of light can occur at the interface of the core-cladding, so that most of the light energy is confined in the core, causing the light signal to transmit along the axial direction of the core.

The core and cladding constitute the fiber. In order to protect the fiber, improve the mechanical strength of the fiber and the resistance to micro-bending, and reduce attenuation, a layer of polymer material is coated on the outside of the cladding. Generally, the coating layer has two layers, the inner layer is a low-modulus polymer material, called the primary coating; and the outer layer is a high-modulus polymer material, called the secondary coating.

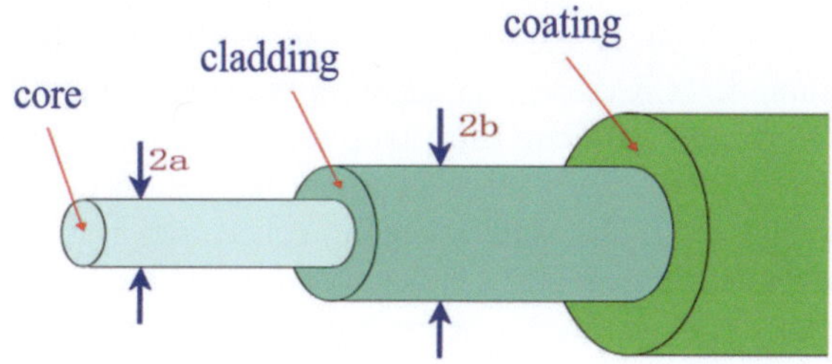

Fig. 2.24 Schematic diagram of fiber structure

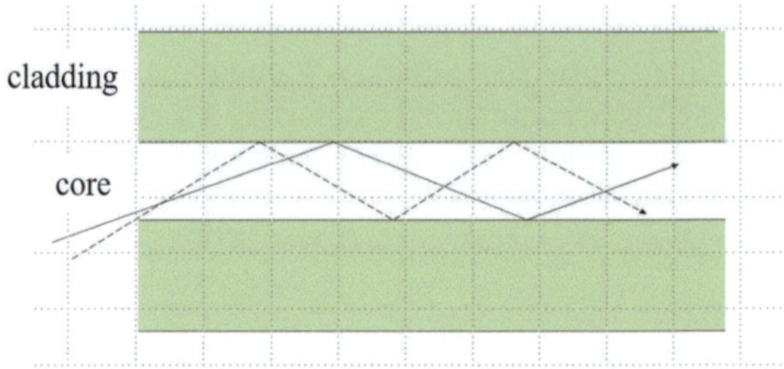

Fig. 2.25 Transmission of fiber

II. **Propagation Characteristics of the Fiber**

The core and cladding of the optical fiber are its essential parts. Using the principle of total reflection, the light signal is confined within the core and transmitted forward. Light enters the fiber from the end face of the fiber, according to the principle of geometric optics, when the angle of incidence of the light is greater than the critical angle θ_c of the fiber, the light is confined within the core. The situation of light propagating in the fiber at different angles of incidence is shown in Fig. 2.25.

The light waves propagating within the core of the optical fiber can be decomposed into plane waves propagating along the axial direction and plane waves propagating in the vertical direction (cross-sectional direction). The plane waves propagating in the cross-sectional direction will reflect at the interface between the core and the cladding. If the phase change in one round trip (incidence and reflection) is an integer multiple of 2π, a standing wave will form. Only those lights that enter the fiber at specific angles that can form standing waves can propagate within the fiber, and these light waves are called modes. The mode with the smallest angle of incidence to the fiber (closest to the fiber axis) is called the fundamental mode, and the modes with larger angles of incidence than the fundamental mode are called higher-order modes.

Only a certain number of modes can be transmitted within the fiber. Usually, when the core diameter is bigger (tens of micrometers and above), hundreds of modes can be propagated, while when the core is very small (a few micrometers), only one mode can be propagated. The former is called multimode fiber, and the latter is single-mode fiber.

III. **Active Ions Doped Fiber**

The fiber whose core is doped with rare earth elements is called doped fiber. The pump light is directly incident on the core of the doped fiber for pumping. After the ground state rare earth ions absorb the pump light, they transition to a higher energy level, and emit light when they return to a lower energy level, as shown in Fig. 2.26. The optimal doping concentration is at the level of 100 ppm. If it is too low, the gain is not enough, and if it is too high, concentration quenching can easily occur between rare earth ions, and the number of particles in the upper laser level will decrease due to non-radiative relaxation, which is not conducive to laser generation. Common doping active ions include: erbium (Er^{3+}), neodymium (Nd^{3+}), praseodymium (Pr^{3+}), thulium (Tm^{3+}), ytterbium (Yb^{3+}), and holmium (Ho^{3+}). For

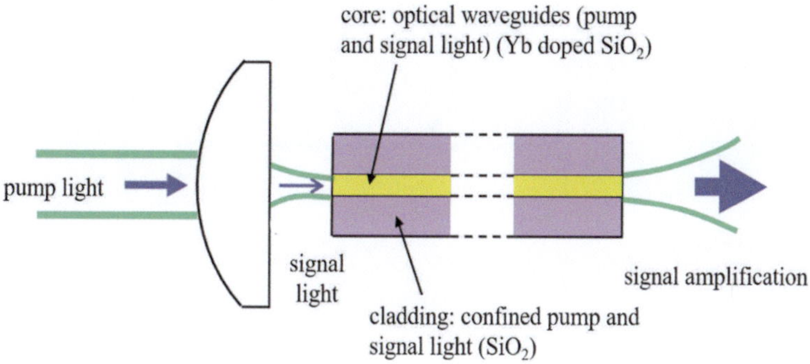

Fig. 2.26 Working principle diagram of fiber laser

neodymium-doped fiber: using pump sources with wavelengths of 800, 900, 530 nm, lasers can be obtained at wavelengths of 900, 1060, and 1350 nm. For erbium-doped fiber: using pump sources with wavelengths of 800, 900, 1480, 530 nm, lasers can be obtained at wavelengths of 900, 1060, and 1550 nm. For ytterbium-doped lasers, using a pump source with a wavelength of 915 nm wavelength pump source, lasers can be obtained at wavelengths from 975 nm to 1.05 μm.

2.6.2 *Basic Structure of Fiber Lasers*

The structure of a fiber laser is similar to other lasers, mainly composed of three parts: a pump source (LD), a gain medium, and a resonant cavity, as shown in Fig. 2.27. In a fiber laser, the gain medium is a fiber doped with rare earth active ions. The pump light is guided into the gain fiber through an optical coupling system, generating stimulated radiation. The light from the stimulated radiation is largely reflected back into the gain fiber after reaching the high reflection mirror on the right, causing the stimulated radiation light to be amplified again in the gain fiber. The light source is amplified again, and the left side of the resonant cavity is a total reflection mirror, which reflects the light source again. Between the two mirrors, each time the light source is reflected, the laser undergoes an amplification process. Only a small part of the amplified light is output through the semi-reflective mirror on the right side after the light intensity reaches the threshold.

With the development of technology, the two resonant mirrors in the fiber laser are directly replaced by coating the fiber surface or creating a grating structure on the fiber. In this way, there is no need to add cavity mirrors outside the fiber, but the total reflection mirror and output mirror are integrated on the fiber. This greatly simplifies the structure and enhances reliability. This type of laser is called an all-fiber laser, which is basically used in the market today.

Fiber lasers have continuous wave (CW) oscillation and pulse oscillation. The former can achieve high power and is often used in cutting and welding; the latter has

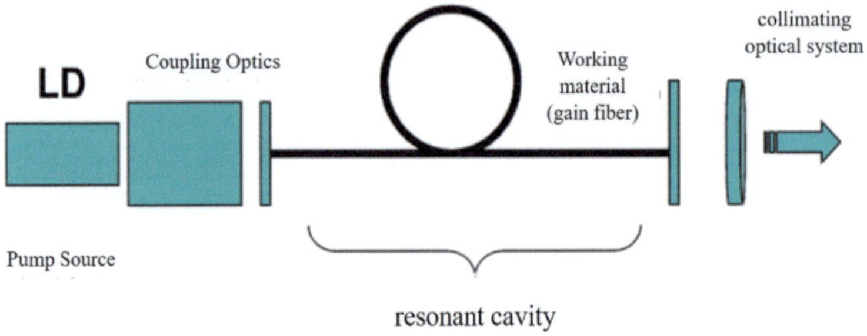

Fig. 2.27 Principle diagram of fiber laser

a smaller average power but a larger peak power, and is often used in fine processing and marking. Because fiber lasers have good beam quality and are easy to transmit through fibers, convenient to use, they are gradually replacing other lasers in industrial applications such as cutting, marking, and remote welding in recent years. In laser cleaning, they are also becoming an increasingly important main force.

2.6.3 Features of Fiber Lasers

Fiber lasers have the characteristics of solid-state lasers, as well as their own unique features.

1. Miniaturization and Lightweight

Fiber is miniaturized, lightweight and can be bent and coiled up. The installation location of the fiber laser can be flexibly determined.

2. Good Stability

All-fiber lasers do not have spatial optical components, so they are not easily affected by dust adhesion and the surrounding environment's heat and mechanical influences, and they have good stability. Almost no maintenance required.

3. High Efficiency, Easy to Achieve High Power, Good Beam Quality, Convenient Transmission

The laser emitted from the fiber has a small NA (numerical aperture), which is easy to focus. The fiber interaction length is long, and the pump light is enclosed in the fiber, so high-efficiency pumping can be achieved (the light-to-light conversion efficiency can reach 70%, and the electro-optical conversion efficiency is about 30%). The pump modules can be connected in series and parallel, so it is easy to increase the output power. A 50 kW ultra-high output power continuous (CW) fiber laser has been put into practical use. The laser output from the fiber laser can be efficiently coupled into the transmission fiber for processing objects at a distance.

4. Easy to Generate Nonlinear Optical Effects

The core diameter of optical fibers is small, with a long interaction length, which can easily produce nonlinear optical effects. Therefore, it is not suitable for high-intensity pulse operation, and the performance of the laser is also somewhat limited.

2.7 Excimer Lasers

Excimer laser is a type of gas laser that uses excimers as the gain medium. Common excimer lasers include ArF lasers, KrF lasers, etc. They are called excimers because these gas molecules are unstable. When an external energy excites a certain medium, this gas undergoes a series of physical and chemical reactions, forming molecules with very short lifetimes (generally tens of nanoseconds), which are called excimers.

Excimer lasers mainly use electron beams or transverse rapid pulse discharges with energy greater than 200 keV for excitation. When the unstable molecular bond of the excited excimer breaks and dissociates into ground state atoms, the released energy is emitted in the form of laser radiation.

2.7.1 Common Types of Excimer Lasers

Since the birth of the first excimer laser in 1972, a variety of gas-phase xenon molecules and other rare gas excimers, rare gas oxide excimers (krypton oxide, xenon oxide, argon oxide, etc.), metal vapor-rare gas excimers (sodium xenon, etc.), rare gas monohalide excimers (xenon fluoride, argon fluoride, krypton fluoride, xenon chloride, xenon bromide, xenon iodide, krypton chloride, etc.), metal halide excimers (mercury chloride, mercury bromide, etc.) and metal excimers (sodium excimers, etc.) lasers have been developed. The emission wavelength, typical output energy, pulse width, efficiency, and other parameters of commonly used excimer lasers can be found in Table 2.6 [17].

2.7.2 Working Principle of Excimer Lasers

Under normal conditions, chemically stable inert gases such as He, Ne, Ar, Kr, Xe and chemically active halogens such as F, Cl, Br, etc. form a mixed gas, which is used as the gain medium and filled into the gas chamber. Under normal circumstances, these inert gas atoms and halogen atoms will not combine to form molecules. If the gas chamber is given appropriate electrical excitation, the two types of atoms can be synthesized into excited excimers. These excited molecules have a very short lifetime and will quickly transition back to the ground state, radiating photons. The molecules

Table 2.6 Main parameters of excimer lasers

	Excimers	Emission wavelength (Å)	Output energy	Pulse width (ns)	Efficiency (%)
Homokaryotic	Xe_2	1730	8 J	20	2.5
	Kr_2	1457		10	
	Ar_2	1261	3 J	50	15
Heterokaryotic	KrF	2485, 2495	117 J		3.5
	XeF	3531, 3511	1 J	50	0.4
	ArF	1932	92 J	55	3.5
	KrCl	2221	60 mJ	15	
	XeCl	3076, 3078 3082, 3084	110 mJ	30	
	ArCl	1750	0.2 mJ	2.5	
	XeBr	2820	60 mJ	20	
	XeO	5300 ~ 5550 four lines	100 mJ	100	

that return to the ground state have an even shorter lifetime and will immediately decompose and revert to the original two types of atoms. Since the lifetime of the molecules in the lower energy level of the laser is much smaller than that in the upper energy level, the lower energy level is almost empty, making it easy to form population inversion. The radiated photons are amplified by the resonant cavity and emit high-energy ultraviolet lasers [17].

2.7.3 Basic Structure of Excimer Lasers

1. Laser Resonator. It is used for storing gas, generating laser through gas discharge excitation, and laser mode selection. It consists of a front cavity mirror, a rear cavity mirror, a discharge electrode, and a pre-ionization electrode, and is connected to the gas storage tank through two rows of small holes for the exchange and replenishment of gain gas. In order to obtain a uniform large-area stable discharge, general excimer lasers will use pre-ionization technology, that is, before the main discharge starts, a high voltage of 20–30 kV is applied between the pre-ionization electrode and the cathode of the main discharge, causing a corona discharge between them, forming a uniform ionization layer near the cathode. When the gas discharges, the pulsed high-voltage power supply is applied to the electrode to discharge the gain gas in the resonator, generating energy level transitions, radiating photons, oscillating through the feedback of the reflection mirror, and finally the laser is output from the front cavity mirror.
2. Laser Discharge Box. It mainly consists of discharge electrodes, pre-ionizers, circulating fans, particulate capture wells, internal box windows, etc. During

operation, the strictly proportioned inert gas and halogen gas are filled into the box through the gas path system and maintained at a certain pressure. The pre-ionizer generates a certain concentration of free electrons in the electrode discharge area, presenting a low impedance state, preparing for subsequent electrical excitation and gas discharge. Since some particles formed during the discharge process will contaminate the box, a circulating fan is used to blow these particles to the particle capture well, and the cleaned gas returns to the discharge area after treatment. The internal box window forms a path for laser resonance.

3. Laser Tuning Module. The wavelength is generally selected by a grating. The beam in the cavity is incident on the grating at a certain angle, diffracts, and only light of a specific wavelength can return to the resonator for resonance, while light of other wavelengths diffracts at different angles, thereby achieving the purpose of selecting the laser wavelength.
4. Laser Energy and Wavelength Sampling and Calibration Module. It consists of a calibrator, a grating, a photoelectric receiver, and an atomic wavelength reference (AWR). The calibrator (Fabry–Perot interferometer) and grating jointly select the laser wavelength, allowing the specific wavelength laser to enter the resonator for resonance, achieving stable laser wavelength output. The atomic wavelength reference module is used to calibrate the laser wavelength. For laser energy, the laser energy information is sent to the main CPU circuit for control and display through the photoelectric receiver.
5. High Voltage Power Supply Pulse Conversion and Pulse Compression Module. It is used to provide excitation pulse electrical output, consisting of a CPU circuit, a high-voltage power supply, a timer, a power switch, a pulse compression module, a coupling module, etc. The emitted electrical pulse outputs alternating pump pulses to the laser discharge box, thereby causing the laser gas in the discharge box to discharge under sufficient electrical excitation, and finally output the laser.
6. Gas Control and Water Cooling System Module. The gain gas is directed to the laser discharge box and other system modules through the busbar, flow meter, and solenoid valve. The proportion of inert gas and halogen gas is strictly controlled by the gas control module. The water cooling system is used to cool the heating components.

2.7.4 Characteristics and Application Fields of Excimer Lasers

Excimer lasers have very special gain media, so they have many unique features:

1. The output wavelength of the excimer laser mainly falls within the ultraviolet band, specifically between 172 and 354 nm. It is characterized by its short wavelength, high frequency, and high energy.
2. Excimers exist in an excited state, with a lifespan on the order of tens of nanoseconds, and the lifespan of the ground state is even shorter, on the order of

10^{-13} s. Transitions between the low excited state and the repelled ground state (or weakly bound) can produce continuous spectral fluorescence, thus enabling wavelength-tunable operation.
3. The quantum efficiency of the excimer laser is very high, approaching 100%, and it can operate at high repetition rates.
4. The number of reciprocations of the excimer laser in the resonant cavity is small, the beam directionality is poor, and the divergence angle is large, generally between 2 and 10 mrad.
5. The peak power of a single pulse is very high, reaching 1–10 GW, and the energy of a single pulse can reach several joules or more.

Excimer lasers have a wide range of applications in agriculture, medicine, defense, biology, and other fields. In medicine, because the specific ultraviolet wavelength of the excimer laser can be almost completely absorbed by certain human tissues, it is very suitable for medical treatment. Excimer lasers are one of the best and most effective methods for treating myopia. In laser processing, the small focal spot and high processing resolution of the excimer laser make it very suitable for high-quality laser processing. In addition, excimer lasers have also been used for isotope separation, high-speed photography, high-resolution holography, material structure analysis, remote sensing, and more. In the field of laser cleaning, especially microelectronic cleaning, it has unique effects.

2.8 Semiconductor Lasers

Semiconductor lasers, also known as laser diodes, abbreviated as LD, use semiconductor materials as the gain medium. Common gain media include gallium arsenide (GaAs), cadmium sulfide (CdS), indium phosphide (InP), and zinc sulfide (ZnS). Excitation methods include electrical injection, electron beam excitation, and optical pumping. Semiconductor lasers can be divided into homojunction, single heterojunction, and double heterojunction types. Homojunction lasers and single heterojunction lasers are mostly pulsed devices at room temperature, while double heterojunction lasers can run in continuous wave operation at room temperature.

Semiconductor lasers were successfully developed in 1962 and achieved continuous output at room temperature in 1970. They are characterized by their small size, long lifetime, high efficiency, and low cost. They can serve as pump sources for solid-state lasers or be used independently. They have been widely used in laser communication, optical storage, laser printing, laser ranging, and other areas [18].

2.8.1 Working Principle of Semiconductor Lasers

Like other lasers, the working principle of semiconductor lasers is also stimulated radiation, which requires particle population inversion. In semiconductor materials, the energy levels of electrons form energy bands, with the high-energy conduction band serving as the upper laser level and the low-energy valence band as the lower laser level. Through electrical excitation, electrons are excited to the high-energy band, and the number of electrons occupying the conduction band exceeds the number of electrons occupying the valence band, forming a population inversion distribution. Then, the electrons transition back to the valence band, and through electron–hole pair recombination, they radiate energy in the form of light. The cleavage planes of the semiconductor crystal are used to form two parallel reflective mirrors as reflectors, forming a resonant cavity, amplifying the radiation of light, and outputting lasers.

The emission wavelength of semiconductor lasers is greatly affected by temperature, generally changing by 0.2–0.3 nm/°C with temperature. The spectral width increases with temperature, and the light intensity correspondingly decreases. The light output of semiconductor lasers increases with the increase of current. At high power, the heat generated is also high, so heat dissipation is very important, which should be fully considered during packaging.

2.8.2 Classification of Semiconductor Lasers

According to different classification ways, semiconductor lasers (LD) can be divided into different types. According to the structure of semiconductor materials, they can be divided into homojunction LD, heterojunction LD (including single heterojunction, double heterojunction); according to specific materials, there are GaAIAs/GaAs lasers, InGaAsP/InP lasers; according to wavelength, there are visible light, far-infrared lasers; according to principles and operating characteristics, there are dynamic single-mode lasers, distributed feedback lasers, quantum well lasers, surface emitting lasers, microcavity lasers.

According to the power size, they can be divided into low-power LD lasers and high-power LD lasers. The former is widely used in the information field, such as distributed feedback and dynamic single-mode LD used for optical fiber communication and optical switching systems, narrow linewidth tunable distributed feedback-LD, red and blue-green LD used for optical disc and other information processing technologies. The development of high-power LD is also very rapid, with output power reaching several kW. The rapid development of solid-state lasers pumped by semiconductor lasers (LDP) benefits from the development of high-power LD.

2.8.3 Features and Application Fields of Semiconductor Lasers

As a special type of laser, semiconductor lasers have the following features:

(1) Small size, light weight. The gain medium and resonant cavity are integrated, eliminating the need for cumbersome water cooling devices.
(2) Low driving power and current, high efficiency. Semiconductor lasers are the most efficient of all lasers, several times or even ten times more efficient than that of conventional solid-state lasers. Therefore, to achieve the same power as other lasers, the required driving current and power are much lower.
(3) Long working lifetime. The device structure is simple and reliable.
(4) Directly electrically modulated. By modulating the driving current, the output laser can be modulated.
(5) Easy to integrate with various optoelectronic devices. Due to its small size, light weight, and simple structure, it is easy to integrate with other optoelectronic components.

Compared to solid-state lasers, the beam quality of LD is much poorer, which affects its applications, such as difficulties in making significant contributions in laser cutting. However, it is widely used in fields such as optical communication, laser indication, laser guidance, and laser radar. In fact, semiconductor lasers have become the best-selling type of lasers. In terms of laser cleaning, semiconductor lasers can also make a difference when the requirement for beam quality is not high.

References

1. Song F, Liu S J, Yan B X. Laser cleaning [J]. Cleaning World, 2004(05): 43–48.
2. Bedair S M, Smith H P. Atomically clean surfaces by pulsed laser bombardment [J]. Journal of Applied Physics, 1969, 40(12): 4776–4781.
3. John F A, Carl G M, Walter H M. Studies on the interaction of laser radiation with art artifacts [J]. Proc Spie, 1974, 41:19.
4. Lazzarini L, Marchesini L, Asmus J F. Lasers for the cleaning of statuary: initial results and potentialities [J]. Journal of Vacuum Science & Technology, 1973, 10(6):1039–1043.
5. Sheng X Z, Lou S Q. Principles of Lasers [M]. Beijing: Tsinghua University Press, 2010.
6. Qing Y B. The development of laser devices [J]. Laser Journal, 1979, 1: 1–7.
7. Zou T C. Advanced Manufacturing in the 21st Century: Laser Technology and Engineering [M]. Beijing: Science Press, 2007.
8. Li G P. Non-equilibrium steady state in three-energy-level laser system [J]. Chinese Journal of Lasers, 1981(07): 3–5.
9. Gu Q. Rate equation of four-level systems depending evidently on geometric parameters of the cavity [J]. Chinese Journal of Lasers, 1986(06): 14–17.
10. Zhou B K. The Principle of Laser [M]. Beijing: National Defense Industry Press, 2014.
11. Sun H B, Pei Y J, Jin K, et al. Calculation of loaded cavity external Q [J]. High Power Laser and Particle Beams, 2006(01): 127–130.
12. Li X. Research on rational harmonic model-locked fiber laser [D]. Beijing: Beijing Jiaotong University, 2013.

13. Zhang J, Guo L, Zhang Q M, et al. Resonant amplification structure of high-power Nd:YAG laser design and analysis [J]. Chinese Journal of Lasers, 2012, 39(4): 7–11.
14. Bo Y, Geng A C, Bi Y, et al. High average power Q-switched quasi-continue-wave Nd:YAG laser [J]. Acta Physica Sinica, 2006, 55(3): 1171–1175.
15. Qiu J L. High power industrial CO_2 laser of a new generation [J]. Chinese Journal of Lasers, 1994(5): 377–381.
16. Digonnet M J F, Gaeta C J. Theoretical analysis of optical fiber laser amplifiers and oscillators [J]. Applied Optics, 1985, 24(3):333.
17. Song J, Zhu P, Zhang W Y, et al. The principle and application of cymer excimer laser [J]. Equipment for Electronic Products Manufacturing, 2008, 7: 60–62.
18. Shan Z G. Semiconductor lasers and their applications [J]. Laser & Optoelectronics Progress, 1990(06): 31–37.

Chapter 3
Physical Basis of Laser Cleaning

3.1 Reflection, Scattering, and Absorption of Laser by Matter

Light is an electromagnetic wave. When a laser shines on a cleaning object, its essence is the interaction between the laser and the material. For most materials, the magnetic field's force on electrons is much smaller than that of the electric field, so the magnetic field's force on electrons can be ignored, and only the effect of the electric field on the matter can be considered.

In a uniform and non-absorbing propagation medium, the electric field of the laser can be expressed as follows:

$$E = E_0 \exp\left[i\left(\frac{2\pi z}{\lambda} - \omega t\right)\right] \tag{3.1.1}$$

where E_0 is the amplitude, z is the direction of laser propagation, ω is the angular frequency, and λ is the wavelength. When the laser shines on the surface of the material, reflection and absorption will occur. The remaining laser energy after reflection and absorption will pass through the material. The total energy is conserved, that is,

$$I = I_r + I_a + I_t \tag{3.1.2}$$

where I is the total incident laser intensity, I_r is the laser intensity reflected (including diffuse reflection, scattering) from the material surface; I_a is the laser intensity absorbed by the finite thickness material; I_t is the laser intensity passing through the material.

F. Song and X. Lin, *Laser Cleaning*, https://doi.org/10.1007/978-981-97-6631-4_3

3.1.1 Reflection and Scattering of Laser by Materials

During the process of laser cleaning, when the laser incident on the surface of the object, the first thing that happens is the reflection and scattering of laser. Studying the reflection and scattering process of lasers during the cleaning process is of great significance. On one hand, when the laser incident on the surface of the object to be cleaned, part of the laser will be reflected and scattered. If this part of the laser energy increases, the laser energy used for cleaning will decrease; on the other hand, different materials have different reflections and scatterings of lasers. By analyzing the reflected and scattered laser, we can judge what changes have occurred in the material during the cleaning process over time, that is, we can perform online real-time detection of the cleaning process by detecting the reflected and scattered laser. This technology has practical application significance for fine cleaning processes and improving cleaning efficiency.

I. **Reflection**

In laser cleaning, apart from some special cleaning mechanisms, such as cleaning at a certain angle and back-incident cleaning, in general cleaning, the laser is vertically incident on the surface of the object to be cleaned. In this case, for a smooth material interface, the reflectivity is:

$$R = \frac{(n_1 - n_2)^2}{(n_1 + n_2)^2} \tag{3.1.3}$$

where n_1 and n_2 are the refractive indices of the two media respectively. The refractive index is related to the wavelength of laser, so the reflectivity of laser on the interface is related to the physical properties of the medium and the wavelength of laser.

In the experiment, the reflectivity can be actually measured by the integrating sphere method. When the material thickness is much larger than the absorption length, the absorptivity A of the material can be obtained through the reflectivity R, that is

$$A = 1 - R \tag{3.1.4}$$

The opposite situation is when the laser-affected material has a thickness less than or on the same order of magnitude as the absorption length, the laser will pass through the absorbing material, so the absorptivity cannot be calculated solely by the reflectivity. In this case, the material's reflectivity and transmittance can be used to calculate the absorptivity. Table 3.1 lists the reflectivity of some metal materials.

II. **Scattering**

When the surface of the material is not smooth, the main factor affecting the attenuation of light intensity is scattering. The intensity of light after scattering can be described by Lambert's law:

Table 3.1 The reflectivity of some metal materials [1]

Material type	Total reflectivity (%)	Regular reflectivity (%)	Diffuse reflectivity (%)	Luminaire absorptivity (%)
Mirror silver	95	90	5	5
Polished oxidized mirror aluminum (purity 99.99%)	90	80	10	10
Polished oxidized mirror aluminum (purity 99.85%)	90	70	20	10
Polished and oxidized coarse grained hammered aluminum (purity 99.85%)	85	60	25	15
Polished and oxidized matter smooth aluminum (purity 99.85%)	85	40	45	15
Bright white polished metal sheet	80	40	40	20
Diffuse white polished metal sheet	80	10	70	20
Polished and oxidized fine grained hammered aluminum	70	20	50	30
Polished and oxidized painted aluminum	70	20	50	30
Polished and oxidized medium smooth aluminum	70	20	50	30
Polished and oxidized aluminum (purity 99.5%)	60	10	50	40
Coated sub-white metal sheet	60	5	55	40
Bright chrome	65			35
Polished stainless steel	55 ~ 65			35 ~ 45

$$I = I_0 e^{-\alpha_s l} \tag{3.1.5}$$

where I_0 is the incident light intensity, α_s is the scattering coefficient, which is related to the properties of the material, surface roughness, and laser wavelength.

1. **Elastic Scattering**

The wavelength of the scattered light is the same as the incident light, that is, the wavelength does not change during the reflection process, similar to the elastic collision process of particles, which is called elastic scattering. In elastic scattering, there are Rayleigh scattering and Mie scattering. When the linear dimension a of the substance interacting with the incident laser is much smaller than the wavelength λ of the incident laser, it is called Rayleigh scattering. The intensity of Rayleigh scattered light is inversely proportional to the fourth power of the wavelength λ of the incident light, which can be expressed as:

$$I(\lambda) \propto \frac{1}{\lambda^4} \tag{3.1.6}$$

This is known as Rayleigh's law, which only holds when $2\pi a/\lambda < 0.3$.

When the linear dimension a of the scatterer in the material is greater than a fraction of the wavelength, or even comparable to the wavelength, Rayleigh's scattering law no longer holds. This is known as large particle scattering, or Mie scattering. The intensity of Mie scattering is almost independent of the wavelength. Unlike the symmetrical distribution of Rayleigh scattering, Mie scattering is stronger in the forward direction of the light than in the backward direction, showing obvious directionality. When the particle diameter is large, Mie scattering can be approximated as Fraunhofer diffraction. Mie scattering can be analyzed through Mie theory, which is a strict mathematical solution of Maxwell's equations under the boundary conditions for a single medium sphere in an isotropic uniform medium with monochromatic parallel light irradiation.

2. **Inelastic Scattering**

When the interaction process between the incident laser and the working material involves quantum processes, the scattering process becomes complex, and the wavelength of the scattered light will be different from that of the incident laser. This scattering process is called inelastic scattering. The elastic scattering of the incident laser with free electrons is called Thomson scattering, while the inelastic scattering of the incident laser with free electrons is called Compton scattering. Compton scattering includes Raman scattering and Brillouin scattering. In the laser cleaning process, it is basically elastic scattering.

3.1.2 Absorption of Laser by Materials

The laser incident on the cleaning object, after being reflected and scattered on the surface, the remaining laser energy will enter the material (contamination layer and substrate) and be absorbed by the material. The laser will first be absorbed by the contamination layer. If the contamination layer is thin or the absorption coefficient of the contamination layer for the laser is small, it will penetrate the contamination layer and enter the substrate to be absorbed.

I. **General Rules of Laser Absorption by Materials**

1. **Lambert's Law**

The absorption of the laser by a material increases as the penetration depth z. The laser intensity I decays exponentially, satisfying Lambert's Law:

$$I(z) = (1 - R)I_0 \exp(-\alpha z) \tag{3.1.7}$$

where R is the reflectivity (including scattering) of the material surface to the laser; I_0 is the incident laser intensity; $(1 - R)I_0$ is the transmitted laser intensity at the surface ($z = 0$), the absorption coefficient α is a physical parameter related to the material properties, and z is the penetration depth.

2. **Absorption Depth (Absorption Length)**

Absorption depth or absorption length refers to the distance the laser travels when the laser intensity deep into the material drops to $1/\mathrm{e}$ of the incident laser intensity I_0. It is generally believed that when the laser reaches the absorption depth, the laser energy is basically absorbed by the material. The absorption depth l_α satisfies:

$$I_0/\mathrm{e} = I_0 \exp(-\alpha l_\alpha) \tag{3.1.8}$$

Therefore, $l_\alpha = 1/\alpha$. As can be seen, after the laser enters the material, it decays exponentially and is absorbed by the material of finite thickness. The larger the absorption coefficient of the material, the smaller the absorption depth, and vice versa. The absorption coefficient α is related to the properties of the material, as well as the laser wavelength, material temperature, and surface conditions, etc.

When the absorption length of the laser is greater than the thickness of the material layer, a portion of the energy will pass through the material layer, and the energy absorbed by the material will depend on the absorption coefficient and thickness of the material layer; when the absorption length of the laser is less than the thickness of the material layer, the laser beam will be completely absorbed by the material layer. Materials with an absorption length much smaller than the thickness of the material layer are called strong absorbing materials, in which the depth that the laser can reach is usually less than the laser wavelength.

In laser cleaning, the contaminants and substrates may be metal or non-metal. Below we will specifically describe the absorption of these two types of materials.

II. Metal Material's Absorption of Laser

1. The Physical Process of Metal Absorbing Laser

In laser cleaning, many substrate materials are metals, for example using lasers to remove paint from aluminum alloy substrates, the substrate is aluminum alloy. Sometimes the contamination layer is also metal, such as rust on the surface of steel, where the contamination layer of rust is a metal oxide. The optical response of metals mainly depends on conduction electrons. When a laser is irradiated onto the metal surface, part of the laser energy is reflected by the metal surface, and another part of the energy is absorbed by the thin layer of the laser action area. For general metals, the absorption length of metal to laser is less than 0.1 μm. After the metal material absorbs the incident laser photons, the lattice point atoms are activated, the lattice vibration amplitude increases, further conducting and diffusing heat to the surface and interior of the metal, forming a surface heating process. In this way, the laser energy is converted into the thermal energy of the metal, causing the temperature of the metal surface to increase rapidly. This completes the process of laser absorption and heat generation [1, 2]. The absorption time of materials for photons is extremely short, the process of absorption and conversion into heat is about $10^{-11} \sim 10^{-10}$ s, much less than the duration of free running or Q-switched pulse laser acting on metal, therefore, in the laser absorption model, the laser beam can be equivalent to a heat source with time and spatial distribution.

When a laser acts on the metal surface, if the energy density is large, an oxide layer can be formed on the surface of the laser action area. When the energy density continues to increase, plasma can be observed in the laser action area, that is, the material layer on the metal surface has been vaporized. If the laser energy density is very high, a melt pool can be seen in the laser action area, forming a surface structure similar to a ring-shaped crater. In the process of the laser beam interacting with the cleaning metal surface, the effect of the laser is mainly thermal. As long as the energy density of the laser beam is not greater than the allowable damage threshold of the substrate, this thermal effect can be used to clean, and meanwhile, it can also achieve surface treatment (known as passivation) of the metal material, playing a protective role against rust.

2. Energy Conversion in the Process of Metal Absorbing Laser

The interaction of the laser with the metal material first triggers energy transfer and conversion. When the laser beam irradiates the metal material, its energy conversion still has to follow formula (3.1.2). When the thickness of the metal material is much greater than the absorption depth, it is a non-transmissive material for the laser, that is to say, $I_t = 0$. Both sides of Eq. (3.1.2) are divided by I_0, then the energy conversion formula of laser acting on metal materials changes to [2, 3]:

$$1 = (I_r/I_0) + (I_t/I_0) = R + A \tag{3.1.9}$$

It can be seen that when a laser irradiates a metal material, its incident laser intensity is ultimately divided into two parts: reflected by the metal and absorbed by

Table 3.2 Absorptivity of metal materials for the wavelength of Nd:YAG laser and CO_2 laser

Metal	YAG:$\lambda = 1.06\ \mu m$	CO_2:$\lambda = 10.6\ \mu m$
Al	0.08	0.019
Cu	0.10	0.015
Au	0.053	0.017
Fe	0.35	0.035
Ni	0.26	0.03
Pt	0.11	0.036
Ag	0.04	0.014
Sn	0.19	0.034
Ti	0.42	0.08
W	0.41	0.026
Zn	0.16	0.027

the metal. For uniform metal materials, the laser intensity incident at a depth of z from the surface still follows

$$F_\nu(z) = F_{\nu 0}(1 - R)\exp(-\alpha z) \tag{3.1.10}$$

where, $F_\nu(z)$ is the power density absorbed by the unit volume of material at the distance z from the surface, $F_{\nu 0}$ is the power density absorbed by the material surface, R is the reflectivity of the material, and α is the absorption coefficient of the material.

As the laser penetrates deeper into the material, the laser intensity will decrease geometrically; in addition, after the laser passes through a thickness of $1/\alpha$, the intensity decreases to $1/e$ of the incident time, that is, the ability of the material to absorb the laser depends on the value of the absorption coefficient α.

3. **Selective Absorption of Metals**

The absorptivity A of metal to laser and the relationship between the laser wavelength λ and the resistivity ρ of the metal can be expressed as

$$A = 0.365\sqrt{\rho/\lambda} \tag{3.1.11}$$

From (3.1.11), it can be seen that the absorptivity A of metal is related to the resistivity ρ of the metal, and the absorptivity A of such metals as titanium, tungsten, and iron are relatively high; the absorptivity of metal is also related to the wavelength, generally for metals, the shorter the wavelength, the higher the absorptivity. Table 3.2 gives the absorptivity of some metal materials. It can be seen that for most metal materials, the absorptivity is low for a CO_2 laser with wavelength of 10.6 μm. For a YAG laser with wavelength of 1.06 μm, the absorptivity is much higher, even an order of magnitude higher, so the material shows a strong selectivity for the absorptivity of the wavelength, and this absorption related to A and λ is called selective absorption.

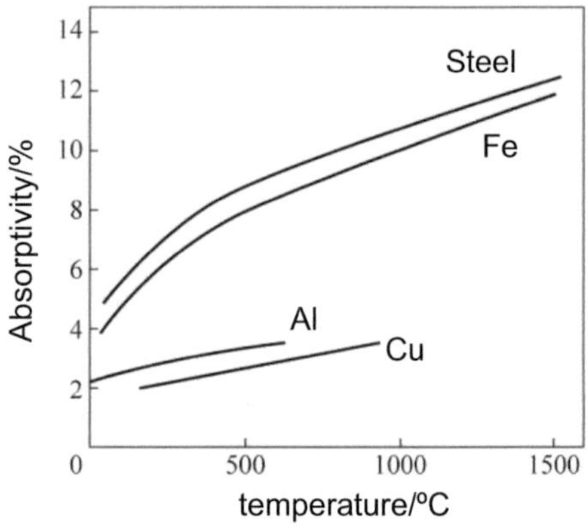

Fig. 3.1 Curves of the absorption rate of some metals with temperature

In addition, the absorptivity of metals to lasers also changes with temperature. The higher the temperature, the greater the absorptivity. At room temperature, the absorptivity is small; near the melting point, its absorptivity can reach 40 ~ 50%; when the temperature is close to the boiling point, its absorptivity can even be as high as 90%. The absorptivity is also related to the laser power density, the greater the laser power density, the higher the metal absorptivity.

Because the resistivity ρ of metals increases with temperature, there is a linear relationship between the absorptivity and temperature t (°C):

$$A(t) = A(20\,^{\circ}\text{C}) \times [1 + \beta(t - 20)] \tag{3.1.12}$$

where β is a constant related to the material.

Figure 3.1 shows the curve of metal absorptivity with temperature.

The above is considering the situation under vacuum conditions. In the actual laser cleaning process, laser cleaning is carried out in the air. Due to the formation of an oxide layer on the metal surface, the laser absorptivity will increase. As the temperature of the metal increases, the oxidation of the metal surface will intensify, and the absorptivity will increase accordingly.

III. **Absorption of Laser by Non-metallic Layer in Laser Cleaning**

In laser cleaning, in addition to metals, the common materials that interact with the laser, whether they are cleaning objects or pollutants, may be non-metallic oxides and organics, such as paint on resin or plastic materials, oil stains on metal surfaces, etc. The laser absorption process in these materials is quite different from the laser absorption process in metals. Usually, without special treatment, an oxide layer will form on the surface of the metal substrate, and this oxide layer plays an important role in the laser absorption process. Existing research shows that laser absorption on the metal substrate is affected by the oxide layer on the material surface.

Unlike metallic materials, non-metallic materials have lower reflection of lasers, correspondingly, their absorptivity is higher. The structural characteristics of non-metallic materials have an important influence on the selective absorption of laser wavelength. Usually, non-metals are insulators. Insulators only have bound electrons when they are not excited by lasers. These bound electrons have a certain inherent frequency, which depends on the energy change of electron transition.

$$\nu_0 = \Delta E/h \tag{3.1.13}$$

where, ν_0 is the inherent frequency, ΔE is the energy change, h is the Planck constant.

When the frequency of the irradiated laser is equal to or close to the inherent frequency of the bound electrons in the material, these electrons will resonate, radiate secondary waves, and form weaker reflected waves and stronger transmitted waves. The absorption and reflection of the material near this resonance frequency are enhanced, showing peaks of absorption and reflection. When the radiation frequency is significantly different from the inherent frequency, the uniform insulator shows transparency characteristics, with lower reflection and absorption characteristics.

For organic materials such as paint coatings, the absorption of lasers not only involves electronic transitions, but also energy coupling through intermolecular vibrations. When a laser beam acts on organic materials, in addition to the energy reflected by the material surface, it will enter the material and be absorbed. Organic materials have lower melting and vaporization points, which can easily form a plasma layer on the material surface, enhance coupling efficiency, further absorb laser energy. Because of the strong instantaneous absorption, the coating is more likely to be cleaned off, ultimately revealing the substrate and completing the cleaning process.

3.2 The Adhesion Force Between Pollutants and the Substrate

3.2.1 Adhesion Force

After understanding the basic principles of the interaction between laser and matter, let's see how laser peels off pollutants from the substrate materials. Although the technical means of laser cleaning are different from traditional physical or chemical cleaning methods, the physical nature is the same, and it is necessary to overcome the adhesion force between the pollutants and the surface of the substrate to which they are attached. Laser cleaning is achieved by the interaction of lasers and matter, absorbing laser energy, overcoming the adhesion force between pollutants and substrate materials, and thus cleaning the pollutants. So, we need to first understand how pollutants and substrates are combined.

The reason that pollutants adhere to the substrate is that there are various forms of force, and the situation is complex which affected by many factors. There are three main adhesion forces, namely van der Waals force, capillary force and electrostatic force [3, 4] (see Fig. 3.2).

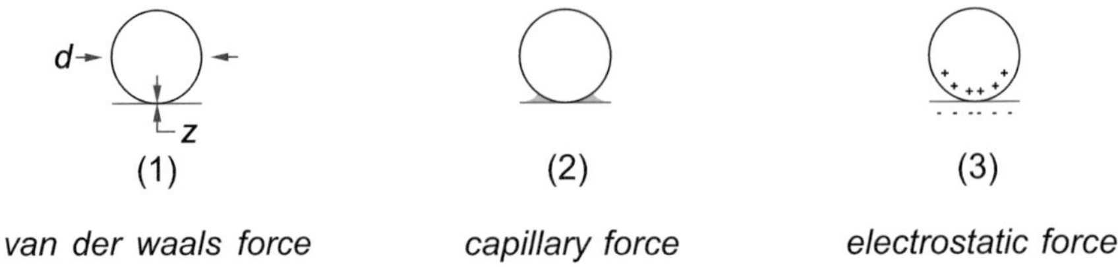

Fig. 3.2 Three types of adhesion forces between particles and substrate

For particles size smaller than a few micrometers, the van der Waals force is the main adhesion force, which originates from the interaction between the dipole moment of one contacting substance and the induced dipole moment of the other contacting substance, manifesting as attraction.

The van der Waals force per unit area between two parallel planes can be expressed as [4–6]:

$$F_v = \frac{h}{8\pi^2 z^3} \tag{3.2.1}$$

where h is the Lifshitz–van der Waals constant, which is related to the material of the substrate and the pollutants. For polymer–polymer, it is about 0.5 eV, and for metal–metal, it is about 10 eV; z is the distance between the pollution layer and the substrate, generally $z = 4 \times 10^{-10}$ m. The van der Waals force of a spherical particle with a diameter of d and a flat plate in point contact (which means the surface is not deformed) is:

$$F_v = \frac{hd}{16\pi z^2} \tag{3.2.2}$$

where z is the distance between the bottom point of the particle and the plane.

If the surface deforms due to strong attraction force, the actual van der Waals force will be much larger than the value given by the above formula. If there is a liquid film on the surface, for example, when water vapor in the air condenses in a humid environment, or when liquid-assisted cleaning is used, where liquid needs to be sprayed before cleaning, capillary forces cannot be ignored. Its expression is

$$F_c = 2\pi\gamma d \tag{3.2.3}$$

where γ is the surface energy per unit area of the liquid film, and d is the particle diameter.

The third type of adhesion force is electrostatic force. When particles come into contact with the substrate, there is a contact potential difference U between them. Under the drive of the electromotive force, charge transfer occurs between the particles and the substrate, forming a double charge layer with opposite charges on both sides of the contact surface, similar to the structure of electrode plates. At this time, the electrostatic attraction between the particles and the surface of the substrate can be expressed as

$$F_e = \frac{\pi \varepsilon U^2 d}{2z} \tag{3.2.4}$$

where ε is the dielectric constant of the air between the particle and the substrate.

The three types of adhesion forces mentioned above are all proportional to the particle diameter d. We know that the gravity of the particle is:

$$G = mg = \rho \cdot \frac{4}{3}\pi \left(\frac{d}{2}\right)^3 = \frac{\pi \rho d^3}{6} \tag{3.2.5}$$

As can be seen, gravity is proportional to the third power of d. Therefore, when the size of the particle is very small, the increase of the adhesion force is much greater than that of gravity. For example, for a particle with a diameter $d = 1\ \mu\text{m}$, the order of magnitude of the van der Waals force is 10^7 times the gravity. Therefore, in the kinetic analysis of particles, gravity can be ignored. From the relationship between gravity and adhesion, the smaller the particle size, the stronger the relative adhesion, and the harder it is to remove. When the diameter of the particles decreases to a certain extent, due to the large adhesion, many cleaning methods are difficult to remove the particles. For continuous pollutants, such as paint and rust, they can be regarded as continuously arranged particles, and their mechanism of action is similar.

3.2.2 Brief Introduction to Principle of Laser Cleaning

Laser cleaning uses lasers as the cleaning medium, transferring laser energy to pollutants and the substrate. By utilizing the interaction between laser and matter, the cleaning force generated by the laser is greater than the adhesion force between the pollutants and the substrate, thereby achieving the purpose of cleaning.

Specifically, when the laser is irradiated on the object to be cleaned, the irradiated area absorbs laser energy, which is converted into heat energy. Due to the different absorptivity of laser and different thermal expansion coefficients between the pollutants and the substrate, different layers will have different temperatures. If the temperature exceeds the melting point or gas point of the pollutants, the pollutants will melt or vaporize, which is the ablation effect. If the temperature is not enough to cause phase change, but cleaning effect is produced due to stress, it is the vibration effect. If the local temperature rises significantly higher than the surrounding temperature instantaneously, there may also be a buckling effect. If a cavity is formed between the pollutants and the substrate, there will be a blasting effect. If there is a liquid film on the surface, the pollutants can be cleaned through the vaporization of the liquid film, through boiling, evaporation and other effects [7].

In fact, there are various different theories about the physical mechanism of laser cleaning. Below, we will introduce base principles of dry laser cleaning and wet laser cleaning, respectively.

3.3 Basic Mechanisms of Dry Laser Cleaning

In dry laser cleaning, according to existing research results, it can be basically determined that the ablation mechanism and vibration mechanism are the two main cleaning mechanisms. When the pollution layer absorbs laser energy, causing the temperature to rise, phenomena such as combustion and gasification occur, which belong to the ablation mechanism. When the pollution layer and the substrate absorb laser energy in turn, convert it into heat energy, raise the temperature, and the heating process satisfies the heat conduction equation, when the thermal stress formed is greater than the adhesion force, it belongs to the vibration mechanism [8].

3.3.1 Ablation Effect

When the laser is irradiated on the pollution layer on the substrate surface, the laser interacts with the pollutants, the pollution layer absorbs the energy of the laser, and converts it into the thermal energy of the system, which is manifested as a temperature rise in material. When the energy density of the laser reaches a sufficiently high level, the temperature of the material will exceed its melting point and even boiling point. The material will therefore undergo combustion, decomposition or vaporization, thereby being removed from the adsorbed substrate surface. According to calculations, after a high-energy laser beam is focused, objects near its focal point can be heated to thousands of degrees. The ablation mechanism is actually using high-energy lasers to act on pollutants, generating thermal effects to destroy the structure of the material itself, thereby eliminating its adhesion force with the substrate and achieving the purpose of cleaning.

Figure 3.3 is a schematic diagram of the ablation effect in laser cleaning. When the laser pulse reaches the surface of the contaminant, the contaminant absorbs the laser energy and converts it into heat, causing the temperature to rise. When the contamination layer absorbs a relatively large amount of laser energy, and the temperature reaches and exceeds the vaporization temperature point of the contamination layer, it will be vaporized, as if it were ablated and stripped away.

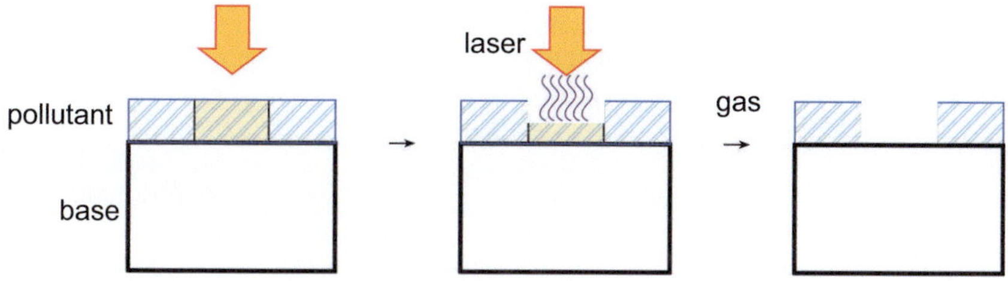

Fig. 3.3 Schematic diagram of ablation mechanism

3.3.2 Vibration Effect

In cleaning, relatively speaking, the contaminant is thin and the substrate is large. Under normal circumstances, we can assume that both are incompressible. We select the irradiation area of the laser on the cleaning object for analyze. In actual laser cleaning, the spot diameter is about 1 mm or so. Under the action of the laser, the substrate and/or contaminant absorbs laser energy, thereby gaining heat and the temperature rises. Different substances have different absorptivity for laser of different wavelengths. For a certain wavelength of laser, if the contaminant absorbs far more laser energy than the substrate, it is mainly the contaminant that absorbs heat. There are two situations: first, after the contaminant absorbs heat, it expands in volume, generating an upward elastic force, causing the contaminant to peel off, as shown in Fig. 3.4a; second, after the contaminant absorbs heat, the heat is not used to expand itself, but is transferred to the substrate, causing the substrate to expand, generating a thrust, causing the contaminant to peel off, as shown in Fig. 3.4b. If for a certain wavelength of laser, the contaminant absorbs far less laser energy than the substrate, such as the case where the contaminant is transparent to the laser (i.e., the contaminant hardly absorbs the laser), it is mainly the substrate that absorbs heat. There are also two situations: first, after the substrate absorbs heat, it transfers the heat to the contaminant, causing it to expand, generating an upward force, causing the contaminant to peel off [6]. As shown in Fig. 3.4c; second, after the substrate absorbs heat, it expands in volume, generating an upward elastic force, and the contaminant in contact with it generates a reaction force in the opposite direction, causing the contaminant to peel off, as shown in Fig. 3.4d.

Due to the heating of the contaminant particles or the substrate, the center of mass of the contaminant particle has moved. We use a one-dimensional model to simply describe this quantitatively.

In the small area of the laser irradiation area, the displacements of the contaminant and the substrate after thermal expansion are: δ_p, δ_s. The total displacement is $\delta = \delta_p + \delta_s$.

Taking the direction perpendicular to the surface of the substrate as the y direction, the linear expansion coefficients of the contaminant and the substrate are α_p, α_s, respectively. After the temperature rises by Tp or Ts (relative to the initial temperatures of the contaminant particles and substrate), the expansion thickness in the

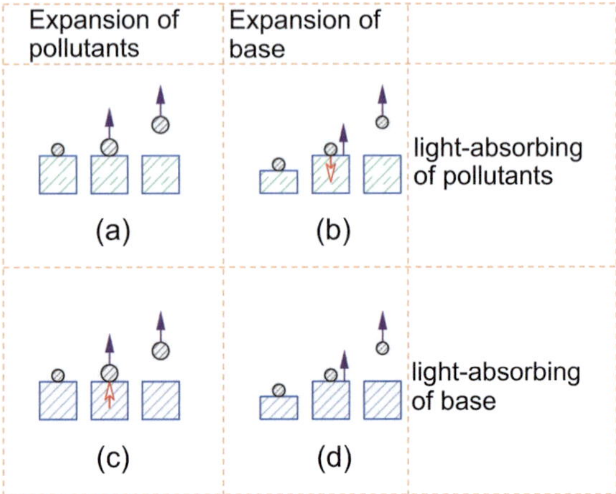

Fig. 3.4 Schematic diagram of laser cleaning mechanism

y-direction is

$$d\delta_i(y) = \alpha_i T_i \tag{3.3.1}$$

where i = p and s representing pollutants and substrates, respectively. So, within the entire range of pollutant or substrate height h_i, the displacement of the center of mass is:

$$\delta_i = k_i \int_0^{h_i} \alpha_i T_i dy \tag{3.3.2}$$

where k_i is a parameter related to pollutants or substrate materials. For the substrate, $k_s = 1$; For particles, k_p depends on the heat absorption and geometric shape of the pollutant, and its value ranges from 0 to 1. For example, for micrometer level metal particles with low heat absorption or small insulating particles, $k_p \approx 0.5$; For particles with high heat absorption rate, $k_p \approx 0.5$; For transparent particles with a high heat absorptivity, $k_p \approx 1$.

According to Newton's second law, the inertial force generated by thermal expansion is:

$$F = m\frac{d^2\delta}{dt^2} \tag{3.3.3}$$

From the above formula, as long as we know the temperature rise of the pollutant and the substrate, we can calculate the force F. If F is greater than the adhesion force, the pollutant will peel off from the substrate.

3.3.3 Thin Film Bending Effect

For thin film pollutants on the substrate material, such as paint, rust, etc., under the condition of large laser spot and large area radiation, there is a bending mechanism. At this time, a large piece (1 mm^2 or even 1 cm^2) of film peels off from the substrate. As shown in Fig. 3.5 [9].

Considering the situation of a large irradiation area in one dimension, the thermal stress in this area is:

$$\sigma_T = E\alpha_f T \tag{3.3.4}$$

where E is the elastic modulus, α_f is the linear expansion coefficient, T is the temperature change of the thin film (relative to the initial temperature before laser irradiation), then the compression energy density per unit volume is:

$$Q = \frac{\sigma_T^2}{2E} = \frac{E}{2}\left(\alpha_f T\right)^2 \tag{3.3.5}$$

It is converted from the surface of the substrate to the kinetic energy of the thin film movement. When this energy is greater than the adsorption energy, the film piece will expand and bend, detaching from the surface of the substrate.

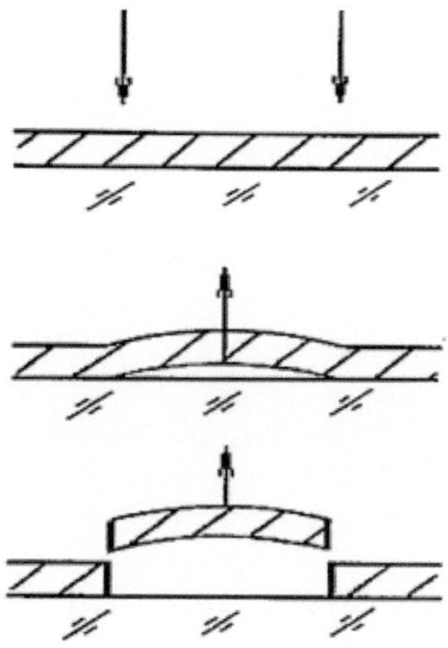

Fig. 3.5 Schematic diagram of thin film bending effect

3.3.4 Blast Effect

There are tiny cavities (bubbles) at the interface between the substrate and the pollutants, or cavities are formed after laser irradiation. The cavities may contain air, gaseous substrate, or gaseous pollutants. When the laser continues to irradiate, the pressure in the cavity rapidly increases, causing an internal explosion in the cavity, which drives the pollutants to peel off from the substrate.

As can be seen from the above laser cleaning theoretical model and related research process, vibration, bending, and blasting effects occur before reaching the melting point and boiling point, and several situations may exist simultaneously [7]. Generally speaking, the vibration mechanism is a common situation. The ablation process depends on the physical and optical properties of the laser and the material absorbing the laser. It only has a satisfactory cleaning effect for specific pollutants and substrates, and there are certain conditions restrictions; compared with this, whether the ablation mechanism works or not, the role of the vibration mechanism is more obvious, that is, the application range is wider. It shows certain advantages, that is, it only requires lower energy compared to the ablation effect, because the cleaning process does not need to absorb energy to raise the temperature of the pollutants above the vaporization point, but uses the vibration effect to make the adsorbate detach, without having to destroy the adsorbate. In addition, the action time is also shorter, and the cleaning process can be completed in nanoseconds, and this instantaneous characteristic can ensure that the residual thermal stress of the substrate is minimized.

3.4 Mechanism of Wet Laser Cleaning

In wet laser cleaning, depending on whether the substrate or the liquid film is easier to absorb laser energy, it can be divided into three situations: the substrate strongly absorbs laser energy, the liquid film strongly absorbs laser energy, or both absorb laser energy at the same time. Below we will discuss the laser cleaning mechanism of these three situations separately [10, 11].

3.4.1 Strong Absorption by the Substrate

The laser energy is irradiated on the substrate and the liquid film, the energy is absorbed and converted into heat, and the heat diffuses in the substrate and the liquid film, which can be represented by the thermal diffusion length. Obviously, the thermal diffusion length is related to the strength of the material's absorption of the laser and the properties of the material itself. For a specific material, if the absorption of the laser is strong, the thermal diffusion length is large. In the case

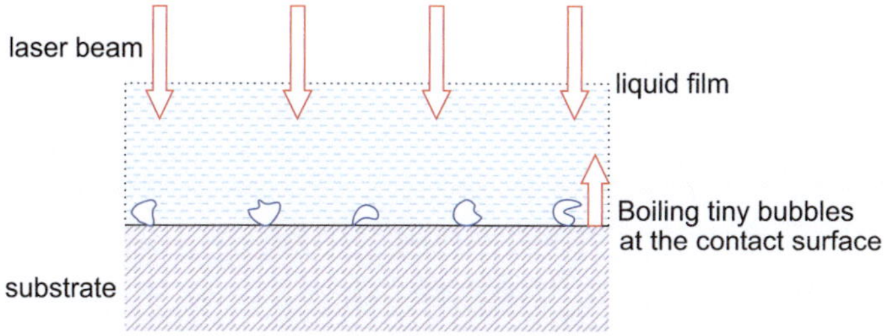

Fig. 3.6 Schematic diagram of wet laser cleaning with strong absorption by the substrate

of strong absorption by the substrate, the absorption of the liquid film layer to the laser is much smaller than the absorption of the substrate. For laser cleaning with a pulse width of about 10 ns, when the pulsed laser is largely absorbed by the substrate material, the resulting thermal diffusion length is about 1 μm in the substrate and about 0.1 μm in the water film, the difference between the two is large, thus a large amount of energy waiting to be dissipated will accumulate at the interface between the liquid and the substrate, and this energy is enough to cause overheating and explosive evaporation of the liquid film covering the substrate interface.

Theoretical calculations show that the shorter the duration of the pulse, the shorter the heat diffusion length in the liquid and the substrate, thus less energy is needed within the heat diffusion range to cause more intense evaporation of the liquid film. A shorter duration means that a stronger overheating phenomenon can occur at the thin liquid interface, resulting in a more intense explosive pressure. When the pulse width exceeds 1 μs, the heat has enough time to diffuse, so a large amount of heat will not accumulate in the liquid film on the contact surface, so the laser pulse width cannot be too long. Experiments have also verified that the efficiency of μs-level pulse width laser cleaning is relatively low. However, very short pulses have a very high corresponding laser peak power, which can easily cause damage to the substrate material.

The physical mechanism of wet cleaning for strong absorption by the substrate can be explained by Fig. 3.6. At the junction of the liquid film and the substrate, a large number of boiling bubbles are generated. When the bubbles escape, they carry away the dirt particles, achieving the purpose of cleaning.

3.4.2 Strong Absorption by Liquid Film

When the liquid film strongly absorbs laser, the laser causes the liquid film to reach a very high temperature instantaneously, forming bubbles and explosions on the surface or inside the liquid, thereby driving the pollutants to vaporize and explode. The schematic diagram is shown in Fig. 3.7. Experiments show that the cleaning effect of strong absorption by the liquid thin film is not as good as that of strong

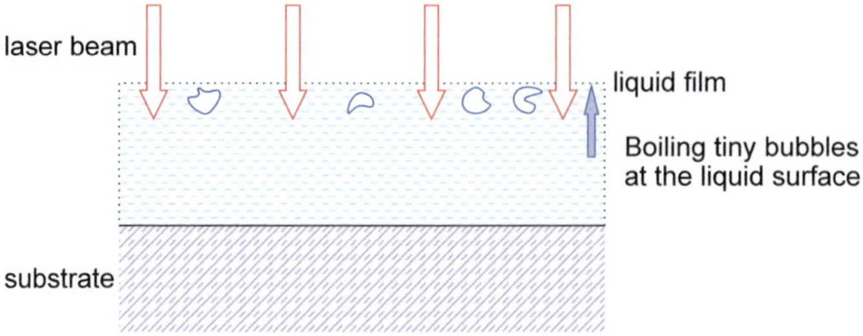

Fig. 3.7 Schematic diagram of wet laser cleaning with strong absorption by the liquid film

absorption by the substrate. This is because the surface temperature of the liquid film reaches a very high temperature under strong absorption, while the temperature at the interface between the liquid film and the substrate reaches a high temperature under strong absorption by the substrate. Therefore, for the case of strong absorption by the liquid, bubbles and intense explosions occur on the surface of the liquid, forming large transient forces on the surface or inside the liquid. For strong absorption by the substrate, bubbles and intense explosions occur at the interface between the liquid film and the substrate, forming large transient forces, which can easily remove the pollutants tightly attached to the substrate, so the cleaning efficiency is naturally higher.

3.4.3 *Joint Absorption by Substrate and Liquid Film*

When the liquid film and the substrate both strongly absorb the laser, vaporization, evaporation, and explosions will occur on the surface and inside the liquid film, as well as at the interface between the liquid film and the substrate. However, the laser beam irradiates the liquid, so a lot of laser energy is first absorbed by the liquid film, and the laser energy reaching the substrate will weaken, which reduces the explosion at the interface. The principle is shown in Fig. 3.8. Shukla's team [10] once used pulsed TEA: CO_2 to irradiate the Si surface covered with thin water film, removing aluminum particles. Although the energy density of the laser pulse used is large, at the level of 10 J/cm^2, the cleaning efficiency is not as high as that of strong absorption by the substrate. Water absorbs 10.6 μm of CO_2 laser. The absorption depth of the laser is only 20 μm, so if the thickness of the water film is a few micrometers, only a part of the laser is absorbed by the water, and the absorbed energy is distributed in a larger volume of water, causing boiling bubbles to form inside the liquid film. The rest of the laser passes through the whole layer of water film to reach the Si substrate, heating the interface. Therefore, more energy is needed to produce evaporation and explosion. This is why this method of cleaning is relatively inefficient.

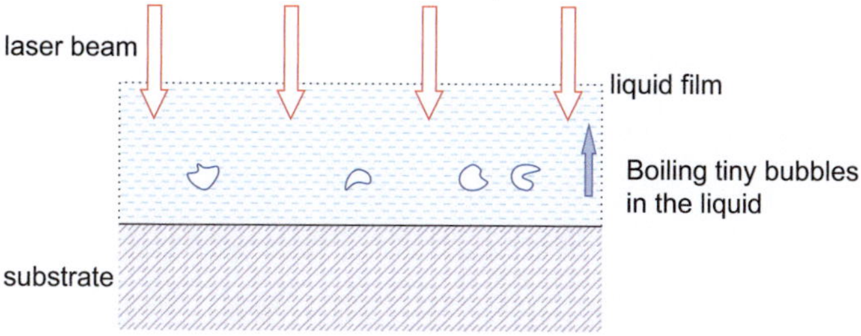

Fig. 3.8 Schematic diagram of wet laser cleaning with simultaneous absorption by the substrate and liquid film

The above three methods are suitable for different situations, and can be selected according to different particle materials and sizes, substrate materials and laser wavelengths in actual cleaning.

References

1. Qiuyu China. Common material reflectance [EB/OL].
2. Fox M. Optical Properties of Solids [M]. Oxford: Oxford University Press, 2001: 143–164.
3. Wang J J. Laser Processing Technology [M]. Beijing: China Metrology Publishing House, 1992.
4. Kelley J D, Hovis F E. A thermal detachment mechanism for particle removal from surfaces by pulsed laser irradiation [J]. Microelectronic Engineering, 1993, 20(1–2): 159–170.
5. Mittal K L. Particles on Surfaces [M]. New York: Marcel Dekker, 1995: 1–33.
6. Lu Y F, Song W D, Ang B W, Hong M H, Chan D S H, Low T S. A theoretical model for laser removal of particles from solid surface [J]. Applied Physics A-Materials Science & Processing, 1997, 65(1): 9–13.
7. Song F, Liu S J, Niu K Z, Li X P. Principle and application research on laser cleaning [J]. Cleaning World: 2005, 21(1): 1–6.
8. Tian B, Zou W F, Liu S J, Song F. Introduction of rust removed by dry laser cleaning [J]. Cleaning World: 2006, 22(8): 33–38.
9. Shi S D. Researches on theoretical model, numerical simulation and application of painting removal by laser cleaning [D]. Tianjin: Nankai University, 2012: 13–70.
10. Shukla S, Kudryashov S, Lyon K, Allen S D. Pulsed laser cleaning of sub- and micron-size contaminant particles from optical surfaces: cleaning vs. ablation and damage [J]. Proc. SPIE, 2005, 5991: 59910N.
11. Lu Y F, Song W D, Zhang Y, Low T S. A theoretical model for steam laser cleaning [J]. Materials Research Society, 1998, 7: 3550.

Chapter 4
Laser Cleaning Equipment

With the continuous improvement of laser cleaning theory and the increasing maturity of various control methods, and the development of related industries and the progress of industrialization, laser cleaning technology is becoming more and more mature. In this context, the related research of laser cleaning gradually shifts from theory and experiment to instrument and equipment development, and there have been many invention patents and utility model patents. Abroad, some professional laser cleaning companies have been established, and large laser production companies have also begun to pay attention to cleaning applications, start to design and produce laser cleaning equipment, and bring certain economic and social benefits. In China, with the increasing requirements for environmental protection, especially since the Ministry of Science and Technology listed laser cleaning research as a key special plan in 2016, many new laser cleaning companies have been established or existing companies have transformed to engage in the cleaning industry.

Specialized laser cleaning companies are basically distributed in several countries such as the United States, Germany, France, Italy, South Korea, and Singapore, and they have laser cleaning equipment for sale. Foreign professional laser cleaning companies include: Adapt Laser (USA), Laser Clean All (USA), Clean Laser (Germany), Norton Sand Blasting (USA), Quanta System (Italy), El En Group (Italy), Rofin (Germany), Quantel Laser (France), IMT—Innovative Manufacturing Technology (Korea), LE Champ (Singapore), etc. These companies mostly use high repetition rate, high power lasers (Nd:YAG) as cleaning lasers, with power ranging from 10 W to thousands of watts. They generally use side-pumped semiconductor modules as pump sources for solid-state lasers. In recent years, fiber lasers have increasingly become commonly used lasers in laser cleaning. Among them, small power models have achieved backpack design, and high-power models can achieve high efficiency cleaning with automatic control systems. The laser Q-switching mode mostly uses active Q-switching, including electro-optic Q-switching and acousto-optic Q-switching. There are also free-running pulsed lasers and long Q-switched lasers.

F. Song and X. Lin, *Laser Cleaning*, https://doi.org/10.1007/978-981-97-6631-4_4

The laser cleaning equipment operates in a handheld manner, with a mechanical arm or fiber optic light guide. The handheld handle can change the spot size and is activated by a switch; the mechanical arm or fiber optic light guide can be connected to a computer via a parallel or serial port, and the various parameters of the laser (start and stop cleaning, single pulse energy, repetition rate, emergency handling, and safety control, etc.) can be controlled directly with software. Additionally, a visible alignment beam, safety coupler, and emergency brake functions can be added at the laser emission point.

4.1 Introduction to the Laser Cleaning Equipment System

The external form of laser cleaning equipment can vary greatly, some use a fixed machine tool, some use a mobile workbench, and there are also small backpack types, etc. Although the external manifestations are different, the internal structure can basically be divided into several subsystems: laser system, beam transmission and shaping system, control system, mobile system, recycling system, and auxiliary system [1–6].

I. **Laser System**

The laser system is the core component of the laser cleaning equipment, mainly composed of several units such as the laser head (optical system), driving power, and cooling. The wavelength of the laser has multiple choices, and in actual work, it can be determined according to different cleaning objects. In Chap. 2, we have already detailed the working principle of the laser and the commonly used cleaning lasers. The commonly used ones on the market are Nd:YAG lasers, CO_2 lasers, and fiber lasers, etc. The main feature of the CO_2 laser is that it has a high output power, high energy conversion efficiency, a wavelength generally of 10.6 microns, and non-metallic materials have good absorption effects at this wavelength. It is cheap, and is commonly used for cleaning paints on non-metallic materials. However, the maintenance cost of the carbon dioxide laser is high, the life span is short, and there is currently no fiber that directly transmits carbon dioxide laser, which is not conducive to long-term operation in complex outdoor environments, affecting its industrial application.

The fiber laser is a laser that directly uses fiber as the gain medium. It has high conversion efficiency, generates less heat, and the fiber output is easy to operate, but the current price of the fiber laser is relatively high, especially the price of the Q-switched pulsed fiber laser is relatively expensive. In addition, the mode of the fiber laser is very good, but it may not be suitable for many cleaning situations, because the peak power of the central area of the basic mode Gaussian laser beam is too high, which can easily damage the substrate, and the peak power of the peripheral part of the Gaussian beam is low, which may make it difficult to clean the pollutants. It is also necessary to shape the beam (such as flattening), which increases the technical difficulty and economic cost, and reduces the efficiency.

Solid-state laser technology is mature and low-cost, especially the technically and industrially mature Nd:YAG laser, which can be transmitted by fiber, is currently the most commonly used laser source for cleaning equipment. Internationally specialized laser cleaning companies basically use this type of laser as the light source for laser cleaning.

Solid-state lasers are mainly divided into flashlamp-pumped and semiconductor-laser-diode(LD)-pumped, with flashlamp-pumped lasers gradually being phased out at this stage. Semiconductor pumping methods are divided into end-face pumping and side pumping, as shown in Figs. 4.1 and 4.2. The advantage of end-face pumping is that the pump light is easy to match with the mode in the resonant cavity, the gain medium absorbs the pump light well, and the output laser beam quality is good. The disadvantage is that when the power increases, the end-face temperature of the pump light irradiation is high, which can cause temperature differences inside the gain medium, forming thermal stress, which is not conducive to stability. Side pumping mainly uses the side of the gain medium to absorb the pump light, and the surface area of the absorbed light is large, the heat dissipation is good. It is conducive to the output of high-power lasers, but the beam quality will be slightly worse. High-power output lasers generally use LD side pumping.

The driving power supply that mainly provides power for the laser has a signal control interface, and can adjust the output current of the power driver (changing the output power of the laser), the form of power output (pulse or continuous wave mode, corresponding to the form of laser output) in real time through the control module,. The cooling system is indispensable in the laser. The power of the laser

Fig. 4.1 Schematic diagram of LD end pumped laser

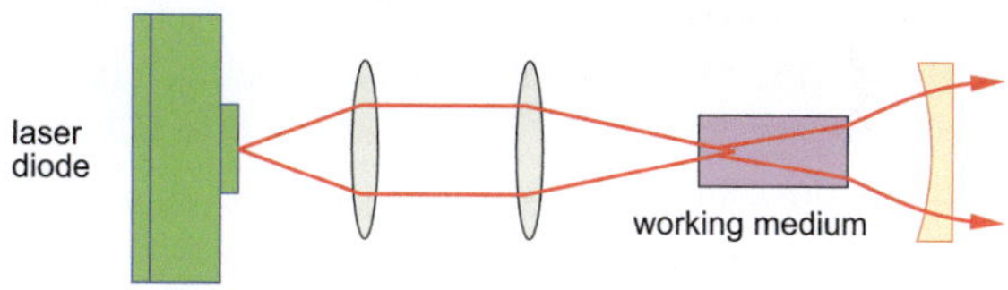

Fig. 4.2 Schematic diagram of LD side pumped laser

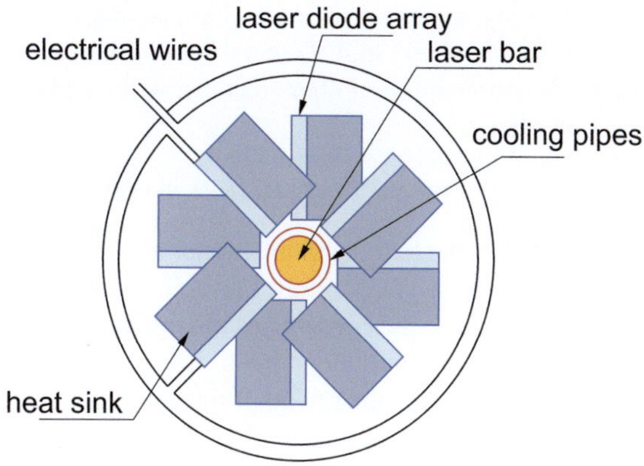

used for cleaning is generally large, so water cooler is mainly used for cooling. The water cooling system includes a water tank, heat pipe, water pump, closed flow pipe (including valves, tees and other structural parts), fans and other devices. The water cooling system should have sufficient cooling capacity and sufficient water pressure and flow rate, and can achieve multi-channel output, providing cooling for multiple components. It has real-time temperature detection and water level detection, flow control and temperature adjustment (completed by the control system).

II. Beam Transmission and Shaping System

1. Three common beam transmission methods

Lasers need a beam transmission system to reach the surface of object to be cleaned. There are basically three ways to transmit lasers, namely free space transmission, multi-joint light guide arm transmission, and high-energy fiber transmission. Free space transmission is to directly irradiate the laser beam from the laser on the surface of the cleaning object. This method has small laser energy loss and low cost, but it is only suitable for short-distance laser operations and the form is fixed and cannot be changed according to actual conditions. Multi-joint light guide arm transmission uses coated mirrors at each joint to reflect the laser to achieve beam deflection, but it has high loss (~ 10–40%). The longer the working distance, the more mechanical arms are needed, and the greater the loss. Figure 4.3 shows the schematic diagram and physical diagram of the multi-joint light guide arm. Another method of beam transmission is to use high-energy fiber transmission. The advantage of the fiber is that it can be bent and is easy to use, with relatively small loss. But during operation, attention should be paid to the bending radius and the damage threshold of the end face of fiber during the coupling process and the deterioration of the output laser quality. The disadvantage of fiber optic transmission is that the coupling requirements are high, the price of the transmission fiber is high, and during use, the damage to the fiber end face during laser transmission under high peak power is a problem that needs to be solved in high-energy fiber transmission. Table 4.1 gives the ability of high-energy fiber transmission of continuous lasers with different core diameters. For laser cleaning, the lasers used are all pulse output, and mainly Q-switched pulses, so the peak power is very high. Generally, continuous laser transmission fibers cannot be used, while transmission fibers that can withstand high peak power and high energy must be used.

2. Coupling of optical fiber and laser in optical fiber transmission mode

Using optical fiber to transmit laser energy is simple and convenient in laser cleaning operations, and it is definitely the mainstream direction of future laser cleaning equipment. In use, the coupling of the optical fiber and the output beam of the laser is very important. On the one hand, it is necessary to ensure the efficiency of light transmission. On the other hand, if the coupling is not good, it is easy to damage the energy transmission fiber.

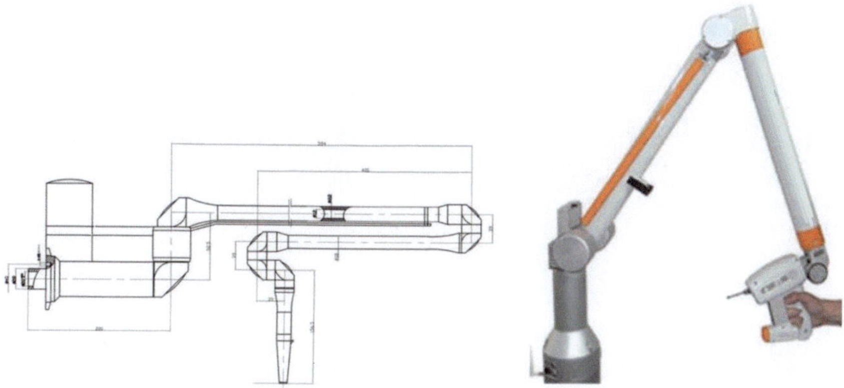

Fig. 4.3 Schematic diagram and physical diagram of multi-joint light guide arm

Table 4.1 The ability of high-energy optical fibers of different diameters to transmit continuous lasers

	Fiber core size	100 μm	200 μm	300 μm	400 μm	600 μm	800 μm	1000 μm	1500 μm
Power	HP SMA (W)	85	340	650	650	650	650	650	650
	MSHP SMA (W)	85	340	650	650	N/A	N/A	N/A	N/A
	HP FD-80 (W)	85	340	750	750	750	750	750	750
	MSHP FD-80 (W)	85	340	750	750	N/A	N/A	N/A	N/A

As shown in Fig. 4.4, when using optical fiber coupling, it is necessary to make the waist radius and far-field divergence angle of the laser beam meet the coupling conditions of the optical fiber at the end face of the optical fiber coupling:

$$d_{in} < d_{core} \tag{4.1.1}$$

$$\theta_{in} < 2\arcsin(NA) \tag{4.1.2}$$

where d_{in} is the diameter of the incident light spot, d_{core} is the fiber diameter, θ_{in} is the divergence angle of the incident light, NA is the numerical aperture of the fiber

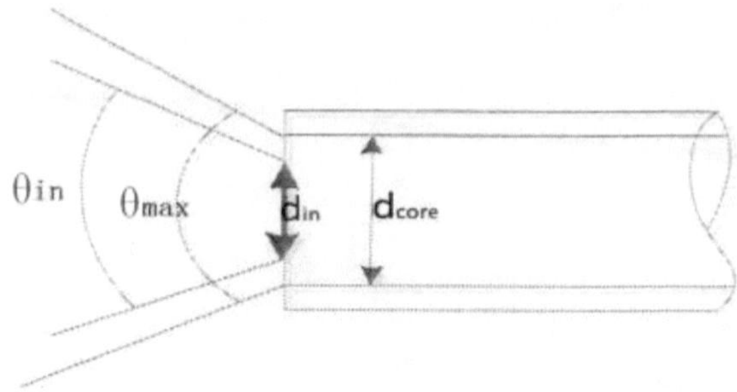

Fig. 4.4 Fiber coupling conditions

core, generally 0.22. Then it can be considered

$$\theta_{in} < 2NA \tag{4.1.3}$$

The quality factor of the beam M^2 is

$$M^2 = \pi d_{in}\theta_{in}/(4\lambda) \tag{4.1.4}$$

By substituting (4.1.1) and (4.1.2) into (4.1.4), we can get

$$M^2 < \pi d_{core}NA/(2\lambda) \tag{4.1.5}$$

According to formula (4.1.5), we can choose the diameter of the fiber to achieve better coupling efficiency. Considering factors such as peak power, if we choose a fiber core diameter of 1000 μm and a length of 5 m, the coupling efficiency at this time is 91%.

3. Beam shaping output

The laser beam has special mode due to different parameters of the laser resonator. During the transmission of the laser, the beam quality will deteriorate (especially when propagating in the fiber). During cleaning, it is hoped that the beam is uniform, and the energy per unit area reaches the cleaning threshold but does not damage the substrate.

Shaping is generally divided into two types, one is energy distribution shaping, and the other is transverse distribution shaping of the output light. In terms of energy distribution shaping, it is to transform the Gaussian distribution of the laser output into a flat-top light with a relatively uniform energy distribution, thereby improving the utilization rate of the laser. The shaping of the shape of the output light is to obtain a point or line spot to obtain sufficient energy density and improve cleaning efficiency. It can be directly focused through a cylindrical mirror to obtain a line spot, or first focused through a convex lens to obtain a point spot, and then scanned into

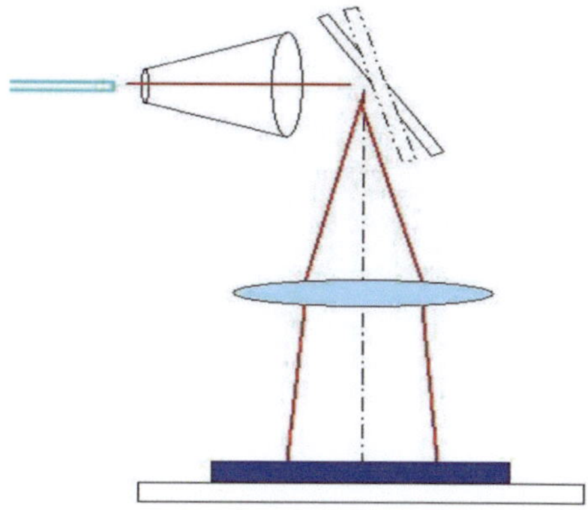

Fig. 4.5 Schematic diagram of the scanning laser output

a line. The method of directly focusing through a cylindrical mirror is suitable for high-power output laser cleaning equipment, because its corresponding peak power density is very large and can reach the cleaning threshold of the cleaning object; the method of focusing through a convex lens and then scanning is suitable for low-power output laser cleaning equipment, which can provide a higher peak power density and a better line spot. The scanning methods of the beam include platform scanning, galvanometer scanning, and rotating mirror scanning. The structure shown in Fig. 4.5 is a schematic diagram of scanning.

Figure 4.6 is a platform scanning device. The platform scanner works in conjunction with a beam shaping device (as simple as a cylindrical mirror) to transform the initial circular light spot into a long strip light spot. Then, through the platform's linear motion, it drives (or moves relative to) the long strip light spot to achieve surface scanning.

The galvanometer scanning device is a common method in laser processing systems. It uses two mutually perpendicular swing mirrors to achieve surface scanning of point-like light spots. The basic structure is shown in Fig. 4.7.

Galvanometers are currently widely used in various types of laser processing equipment, such as in laser marking machines. The galvanometer mainly consists of

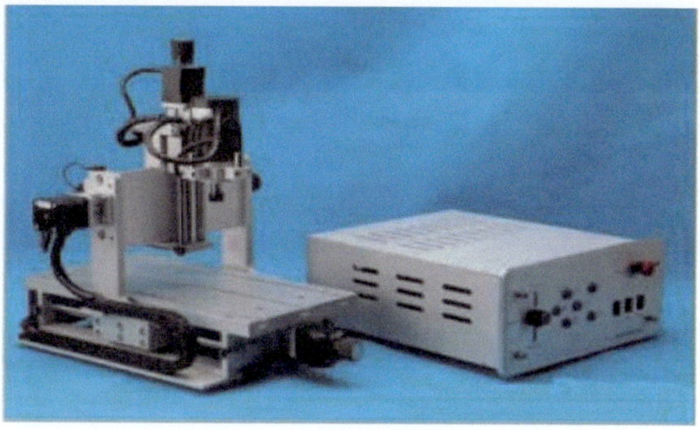

Fig. 4.6 Platform scanning device

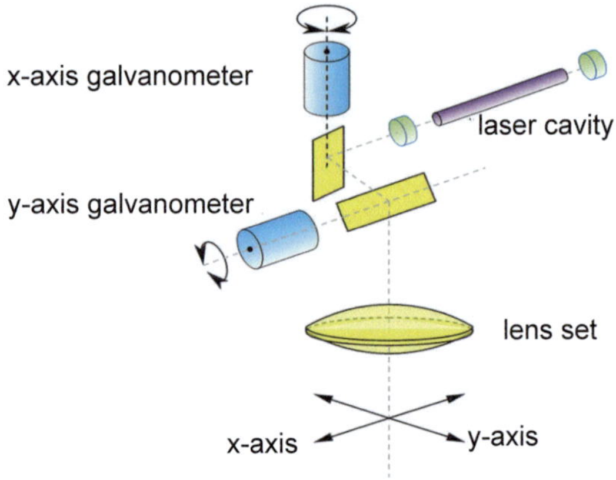

Fig. 4.7 Galvanometer X–Y scanning structure

XY scanning mirrors, field mirrors, and control software, with the scanning mirror driven by a motor. As shown in Fig. 4.7, the laser beam output from the laser is incident on two reflecting mirrors (X and Y direction galvanometers). The signal provided by the computer drives the motor through the driving amplification circuit, controlling the reflecting mirror to swing at a certain angle according to a certain voltage and angle conversion ratio. This controls the laser beam to scan along the X and Y axes separately, achieving the deflection of the laser beam. It then focuses through the field mirror and shoots onto the cleaning object. The entire process uses closed-loop feedback control, with the position sensor, error amplifier, power amplifier, position discriminator, and current integrator controlling the entire process.

Since the galvanometer swings back and forth, its speed is zero when it swings to the maximum value on both sides. It is not uniformly swinging. Therefore, when the galvanometer swings to the maximum and needs to move in the opposite direction, there is a pause (mechanical pause), causing excessive exposure of the laser, leading to uneven power density. This is an inherent problem with galvanometer scanning. To overcome the mechanical pause of the galvanometer swing, rotating mirror scanning is used in some laser processing devices that require high power density uniformity. Its basic structure is shown in Fig. 4.8. Since the rotating mirror rotates uniformly in one direction, there is no problem of mechanical pause, so the uniformity of the beam energy density can be better guaranteed. Figure 4.9 shows a schematic diagram of the application of the rotating mirror in the cleaning device.

III. **Mobile System**

The mobile system completes the relative movement of the cleaning laser and the cleaning object during the cleaning process. This can be achieved by moving the fiber optical head or other beam transmission components, or by moving the cleaning workpiece. Common methods include handheld heads, robotic arms, and three-dimensional displacement platforms. The handheld head is easy to operate, but manual operation inevitably lacks uniformity, and the requirements for the operator are relatively high. At the same time, the safety protection of the operator must be

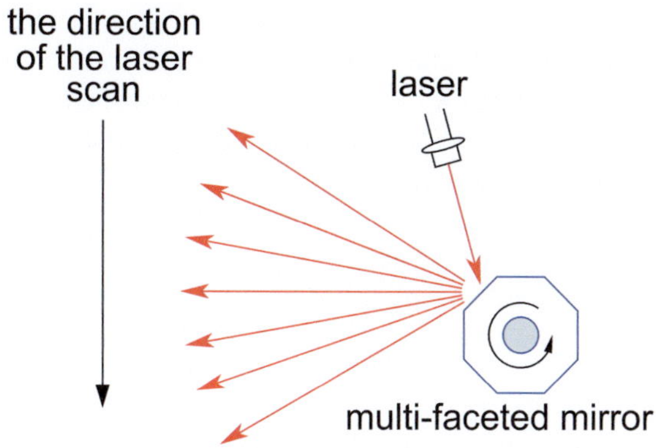

Fig. 4.8 Basic mode of rotating mirror scanning

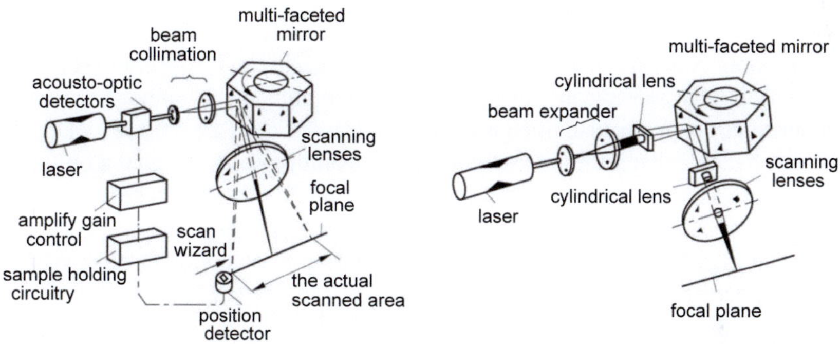

Fig. 4.9 Application of rotating mirror in cleaning device

taken into account. The robotic arm fixes the laser head on the mechanical hand and cleans through mechanical movement. The operation is more stable, and the uniformity of scanning is also improved. However, the flexibility of the mechanical hand's movement is poor, and there may be dead angles. The three-dimensional displacement platform can achieve x–y–z axis cleaning by fixing the laser cleaning head, which is suitable for cleaning flat or regular surface objects and can effectively improve cleaning efficiency. If the cleaning object is small and relatively regular, the cleaning object can be placed on the moving platform, and the laser beam remains stationary.

IV. **Recycling System**

During the cleaning process, contaminants that are stripped from the base material must be recycled. The stripped contaminants may be gaseous or solid (powder or fragments). Generally, because the vibration mechanism dominates in the laser cleaning process, the contaminants that are cleaned off are mainly solid. A vacuum device can be installed for recycling. The sealing of the vacuum device should be noted, and it should not affect the cleaning system. The vacuum device can be kept turning on all the time, or it can be turned on at the appropriate time as needed.

V. **Monitoring System**

When cleaning it is necessary to detect whether laser cleaning has achieve the expected effect. In addition to adopting offline detection methods (that is, after the cleaning is completed, the cleaning effect is checked in some way), online monitoring can also be carried out, that is, the cleaning effect is monitored in real time during the cleaning process. The monitoring system generally includes: observation windows, signal measurement devices. Signals include vibration signals, spectral signals, sound wave signals, etc. Among them, the observation window is a basic structure, and the operator can directly observe the surface of the cleaning object to determine the degree of cleaning. The basic structure of the observation window is shown in Fig. 4.10. The human eye can observe the surface of the cleaning object while the laser is acting through the observation microscope. Eye safety must be paid attention to. If safety cannot be guaranteed, the human eye cannot be used for direct observation, but instead, imaging devices such as CCDs are used to replace the human eye to achieve real-time recording of the cleaning process. Signal monitoring is carried out in an automatic manner, which is detailed in the fourth section of this chapter.

The cleaning effect can be detected offline and monitored online through various methods. Spectroscopy is an effective and relatively accurate method.

VI. **Control System**

The control system is an indispensable part of equipment with a high degree of automation. It can be a simple knob + digital tube form or a fully automatic touch form. The control module should have the following functions: (1) Control the operation of the entire laser cleaning equipment; (2) Monitor various parameters, including the temperature (water temperature in multiple places, temperature of the

Fig. 4.10 Observation window

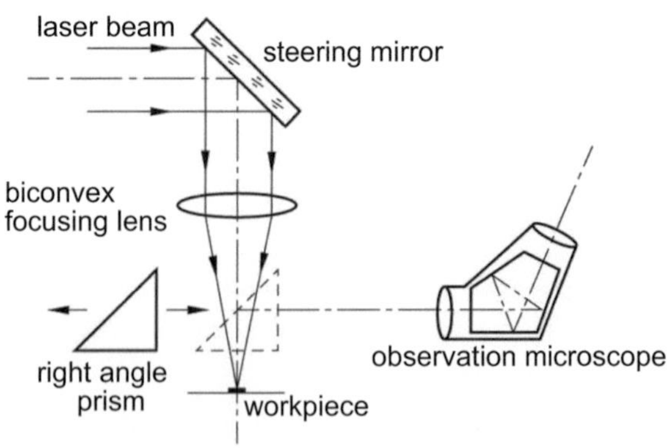

laser cavity), humidity, vibration, power-on situation, etc., with a complete alarm and protection device; (3) Control the parameters of laser output by outputting signals, adjust the speed of the scanning device during output of laser (and the moving speed of the displacement device), display and record the equipment status in real time, and have remote communication functions. Through the above functions of the control system, the laser cleaning equipment can work methodically, help staff to complete work efficiently, deal with possible failures in time, and reduce maintenance workload and operation costs.

VII. **Other Systems**

Other systems include power supply modules, protection modules, cooling modules, etc. In addition to supplying power to the laser, other systems of the laser cleaning equipment including the control system, beam output shaping system, and Q-switching system, also need power supply. The power supply module mainly has output of AC 220 V and DC 24 V, 5 V, 12 V, etc.

For wet laser cleaning equipment, there is also a liquid spray module, which sprays a certain amount of liquid on the object to be cleaned according to actual needs.

For safety reasons, the entire equipment should have water cut-off, power cut-off, overcurrent, overheating, and leakage protection, and the signals of various protections are connected to the control module.

4.2 Example of Nd:YAG Pulsed Laser Cleaning Machine

Nd:YAG Q-switched laser cleaning machine is currently the most commonly used laser cleaning machine, and there are already many commercial devices. In this section, we will introduce the structure and main technical parameters of the laser cleaning machine using Nd:YAG Q-switched lasers we made by ourselves as an example. From Chap. 5, we will introduce specific applications.

4.2.1 Electro-optic Q-Switched Nd:YAG Laser Cleaning Equipment

The pulse-pumped electro-optic Q-switched Nd:YAG laser is a commonly used laser source, with a pulse width of about 10 ns and a pulse peak power density of about 10^7 W/cm^2. The single pulse energy can easily reach 0.5–1 J, which is the ideal light source parameter for laser cleaning. We have conducted a large number of cleaning experiments with this type of laser, accumulated some experience, and developed a laser cleaning machine based on this. This machine, as the first-generation prototype, mainly explores some technical issues of laser cleaning equipment, including stability, beam transmission, beam shaping, machine packaging, device interconnection, cleaning operations, etc., and is convenient for cleaning experiments in front-line factories. Although its performance is limited, many problems were encountered and solved during the production process, and relevant experience was accumulated, laying the foundation for the development of the second-generation machine.

I. **Main Technical Indicators**

1. Overall rated parameters

   Input voltage: AC 220 V, allowing ± 10% fluctuation, input power: 1000 W.
2. Nd^{3+}:YAG pulsed laser

   Laser wavelength: 1064 nm, laser pulse width: ~ 200 μs (free running), ~ 10 ns (Q-switched), single pulse energy (at 2 Hz): 820 mJ (free running), 560 mJ (Q-switched), laser working frequency: 2, 10, 30, 60 Hz, indicating laser: wavelength 650 nm (red light), cooling: closed-loop cooling with water tank.
3. Transmission system

   Seven-joint light guide arm: Total length 1650 mm, 1064 nm laser total transmission rate of 80%.
4. Spot shaping output head

   Contains a 25 mm focal length cylindrical lens to achieve linear spot output.
5. Three-dimensional translation stage (controlled by Mach2 software)

   The range of movement in three directions: 400 * 300 * 200 (unit: mm).
6. Control system

   Power supply, temperature monitoring, power monitoring.

II. **Structure of the Laser Cleaning Machine**

The main part of this machine is the laser. The laser head and the main box are designed separately. The light guide arm is installed on the laser head to transmit the laser beam, and the cleaning head at the end of the light guide arm is clamped on a three-dimensional moving platform to clean some smaller samples.

1. Overall structure

The structure is shown in Fig. 4.11. It contains 1—the main machine box (power supply system, control system, cooling system), 2—laser cavity, 3—light guide arm, and 4—three-dimensional translation stage.

2. Control panel

The laser cleaning machine is controlled through the buttons on the control panel. The control panel is shown as Fig. 4.12.

(1) Power section (POWER): 1—Main power switch (two-stage key switch), 2—Power indicator (POWER ON) 3—Emergency brake switch (EMERGENCY)

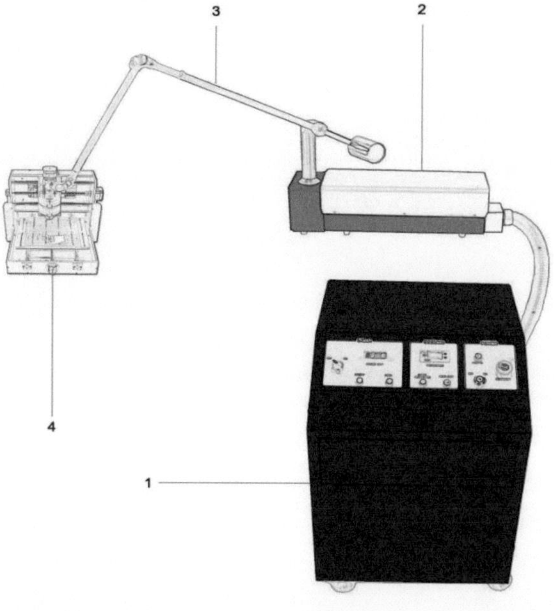

Fig. 4.11 Schematic diagram of the whole machine structure (1–4 are respectively the main machine box (power supply system, control system, cooling system), laser cavity, light guide arm, three-dimensional translation stage)

(2) Cooling system (COOLING): 4—Water flow indicator light (CIRCULATION), 5—Overheat buzzer alarm (OVER HEAT), 6—Temperature display and setting (TEMPERATURE)
(3) Laser power switch section (LASER): 7—Laser power switch, 8—Pre-ignition button (SIMMER), 9—Work button (WORK), 10-Xenon lamp charging voltage display (CHARGE VOLT).

3. Internal Structure of the Laser Cavity.

The internal structure of laser cavity is shown in Fig. 4.13, the main parts of which are: 1—pumped condensing cavity (with full cavity water-cooled diffuse reflection), 2—total reflection mirror, 3—partial reflection mirror, 4—Brewster window, 5—KD * P electro-optic Q-switcher, 6—650 nm LD indicator laser, 7—two 45°/1064 nm total reflection mirrors (used to reflect the laser into the light guide arm).

4. Cooling System

Due to the low power of the laser, the deposited thermal power is also low, so the cooling system uses a water circulation and water–air heat exchange method. The heat

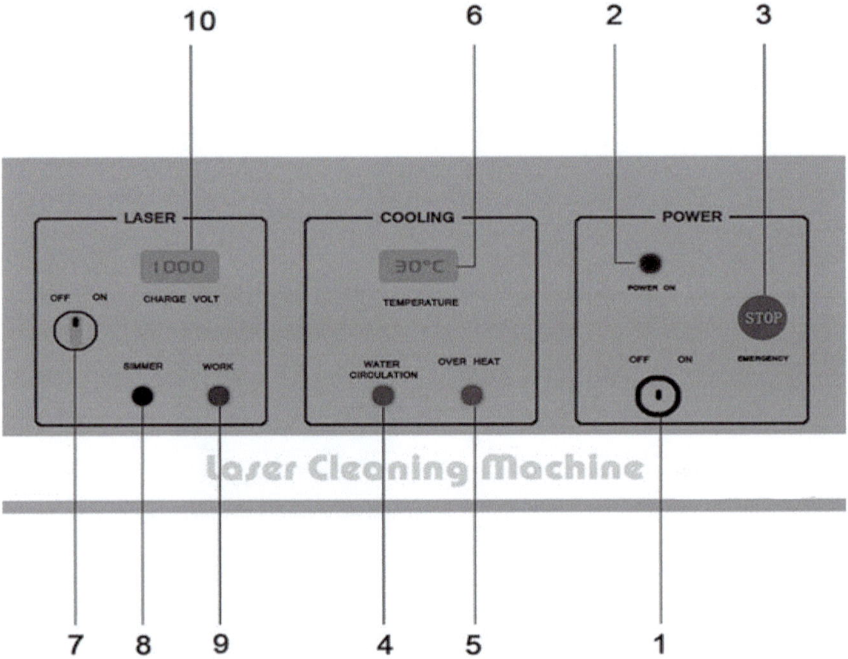

Fig. 4.12 Schematic diagram of front panel

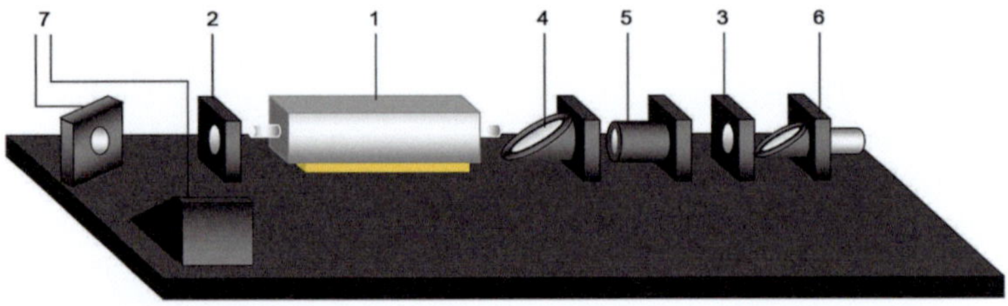

Fig. 4.13 Internal structure of the laser cavity

exchanger reduces the temperature of the circulating water to the ambient temperature, as shown in Fig. 4.14. The entire cooling system has water flow protection, overheat protection, and liquid level protection. If any of these are abnormal, the laser will automatically adjust to a dormant state, the flash lamp will stop working, and an automatic alarm will be triggered. The alarm setting for the water flow switch is set at a flow rate below 2 L/s; the alarm temperature for the thermostat is set at 35 °C; the alarm setting for the water level switch is when the water storage is below 8 L.

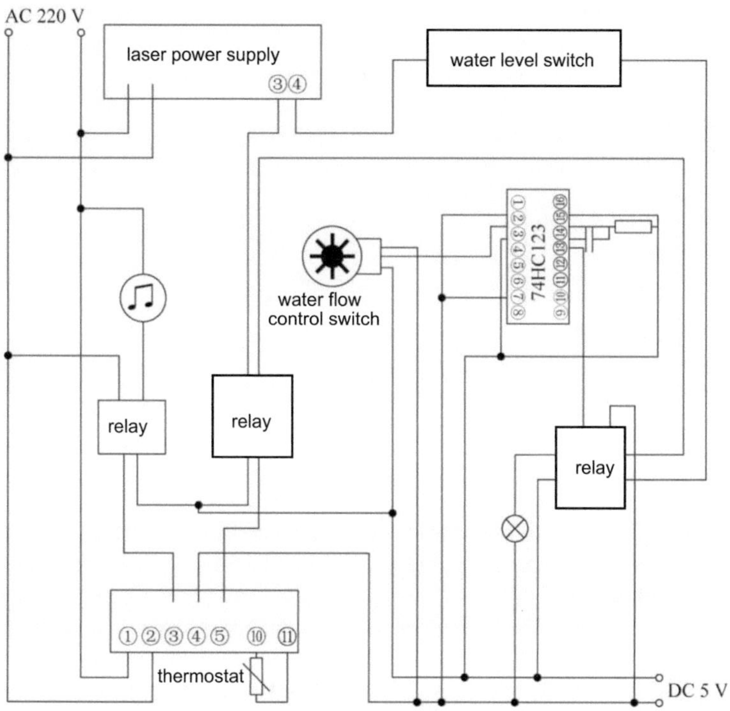

Fig. 4.14 Schematic diagram of water cooling system wiring

5. Light Guide Arm

We use a 7-joint counterweight light guide arm, as shown in Fig. 4.15. Each joint is a 45°/1064 nm total reflection mirror. An external laser output head for shaping is attached to the output part of the existing light guide arm, which contains a cylindrical mirror and a protective lens, capable of shaping the circular laser spot into a linear spot.

III. Features of the Cleaning Equipment

1. Selectable parameters. According to specific cleaning needs, by selecting the knob on the control panel or entering relevant parameters, you can choose the output laser energy, repetition rate, peak power, average power, etc.
2. Safety. The device has a thermostat displaying the current cooling water temperature, a water flow indicator light (water circulation, blue light), and a buzzer alarm for setting the upper limit of water temperature (over heat,

Fig. 4.15 Light guide arm

red light). If there is a problem with the water circuit, electrical circuit, or monitoring parameters, the laser cleaning machine will automatically enter a dormant state.

3. Controllability. The three-dimensional translation stage is controlled by the Mach2 software on the computer, loading the scanning program; setting the starting position and height, determining the scanning speed of the translation stage based on the thickness of the sample paint or rust layer, changing the F factor on the right side of the software interface (can be set to 100), executing the program scan, and stopping the program operation after the sample cleaning is finished.

4.2.2 *Quasi-continuous Laser Cleaning Machine*

Under the existing technological conditions, it is difficult to increase the pulse repetition frequency of the electro-optically Q-switched Nd:YAG laser to the kHz level, which greatly limits its cleaning efficiency. At the same time, the practical application of high peak power lasers with a pulse width of 10 ns is affected due to the lack of practical transmission fibers. Using a continuously pumped acousto-optic Q-switched Nd:YAG laser as the laser source for the cleaning machine can solve these two problems. After adopting the acousto-optic Q-switching technology, the single pulse energy of the laser is lower, the pulse width becomes wider, and the peak power of the pulse is two orders of magnitude lower than that of the electro-optic Q-switched laser, but it can be transmitted by optical fiber. The output spot of the laser must be focused to reach the cleaning threshold. For the convenience of cleaning operations, we hope to obtain a linear spot. The method of directly shaping the spot like the aforementioned electro-optic Q-switched laser is not feasible. It is necessary to introduce a spot line scanning system, and perform one-dimensional scanning on the focused pulsed laser through a galvanometer, so that the pulsed laser with a repetition frequency of kHz level forms a quasi-continuous linear spot. Based on this design idea, we have developed a quasi-continuous laser paint removal test machine.

We have adopted an all-in-one structure, which is more practical and easier to use. Since the laser head and cooling system are packaged together, this requires the laser cavity to have extremely high stability and shock resistance, so we have also made special designs for the optical path components. In terms of cleaning operations, the use of optical fiber to transmit lasers greatly enhances operational flexibility. We have also designed a handheld cleaning head that integrates focusing, scanning, and pollutant recycling, which is very compact and makes cleaning operations more convenient.

I. **Main Technical Parameters**

1. Rated parameters of the whole machine Input voltage

   AC 220 V, allowing ± 10% fluctuation, input power: 5000 W.
2. Cleaning laser source: Nd:YAG laser

   Laser wavelength: 1064 nm, pump source: continuous krypton lamp, Q-switching mode: acousto-optic Q-switch, laser pulse repetition frequency: 5 MHz, laser pulse width: 350 ns, average laser power: 45 W, indicating laser: wavelength 650 nm, output power 100 mW, cooling method: water circulation cooling.
3. Beam transmission

   Fused quartz energy transmission fiber, length: 5 m, core diameter: 800 μm, connector model: SMA05.
4. Spot shaping Focusing + galvanometer line scanning method, realizing quasi-linear spot output
5. Operation mode

   Handheld cleaning/mechanical clamping cleaning.
6. Control system and auxiliary devices

   Temperature monitoring, power supply, vacuum device, etc.

II. **Structure of the Cleaning Machine**

Figure 4.16 shows the top and side views of the laser cleaning machine. It adopts an integrated structure, organically encapsulating the laser, driving power supply, cooling system, control system, energy transmission fiber, and handheld cleaning head within a mobile chassis. When not in operation, the fiber optic cable can be wound at the front of the chassis (equipped with a dedicated cable winding device), and the handheld cleaning head at the tail end can be stored in the storage box at position 4 in the figure, thereby improving the integrity of the equipment.

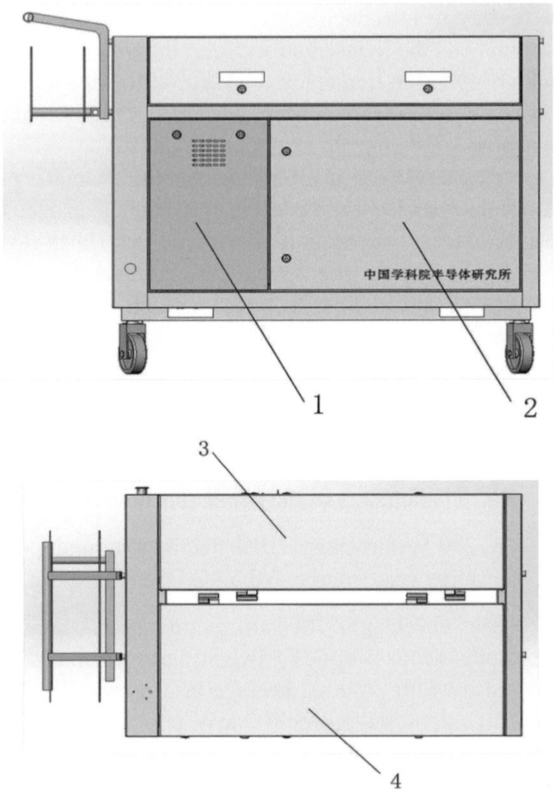

Fig. 4.16 Overall structure diagram. 1—Lower part of the chassis control system, 2—Lower part of the chassis cooling system, 3—Upper part of the chassis cleaning head and fiber placement, 4—Upper part of the chassis laser placement

1. Control panel

The control panel is shown in Fig. 4.17, including power setting, laser standby and emission buttons, indicator, and so on.

2. Internal structure of the laser head

As shown in Fig. 4.18, the laser head adopts a fully enclosed structure, which can keep the internal environment clean and dry. The bottom is a grid-shaped steel plate profile, which has strong stability. The internal components mainly include a focusing cavity, a resonant cavity, a Q switcher, an indicating laser, a fiber coupler, etc. The resonant cavity and the elements inside the cavity are connected by steel pipes to enhance the relative stability between each adjustment frame and reduce the vibration impact of the cooling system. The specific structure of the laser head is shown in Fig. 4.19, including the laser resonant cavity, acousto-optic Q switcher, fiber coupler, indicating laser, cooling, and other parts.

3. Handheld cleaning head

As shown in Fig. 4.20, the handheld cleaning head is the output end of the working laser and contains a focusing and scanning mirror system. The laser transmitted

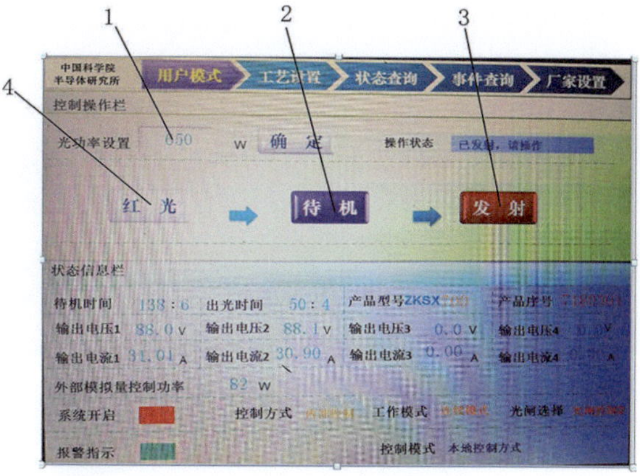

Fig. 4.17 Control panel schematic: 1—Power setting; 2—Laser standby button; 3—Laser emission button; 4—Red light indicator button

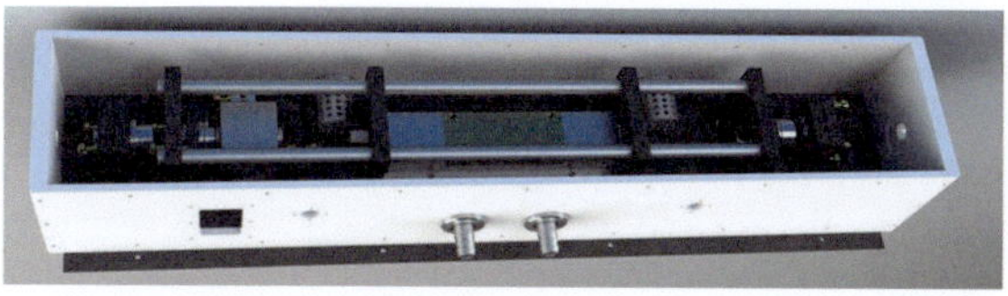

Fig. 4.18 Internal components of the laser head

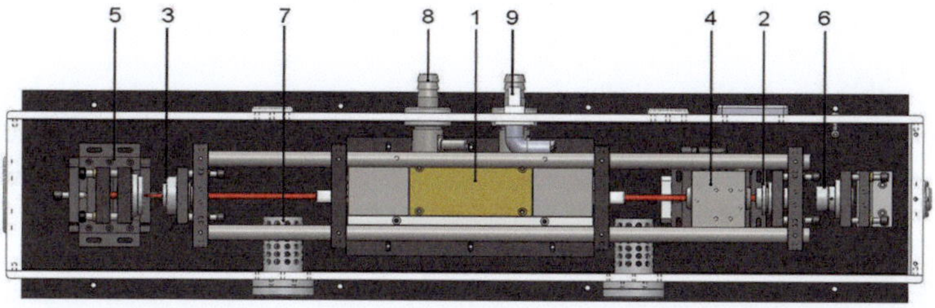

Fig. 4.19 Laser head optical path structure diagram: 1—Focusing cavity (tightly wrapped gold-plated cavity), 2—Total reflection mirror, 3—Partial reflection mirror + coupling focusing lens, 4—Acousto-optic Q switcher, 5—Fiber coupler, 6—650 nm LD indicator laser, 7—Dryer, 8—Focusing cavity cooling water inlet and outlet, 9—Q switcher cooling water inlet and outlet

through the fiber is first focused. This focusing lens and the focusing lens at the fiber head form a shaping system, making the single pulse spot diameter at the cleaning surface about 0.5 mm after the total optical path; the converged laser will be scanned by the scanning galvanometer and reflected to the reflector at a 60° incident angle. After the reflector, the laser will be emitted horizontally and form a scanning line in the direction perpendicular to the paper, ultimately achieving the purpose of shaping and scanning the beam.

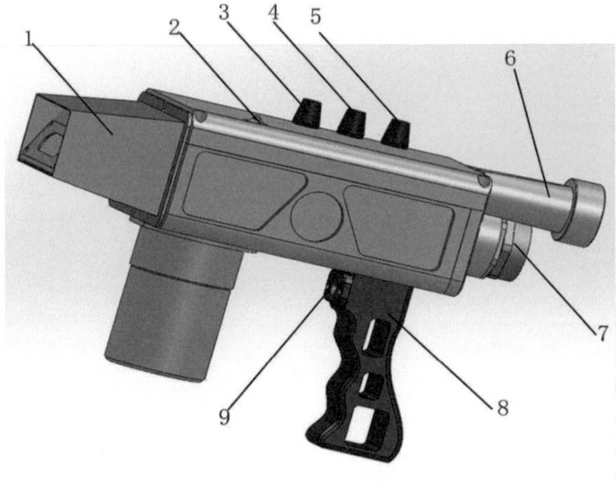

Fig. 4.20 Handheld cleaning head structure diagram (mainly includes: 1—Field lens protective cover, 2—Indicator light, 3—Power adjustment knob, 4—Line width adjustment knob, 5—Speed adjustment knob, 6—Dust suction device interface, 7—Energy transmission fiber end, 8—Handle, 9—Operation switch)

III. **Features of the Instrument**

Like the electro-optic Q-switched laser cleaning machine, this machine also has selectable parameters and safety. In addition, this machine adopts an integrated structure and pre-optical transmission.

1. Selectable parameters. According to specific cleaning needs, by selecting the knob on the control panel or inputting related parameters, you can choose the output laser energy, repetition rate, peak power, average power, and other parameters.
2. Safety. It is controlled by the overall control system, with monitoring of water, electricity, mechanical parameters, etc. For example, if the current is abnormal, the temperature reaches the set upper limit, there is insufficient cooling water, or the work switch of the handheld cleaning head is not closed, the corresponding warning lights will light up, the buzzer will alarm, and the RF output and laser output will stop, with the whole machine in standby mode.
3. Convenience. The laser head and cooling system are encapsulated together, transmitted through fiber optics, and the galvanometer focuses into a quasi-continuous line spot. The structure is more compact and easier to use.

4.2.3 Acousto-optic Q-Switched Laser Cleaning Machine Pumped by Semiconductors

The Nd:YAG laser with acousto-optic Q-switching pumped by semiconductors is currently widely used in the cleaning field. Compared with flash lamp pumping, it has higher efficiency, a more compact laser head structure, and lower cooling requirements. Common models of 50 W, 100 W, 200 W, 500 W (requiring amplification stage) have been designed. First, the beam is shaped by scanning with a galvanometer and then focusing with a field lens, ensuring a sufficiently large laser pulse energy density and high cleaning efficiency. Considering the convenience of the whole machine, an all-in-one structure is adopted, and each area is divided to install its own modules, which is conducive to modularization. In operation, the use of fiber optic transmission and handheld heads greatly increases operational flexibility, and the compact structure is suitable for different laser cleaning occasions. A touch screen is used to display and adjust parameters, making the panel simple and easy to operate. The central processing unit uses a PLC to timely collect data analysis and control the normal operation of each module, and sets up various protection and alarm devices. The cooling device uses compressor refrigeration to ensure the normal operation of the side pump module and the acousto-optic Q switch.

I. **Technical Indicators**

1. Overall machine rated parameters

   Input voltage: AC 220 V, allowing ± fluctuation, input power: 3–10 kW.
2. Laser parameters Laser type: Nd:YAG laser

   Pumping method: semiconductor side pump, Q-switching method: acousto-optic Q switch, laser pulse repetition frequency: 3–10 kHz, pulse width: 150–400 ns, average laser power: 50 W (min)–500 W (max), indicating laser: wavelength 635 nm (5 mW), cooling method: compressor refrigeration, full module water cooling.
3. Beam transmission method

   Fused quartz energy transmission fiber, length: 5 m, fiber core diameter: 1000 μm, connector type: SMA05.
4. Beam shaping

   Galvanometer scanning + field lens focusing method, resulting in linear spot output.
5. Operation method

   Handheld cleaning or mounted on a robot arm.
6. Display control method

   PLC plus touch screen, laser control signal: 0–7.5 V (corresponding to current 0–30 A), Q drive signal: TTL signal, galvanometer signal: 0–5 V triangle wave (0–50 Hz).

IV. **Machine Structure**

1. Overall machine layout structure

This machine adopts an integrated layered structure, divided into three layers. The bottom layer is the cooling module, the middle layer is the power supply module, and the top layer is the laser and control system, with external energy transmission pipelines and laser cleaning heads, achieving overall machine encapsulation and good mobility. The diagram is shown in Fig. 4.21. The handheld head is fixed on the side handle, and there is a fiber winding disc on the side, which usually winds the fiber protective cover, offering a more convenient and swift experience for users.

Since the compressor will vibrate during operation, placing the cooling system at the bottom will minimize the side effects of vibration and facilitate water injection and drainage, making it easier to operate. To prevent interference between the control module and the power supply module, all power supplies are placed on the second layer. The third layer mainly places the laser and control module, with a parameter display panel on top. Figure 4.22 shows the distribution of modules in the second and third layers.

2. Control Panel

As shown in Fig. 4.23, the control panel mainly consists of a display screen, signal generator, laser emission button, and emergency stop switch.

The display panel includes four interfaces: the main control interface, alarm records, temperature setting, and operating parameters. The status of each interface is shown in Fig. 4.24.

3. Internal Structure of the Laser

The laser adopts a fully enclosed aluminum alloy structure, with a shock-absorbing pad added at the bottom, providing good stability. The internal components are fixed

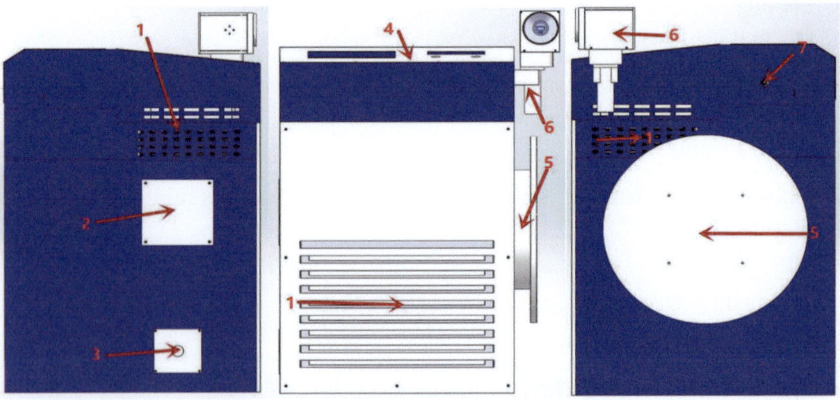

Fig. 4.21 Overall structure diagram. 1—Heat dissipation hole (to ensure the normal operation of the internal module), 2—Water injection side panel (after opening, there will be a water injection port for the chiller inside), 3—Drainage port, 4—Main control panel, 5—Fiber winding reel (the usual placement of the fiber protective sleeve), 6—Handheld head (laser focusing and scanning device, containing galvanometer and field lens), 7—Fiber outlet (there is a laser inside, external fiber)

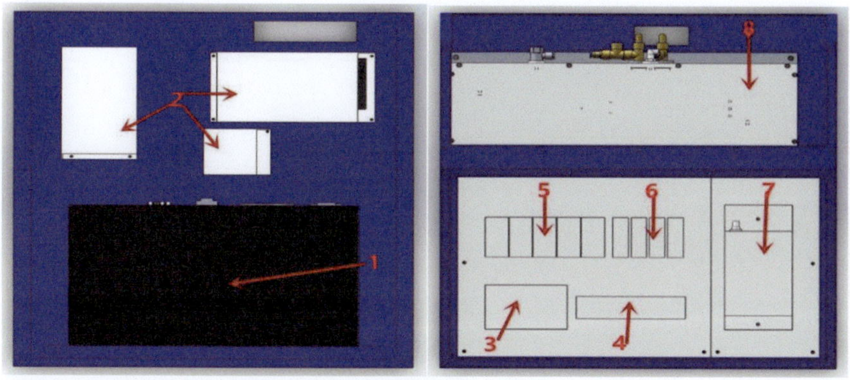

Fig. 4.22 Second and third layer module distribution diagram. 1—Laser driver power supply, 2—Switching power supply (provides power to Q driver source, indicator light source, processing unit, etc.), 3—Processing unit (PLC, generates 0–7.5 V control voltage of laser driver power supply and on–off of each relay), 4—Wiring terminal (connects various components, using wiring), 5—Air switch (the main switch of each module, to prevent overcurrent), 6- Relay (realizes the weak electric control of the control module), 7- Q driver (acousto-optic Q switch driver module, provides RF signal)

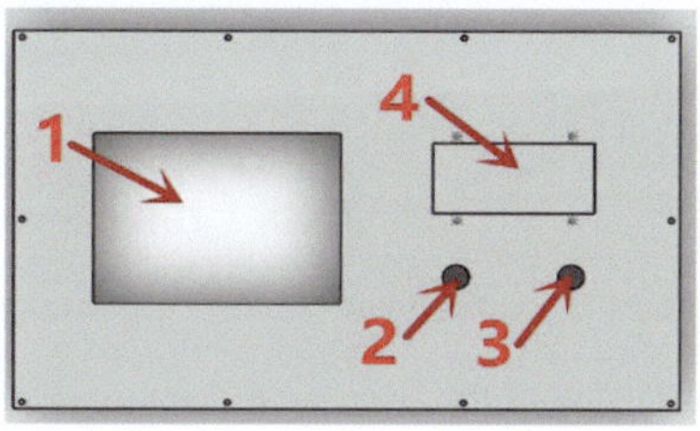

Fig. 4.23 Display control panel. 1—Display touch screen (displays parameters, error records and output control signals, adjusts working status), 2—Laser emission switch, 3—Emergency stop button, 4—Signal generator (controls the repetition rate of laser pulses, scanning frequency of galvanometer)

to the base plate, and the upper part uses an Invar rod for reinforcement (Invar has an extremely low expansion coefficient), ensuring the stability of the laser module, cavity mirror, acousto-optic Q switch, and coupling module. The structure is shown in Fig. 4.25.

4. Handheld Cleaning Head

The handheld head is implemented by galvanometer scanning and field focusing. The optical path diagram is shown in Fig. 4.26. The laser is incident from the bottom, reflected by a 45° mirror to become horizontal, and after passing through

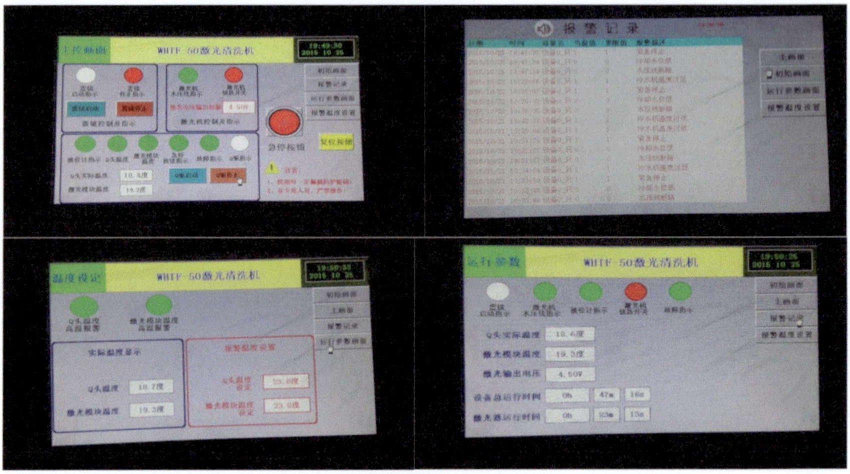

Fig. 4.24 Four interfaces of the display panel

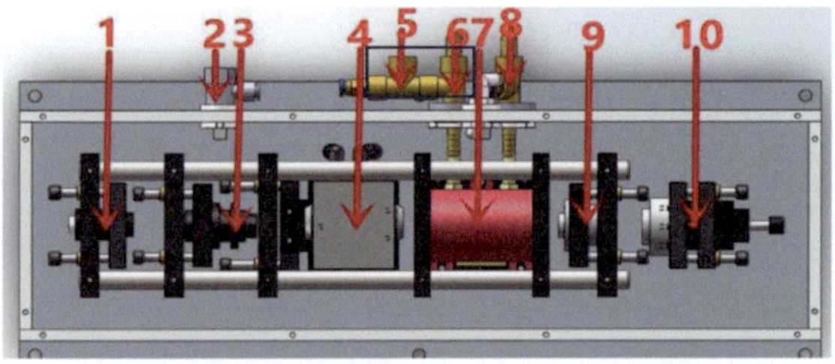

Fig. 4.25 Internal structure diagram of the laser. 1—635 nm semiconductor indicator laser, 2—Acousto-optic Q switch outlet (with external pt100 temperature sensor), 3—Total reflection mirror, 4—Acousto-optic Q switcher, 5—Total cooling water outlet, 6—Semiconductor side pump module outlet (with external pt100 temperature sensor), 7—Semiconductor side pump module, 8—Total cooling water inlet, 9—Partial reflection mirror + coupling focusing lens, 10—Fiber coupling device

the galvanometer and field lens, it forms a focused linear light spot. The schematic diagram of the structure is shown in Fig. 4.27.

V. **Features of the Laser Cleaning Instrument**

The instrument uses a microcomputer as the control system and is equipped with an LCD screen for visibility. Indicators such as laser water pressure line, liquid level gauge, Q-head temperature, laser module temperature, emergency stop button, fault indicator, etc., can all be displayed on the LCD screen. If a problem occurs, all the buttons on the panel will be locked by the program, followed by automatic power off, making it inoperable. The machine's control module will record the type and time of the fault in real time for daily maintenance. Additionally, after connecting the GPRS module, the machine's working status can be remotely viewed.

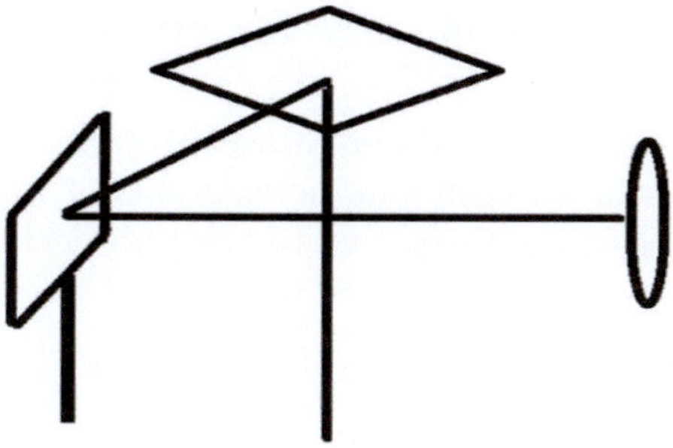

Fig. 4.26 Handheld head optical path schematic

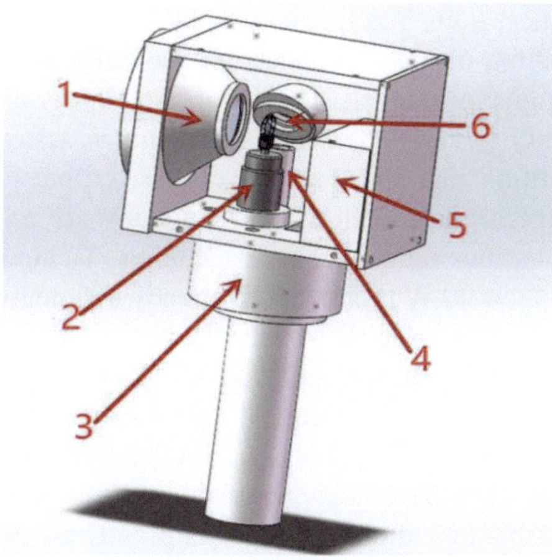

Fig. 4.27 Handheld head structure schematic. 1—Field lens (for focusing, aspherical lens group), 2—Scanning galvanometer (X-axis), 3—Wire storage location, 4—Laser fiber connection post, 5—Scanning galvanometer drive board, 6—45° reflection mirror

In addition to our research and development work on laser cleaning equipment, many domestic institutions and companies have also conducted research in this field. As early as the Anhui Institute of Optics and Mechanics of the Chinese Academy of Sciences, Huazhong University of Science and Technology, Institute of Engineering Thermo-physics, Changchun Institute of Automation, and other companies have also conducted research, some of which have developed laser cleaning equipment.

4.3 International Status of Laser Cleaning Equipment

In the field of laser cleaning, the United States and Europe have always been in a leading position. With the continuous accumulation of technology, they have moved from laboratory research to equipment development and practical application, achieving certain social and economic benefits.

4.3.1 Current Research Status of Laser Cleaning Equipment

Cleaning companies are mainly distributed in several countries such as the United States, Germany, France, Italy, South Korea, and Singapore, as shown in Table 4.2. At present, laser cleaning machine products are basically divided into low power, medium power, and high power levels. The mainstream lasers all use Nd:YAG lasers, combined with active Q-switching technology (mainly acousto-optic Q-switching), to achieve high repetition rate and high power laser output to meet the cleaning conditions. Among them, the products from Germany are relatively mature, and they have developed various models of laser cleaning machines. There are several domestic agents, and their 20 W products can already be integrated into backpacks. The workers can operate and use them very conveniently. They have also developed a variety of handheld heads and operating platforms. For high-power products, they have integrated robotic arms, combined with automatic operating systems for remote control. The products of Italy's EL.En.SPA company are aimed at the field of cultural relic cleaning. They have manufactured short free-running lasers and long Q-switched lasers suitable for cultural relic cleaning, and have achieved good results. Table 4.3 lists the performance parameters of some laser cleaning machine products developed abroad.

Table 4.2 Some laser cleaning companies

S. No.	Company	Country
1	Adapt laser	USA
2	Laser clean all	USA
3	Clean laser	Germany
4	Norton sand blasting	USA
5	Quanta system	Italy
6	El En group	Italy
7	Rofin	Germany
8	Quantel laser	France
9	IMT—innovative manufacturing technology	South Korea
10	LE champ	Singapore

Table 4.3 Some laser cleaning equipment and performance parameters

Country	France	Italy	Germany	Poland	Netherlands	Germany
Company name	QUANTEL	EL. EN. SpA	JET	Military Univ. of Technology	Art Innovation	Rofin
Model	LASERBLAST 60/50001000/ 2000	EOS1340	TMCS	ReNOVALaser	LCS	Rofin DQ10/50/ 80
Laser type	Nd:YAG	Nd:YAP	TEA–CO_2	Nd:YAG	Excimer	Nd:YAG
Wavelength	1064 nm	1341 nm	10.6 μm	1064/532 nm	248 nm	1064 nm
Pulse width	10–12 ns	60–120 μs	μs-level	8 ns	20 ns	38 ns
Power	8–72 W	14 W	kW-level	5.5 W	30 W	450–900 W
Transmission mode	Optical fiber	Optical fiber	Optical fiber	Light Guide Arm	Light Guide Arm	Optical fiber
Application	Buildings	Sculptures	Tire Molds	Tire Molds	Paintings	Paints/ Greases

4.3.2 Laser Cleaning Machine Using Nd:YAG Laser

I. **LaserBlast Series (Quantel, France)**

The LaserBlast series includes four models: LaserBlast60, LaserBlast500, LaserBlast1000, and LaserBlast2000. They use lamp-pumped electro-optically Q-switched Nd:YAG lasers as the cleaning laser source, with output powers of 8 W, 20 W, 40 W, and 72 W respectively. They use optical fibers with larger core diameters for beam transmission. The lasers of the LaserBlast series run in the fundamental mode, and the laser beam is shaped in two ways: Gaussian laser beam is shaped into flat-top shape, and circular light spots are shaped into rectangular light spots. This series of laser cleaning equipment has high pulse energy and narrow pulse width, which can easily reach the cleaning threshold of pollutants. The cleaning effect is good, but the repetition frequency is low, and the cleaning speed is not high. Figure 4.28 shows the LaserBlast500 and LaserBlast1000 cleaning equipment [7].

II. **DQ Series**

The DQ series from Rofin uses high-frequency Q-switching technology to output high repetition frequency, narrow pulse width lasers, with optical fibers for beam transmission. It has launched three models: DQx50, DQx80, and DQ010, with output parameters shown in Table 4.4. It can be seen that when the repetition frequency is 6 kHz, the pulse width is 38 ns, and the power range is 500–1000 W, ensuring sufficient peak power and average power, guaranteeing the cleaning effect and sufficient cleaning efficiency. It is an ideal tool for laser ablation, laser cleaning, laser insulation, and surface treatment, and can be used for laser film removal of splicing plates,

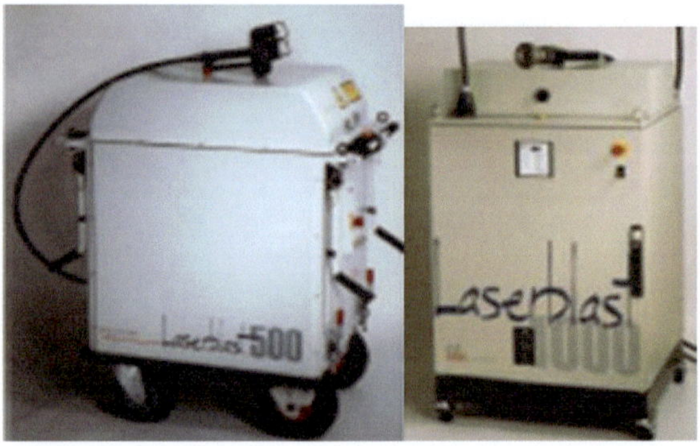

Fig. 4.28 LaserBlast series (Quantel, France)II. DQ Series (Rofin, Germany)

solar cells or LCD display screens, color ablation of glass and other materials, and cleaning of tools and molds. Figure 4.29 shows the appearance of DQx80 [8].

III. **EOS Series (En.El.S.P.A, Italy)**

The Italian company En.El specializes in the development of laser cleaning machines for different types of cultural relics, mainly focusing on the cleaning of stone artifacts. The appearance of the EOS-1000-LQS, EOS-1000-SFR, and EOS COMBO models are shown in Fig. 4.30, and their relevant parameters are listed in Table 4.5 [9].

IV. **CL Series from Clean Laser Company**

"Clean Laser" is a German company specializing in laser cleaning equipment. Its laser cleaning equipment power ranges from 20 to 1000 W, and it has developed a variety of laser cleaning handheld heads and mechanical arms, with a variety of laser cleaning solutions. Among them, the 20 W product is highly integrated, made into a backpack style, easy to use, the 20–100 W product has a small desktop type

Table 4.4 DQ series related parameters

Output power	500 W, 800 W, 1000 W
Wavelength	1064 nm
Maximum pulse energy	75 mJ@6 kHz (500 W) 120 mJ@6 kHz (800 W) 150 mJ@6 kHz (1000 W)
Pulse repetition rate range	6–15 kHz
Pulse width	38 ns@6 kHz (± 10%)
Pulse stability	± 3%
Beam quality at collimation	< 25 mm mrad (500 W) > 30 mm mrad (800 W and 1000 W)
Fiber diameter	600 μm(500 W), 800 μm(800 W and 1000 W)

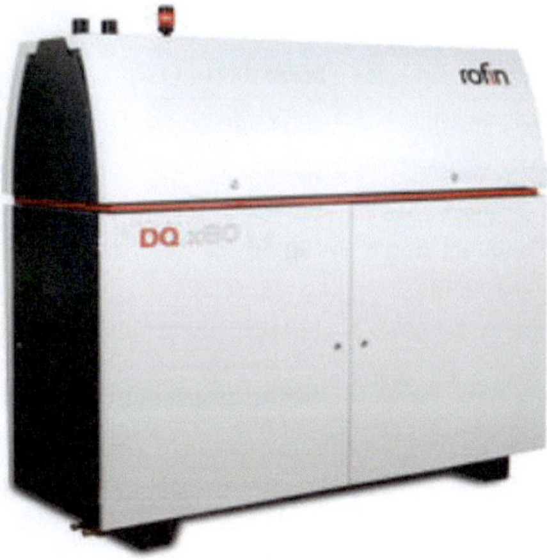

Fig. 4.29 Appearance of DQx80

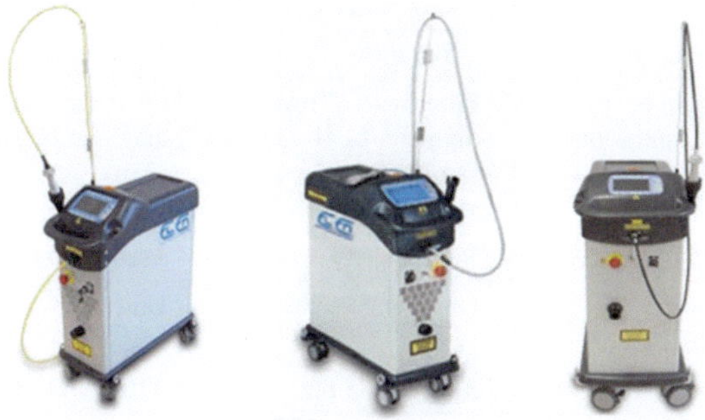

Fig. 4.30 Appearance of the EOS series (from left to right are EOS-1000-LQS, EOS-1000-SFR, and EOS COMBO)

with a small footprint, and the 100–1000 W product is a trailer-type workstation, convenient for mobile work. Figure 4.31 lists the representative products within each power range [10, 11].

In addition to the models of various powers, the Clean Laser Company has also developed different handheld heads according to different products to adapt to different working environments, integrating various sensors and cooling devices, with optical shaping functions (line-shaped flat-top light). In addition, various auxiliary brackets have been developed, which can clean paint, rust, dust, oil stains, fingerprints, etc. In recent years, Clean Laser Company has been committed to researching laser cleaning equipment that can realize online and high-temperature cleaning of plastic injection molds and rubber vulcanization molds. It can be used for cleaning different items in different environments, greatly improving the cleaning efficiency and reducing the limitations of laser cleaning. Figure 4.32 shows the appearance of different model cleaning heads.

The company's laser cleaning equipment is more suitable for industrial applications, among which the 150 W product is suitable for various cleaning fields. Its internal structure diagram is shown in Fig. 4.33, which includes all modules.

Table 4.5 Relevant parameters of EOS-1000-LQS, EOS-1000-SFR, and EOS COMBO

Parameter	EOS-1000-LQS	EOS-1000-SFR	EOS COMBO
Wavelength	1064 nm	1064 nm	1064 nm
Pulse width	100 ns	50–130 μs	SFR mode: 30–110 μs LQS mode: 100 ns
Maximum single pulse energy	130 mJ	1 J	SFR mode: 2 J LQS mode: 150 mJ
Energy	130 mJ (single pulse), 250 mJ (double pulse), 380 mJ (triple pulse)	50–500 mJ (50 mJ per step); 600–1000 mJ (100 mJ per step)	SFR mode: 200-1400 mJ (100 mJ per step), 1600 mJ, 1800 mJ, 2000 mJ LQS mode: 150 mJ (single pulse), 300 mJ (double pulse), 450 mJ (triple pulse)
Repetition rate	Single pulse, 1–10 Hz, 15 Hz, 20 Hz	Single pulse, 1–10 Hz, 15 Hz, 20 Hz	Single pulse, 1–10 Hz, 15 Hz, 20 Hz
Light spot	1.5–6 mm	1.5–6 mm	1.5–6 mm
Light beam transmission	Fiber with a core diameter of 1000 μm (3 m)	Fiber with a core diameter of 1000 μm (3 m)	Two fibers with a core diameter of 1000 μm (respectively 3 m and 10 m)
Handheld head	Variable focus	Variable focus	Variable focus
Energy distribution	Uniform	Uniform	Uniform
Indicator light	3 mW (635 nm)	3 mW (635 nm)	3 mW (635 nm)
Power supply	230 V-50/60 Hz, 8.5 A	230 V-50/60 Hz, 8.5 A	230 V-50/60 Hz, 12 A
Size	$23 \times 65 \times 68$ cm^3	$23 \times 65 \times 68$ cm^3	$33 \times 95 \times 75$ cm^3
Weight (kg)	40	40	80
Control method	Operation panel	Operation panel	Operation panel
Cooling circuit	Closed heat exchanger (air/water)	Closed heat exchanger (air/water)	Closed heat exchanger (air/water)

Figure 4.34 is a schematic diagram of the user operation display interface. Figure 4.35 is the cooling system of this model product, the left side is compressed air cooling, and the right side is compressor cooling. Compressed air is mainly used for internal cooling of the handheld head and blowing away the residues of pollutants to keep the cleaning surface clean. The compressor cooling mainly cools the water flow in the laser module to ensure the normal operation of the laser module.

Fig. 4.31 CL series products: **a** 20 W backpack type, **b** 100 W desktop type, **c** 150 W trolley type, **d** 1000 W trolley type

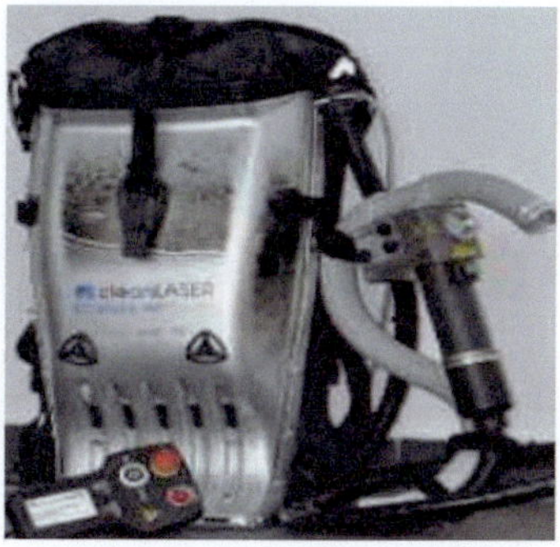

(a)

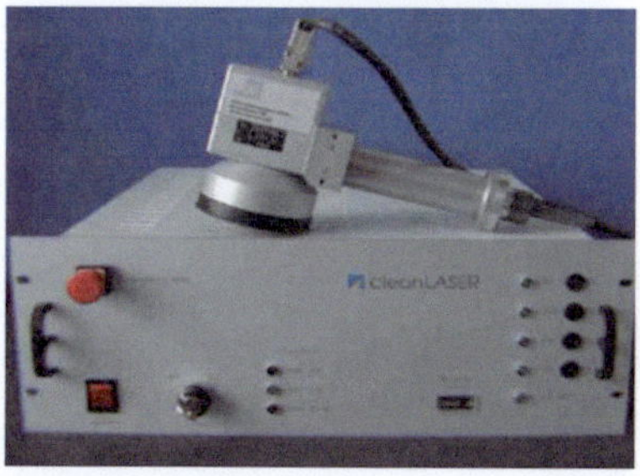

(b)

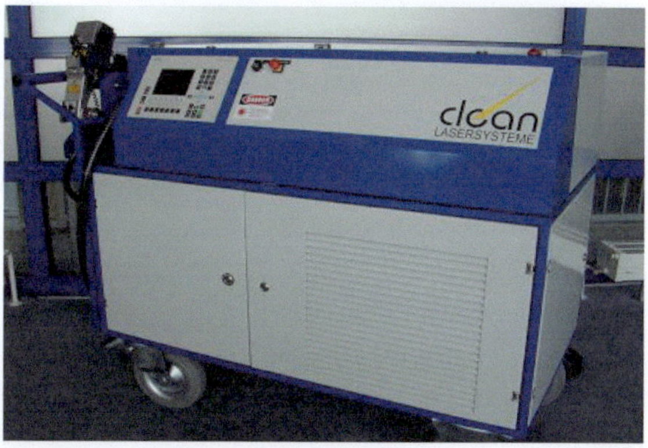

(c)

Fig. 4.32 CL series handheld heads (from left to right are OSH20, OSH50, OSH80)

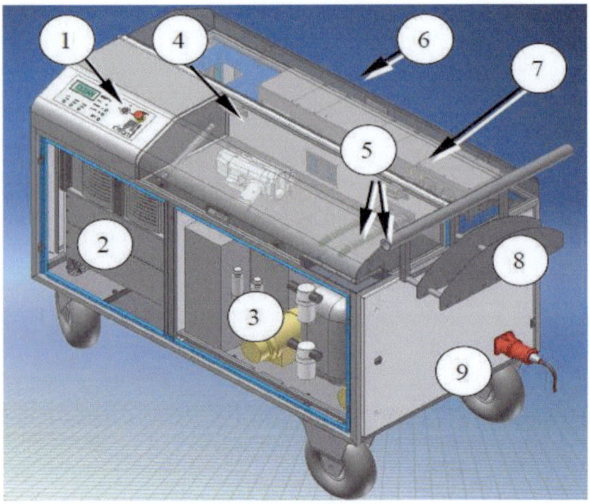

Fig. 4.33 Internal structure diagram of the laser cleaning machine. 1—User operation display interface, 2—Central control unit, 3—Cooling system, 4—Laser output device storage room, 5—Fiber optic signal line protective cover, 6—Laser resonator protective cover, 7—Laser resonator, 8—Protective cover hanging place, 9—Main power cord)

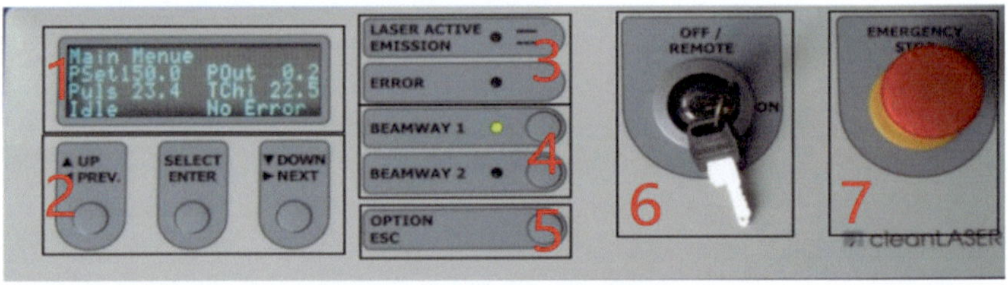

Fig. 4.34 User operation display interface device. 1—LCD display panel, 2—Menu navigation buttons (up, select, down), 3—Status indicator lights (laser emission, error), 4—Optical path switch button, 5—Select exit button, 6—Switch key, 7—Emergency stop button (only for emergency use)

Fig. 4.35 Cooling system

4.3.3 CO_2 Laser Cleaning Machine

I. **JET Company's TMCS Tire Mold Laser Cleaning System**

The TMCS (Tire Mode Cleaning System) laser cleaning system, shown in Fig. 4.36, is a laser cleaning machine produced by Germany's JET Laser System GmbH Company for cleaning the residual rubber on the surface of tire molds. The TMCS Laser Cleaner uses a CO_2–TEA laser, with pulse width of ms–μs level, radiation range of 1 cm^2 laser, and the cleaning depth of 0.01 mm. This instrument has extremely high cleaning quality, which can automatically clean every part of the mold, and the precision five-dimensional coordinate scanning device under the groove guides the laser output. It can thoroughly clean every corner of the mold side without damaging the mold, and timely remove and filter the removed residues. The movable body can achieve online or offline cleaning, eliminating the trouble of disassembling and pulling the mold. Depending on the size, complexity, and hardening cycle of the mold, the time to clean a set of molds is generally 45–90 min. At the same time, the convenient and flexible cleaning method can clean molds composed of two parts pressed by ordinary methods. The standard JET cleaning unit can clean tire molds with a rim diameter of 13–21 inches. For larger molds, they need to be disassembled into appropriately sized parts for offline cleaning. The entire cleaning process is fully automated, without any manual cleaning process, providing good safety protection for cleaning personnel.

II. **Other JET's Laser Cleaning Equipment**

The Flex JET Cable Stripping System and Bless Tube Decoating System laser cleaning machines produced by JET can be used to strip various coatings or claddings (such as PA, PVC, PTFE) on objects such as metal tubes and optical cables. The system uses a CO_2 laser, with a clever structural design, fully automated operation, and can strip various shapes and sizes of patterns, accurately removing the cladding along the edge. Its accuracy can reach 100 μm, and the typical cleaning speed can reach 50 mm^3/s.

Fig. 4.36 Appearance of CO_2–TEA laser cleaning machine

4.3.4 Laser Cleaning Machine Using Excimer Laser

I. **JET's Laser Cleaning System**

The MoldJET laser cleaning system (Fig. 4.37) is a product of the German JET company. It is mainly used to clean all types of molds other than tire molds. Similar to the previously introduced TMCS system, MoldJET has computer numerical control devices, filtering devices, cooling devices, and other auxiliary systems. However, because the MoldJET laser cleaning machine uses a shorter pulse width excimer laser, it can be used to remove some fine particles that are more firmly combined with the mold. The laser cleaning head of the MoldJET system consists of a four-dimensional coordinate scanning device, which can clean every part of the mold surface at different feed angles. The entire instrument, while facilitating the cleaning of other types of molds, also has the excellent characteristics of TMCS.

II. **USHIO's Excimer VUV/O_3 Laser Cleaning System**

The Excimer VUV/O_3 laser cleaning system produced by the American USHIO company can meet the high-precision cleaning operations of the microelectronics industry. Its laser system uses a specially designed excimer laser, which has the characteristics of single-mode radiation and extremely narrow bandwidth. It can radiate a vacuum ultraviolet pulse laser with a wavelength of 172 nm different from the general excimer laser ultraviolet pulse laser. Its single photon energy can reach 166.7 kcal/mol, which is greater than the binding energy of common organic chemical bonds (such as O–O, O–H, C–H, etc.). At the same time, its extremely high energy conversion efficiency (10%) can ensure to efficiently remove various tiny organic pollution particles on the semiconductor chip. This device can also control the energy

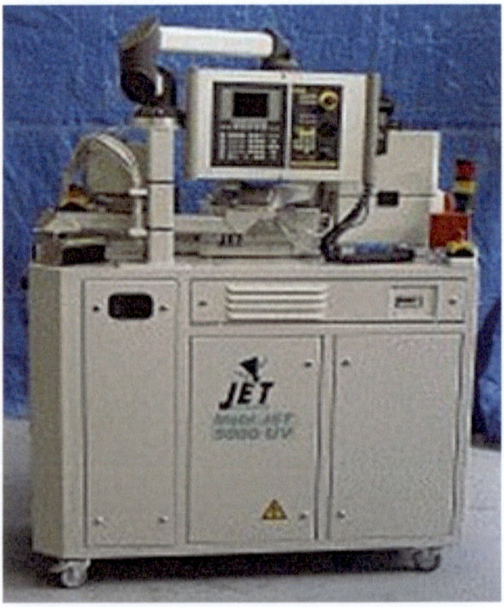

Fig. 4.37 Appearance of excimer laser cleaning machine

Fig. 4.38 Appearance of UVS-4200 laser cleaning system developed in the United States

of each output pulse, used to remove particles of various properties without damaging the substrate. In addition, the direction of laser radiation of this system is quite free, which can keep the output light direction perpendicular to the plane of the substrate to be cleaned on the production line. This product can meet the cleaning requirements of various electronic components such as liquid crystal units and semiconductor units. Among them, the UVS-4200 model is used for semiconductor surface cleaning (Fig. 4.38) before the deposition, improving the wettability and adhesion with solid aerosols and photoresist, etching line cleaning and other electronic processes [12].

4.3.5 Laser Cleaning Machine Using Fiber Laser

IPG, the leader of high-power fiber lasers and amplifiers, has also developed fiber lasers for laser cleaning. In recent years, IPG has completed the full-body paint removal using a 6 kW continuous fiber laser. In response to different market demands for laser cleaning, IPG has innovatively launched new pulsed fiber lasers. They have round/square spots, fixed/adjustable pulse widths, higher pulse energy (up to 100 mJ), and higher output power (peak power can reach MW level). Compared with round fibers, square fibers can obtain the maximum cleaning area with the smallest overlap area, improve cleaning efficiency, and achieve higher cleaning efficiency. The typical appearance of a fiber laser cleaning machine is shown in Fig. 4.39.

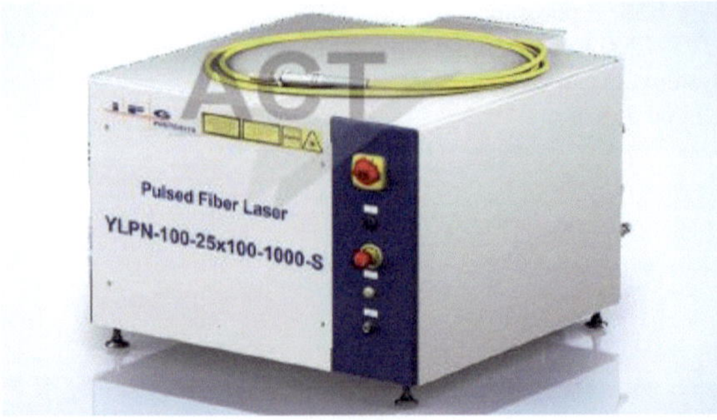

Fig. 4.39 Appearance of YLPN-100—25 × 100-100-S model fiber laser cleaning machine

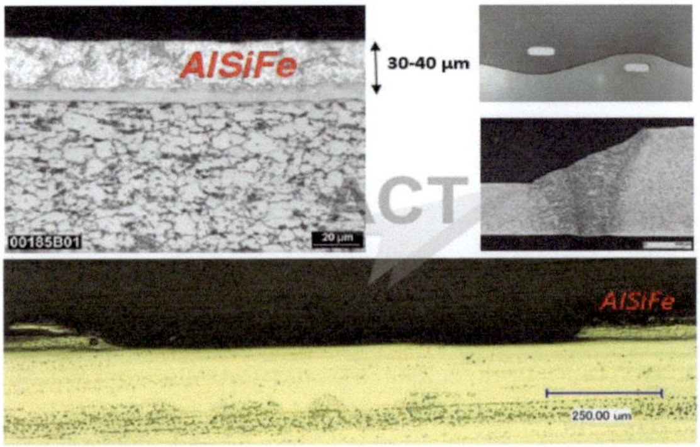

Fig. 4.40 IPG 1 kW pulsed laser cleaning of high-strength steel welded plate coating

The coating of high-strength steel welded plates for cars is removed with IPG's fiber laser cleaning machine, as shown in Fig. 4.40. A 600 × 600 μm square fiber, with an average power of 1000 W and a pulse energy of 100 mJ, is used. It synchronously removes the coating with the welding at a speed of 8–10 m/min, leaving a smooth surface, improving the stability of subsequent welding processes.

4.4 Online Monitoring Methods of Laser Cleaning

During the operation of laser cleaning, when the pollutants detach from the surface under the action of the laser, the substrate layer under the pollutants will continue to be irradiated by the laser. At this time, depending on the optical properties, thermal properties, and surface structure of the substrate material, there are generally two different manifestations. The first is the occurrence of self-termination phenomenon, where the substrate basically does not change when it continues to be irradiated with the same laser parameters. This phenomenon generally occurs on the surface of a smooth, highly reflective metal substrate or on materials that are transparent to the incident laser. However, the reflected laser may cause damage to the light path or other optical components, or even the surrounding operators. The second is that the laser will interact with the substrate material, resulting in physical changes such as melting and vaporization, or chemical changes such as decomposition, crystallization, amorphization, and passivation. This situation can cause damage to the substrate. The difference between the two phenomena lies in whether the laser energy density absorbed by the substrate surface has reached the threshold conditions (minimum laser energy density) required for the material itself to undergo corresponding physical or chemical changes.

During the laser cleaning process, changes in the characteristics of the substrate surface (microscopic images of the surface, reflectivity, hardness, etc.), will cause the changes of the acoustic waves or changes in the parameters of the laser, which can be monitored by acoustic signals and light wave signals to determine whether the surface is clean. For example, the invention patent with the application number 201210582928.6 proposed using chroma to distinguish between the pollution layer and the substrate layer of the cleaning object. During the laser cleaning process, in order to achieve the purpose of non-damaging cleaning, a monitoring system is introduced to monitor the laser cleaning effect in real time, measuring the parameters of the cleaning area, and modifying the output laser parameters and scanning speed for feedback compensation based on reference values. This so-called online monitoring or in-situ monitoring is of great significance for the automation of laser cleaning [13–15].

4.4.1 Laser Cleaning Monitoring Methods

With the development of laser cleaning technology, various monitoring methods have been proposed. Depending on whether the cleaning parameters can be obtained during the laser cleaning process and the cleaning effect can be evaluated, these monitoring methods can be divided into two categories: offline monitoring methods and online monitoring methods. The detection methods used in monitoring are mainly divided into four categories. The first is surface imaging detection of the cleaning object, which is to observe the surface of the cleaning object through the lens of a camera or video camera, and the technician personally evaluates the cleaning effect and adjusts the laser parameters and cleaning methods accordingly. This method is most commonly used in practical applications. The second is to measure some parameters of the surface of the cleaning object, such as surface hardness, roughness, reflectivity, and surface potential. The third is to collect the vibration signals generated during laser cleaning, and analyze the intensity and changes of the vibration signals (sound waves) to determine the cleaning stage and cleaning effect. Typical examples include the measurement of sound wave intensity in laser rust removal and more widely used flight time measurement. The fourth is to collect the light signals generated during laser cleaning, and determine the material composition of the current cleaning layer through the spectral characteristics, thereby determining the cleaning effect, stage, efficiency, etc. Typical examples include XRD spectra, laser-induced fluorescence spectra, and laser-induced breakdown spectroscopy (LIBS). These detection methods can be used for both offline and online monitoring. The various detection methods used in laser cleaning are summarized in Table 4.6.

Considering the two main removal mechanisms currently used in laser cleaning, one is below the material vaporization threshold, which relies on thermal elastic expansion or the pressure difference at the interface to achieve cleaning, and the other is above the material vaporization threshold, where the material undergoes phase change and ionization through ablation, generating plasma, and then detaching from the substrate surface. The latter removal mechanism has a wide range of applicability. As long as the laser energy density can reach the material's vaporization threshold, the removal of various contaminants can be carried out under this mechanism. During the laser ablation removal process, both vibration and light information are generated, so acoustic and optical monitoring are the two most commonly used methods. These will be introduced in details below.

4.4.2 Real-Time Monitoring of Acoustic Intensity Measurement

In the mechanism of thermoelastic vibration, the sample surface will excite sound waves in the air due to heat expansion, elastic vibration, and the detachment of pollutants from the substrate. Therefore, sound waves or even ultrasonic waves will be generated during the cleaning process. In the laser rust removal experiment, rust is removed by pulsed laser, which also generates sound waves. The frequency range of these sound waves may range from a few hertz to the ultrasonic band. Actual monitoring does not need to cover the entire frequency range, but only needs to monitor the intensity of sound waves at certain specific frequencies. The specific method is to use sound recording devices (microphones, recorders, etc.) to collect signals during the cleaning process, then perform Fourier transformation to convert the intensity signal in the time domain into spectral features in the frequency domain. The changes in specific spectral peaks indicate certain specific stages of laser cleaning, thereby achieving the acquisition of cleaning information.

As shown in Fig. 4.41, the AESM-98 acoustic emission meter produced by Vallon, Germany, was used for experimental research. A Q-switched Nd:YAG laser with a wavelength of 1.06 μm, pulse width of 20 ns, and pulse repetition rate of 5 Hz was used. After being focused and homogenized by an aperture, it was irradiated vertically onto the surface of the workpiece with a power density of 9.8×10^7 W/cm^2. The acoustic emission signal was detected by a broadband sensor (10 Hz–15 kHz), and the signal was amplified by a preamplifier and displayed by the acoustic emission meter. A few pulses were used to act on a point in succession to study the relationship between the change in the intensity of the acoustic emission signal generated during

Table 4.6 Detection methods used in laser cleaning

English name	Abbreviation
Auger electron spectroscopy	AES
Atomic force microscopy	AFM
Cleaning efficiency = clean area/laser spot area	CE
Acoustic wave monitoring	AWM
CCD camera	CCD
Dektak3 surface profiometer	Dektak3
Dark field microscopy	DFM
Diffractive optical element (intensity projection)	DOE
Emission current monitoring	ECM
Energy dispersion x-ray analyzer	EDXA
Electron probe microanalysis	EPM
Photodiode monitoring of laser induced fluorescence	Fluor
Fourier transform infrared spectroscopy	FTIRS
Fuzzy rule system	Fuzzy
Gamma spectrometry (measure sample activity)	Gamma
Heterodyne interferometer (surface displacement)	HI
Image processing	IP
Low-energy diffraction spectroscopy	LEDS
Laser induced plasma spectroscopy	LIPS
Light scattering	LS
Optical/chromatic monitoring	OCM
Optical microscope/optical micrographs	OM/OMG
Optical reflectance	OR
Optical surface analysis	OSA
Particle counter/censor ANS100 light scattering particle counter	PC/CANS
Particle densities	PD
Probe deflection technique	PDT
Polarized light microscopy	PLM
Photomultiplier	PM
Particle measuring system	PMS
Polaroid camera	Pol
Profilometer	Profile
Scanning electron microscopy	SEM
Secondary ion mass spectrometry	SIMS
Spectrocolorimeter	Spect
Surface photo voltage	SPV
Specular reflection	SR

(continued)

Table 4.6 (continued)

English name	Abbreviation
Surface roughness	Surf
Video-based PC driven system counts particles > ~75 μm diameter	VB
Visual observation of contaminant removal	Visual
Video monitor	VM
Voltage readings	VR
X-ray photoelectron spectroscopy	XPS

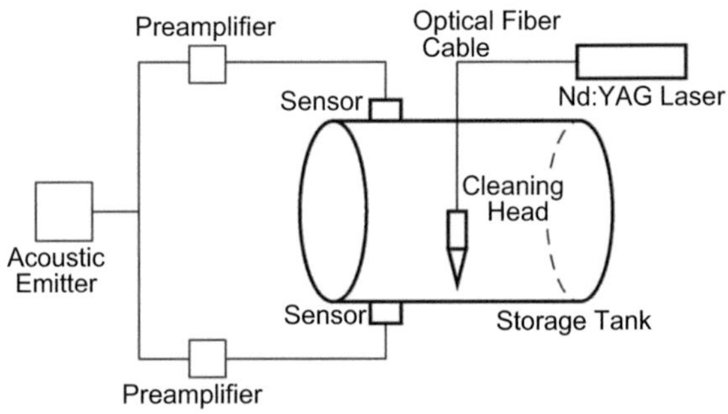

Fig. 4.41 Schematic diagram of the working principle of ultrasonic monitoring laser cleaning figure

the laser cleaning process and the cleanliness of the substrate surface. Through this method, the cleaning threshold can be easily measured.

4.4.3 Shock Wave Monitoring

During the process of removing surface pollutants with laser cleaning technology, shock waves will be generated. After the surface pollutants are removed, the laser continues to irradiate the underlying substrate. Due to the differences in optical and thermal properties between the substrate and the pollutants, the frequency of the generated acoustic waves is different. The acoustic waves propagate outward, causing periodic changes in the surface air density, forming a phase grating, and deflecting the grazing incident beam. However, since the propagation speed of the acoustic wave is fixed at the speed of sound in the air, the deflection angle is a constant value, different from the deflection angle of the shock wave. This transition between shock waves and acoustic waves is reflected as a change in the deflection angle, corresponding to the transition of the laser action object from the pollution layer to the substrate layer, marking the change in the cleaning stage, and also achieving the quantification of the cleaning stage.

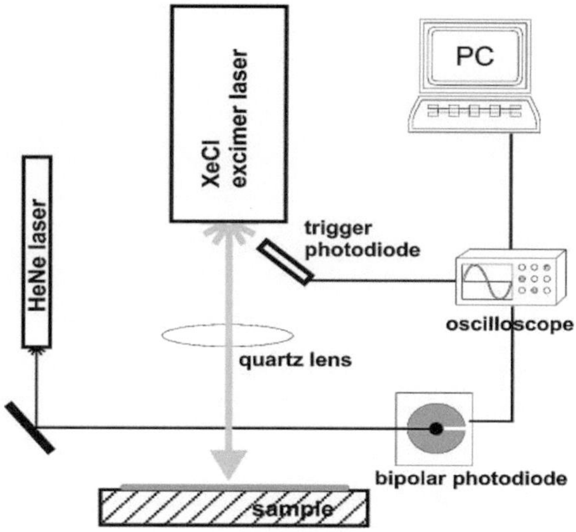

Fig. 4.42 Time of flight measurement monitoring device

Shock waves are medium waves that propagate faster than sound, not acoustic waves with frequencies in the ultrasonic band. The phase grating formed in the air is different from that of sound waves, so the Time of Flight (TOF) measurement method can be used to detect the transition point of these two waveforms. The flight times of shock waves and sound waves have obvious changes. This transition from shock waves to acoustic waves indicates that the mechanism and stage of laser action have changed. For ablation cleaning, this transition is a sign of the end of laser cleaning. The measurement principle of the flight time measurement monitoring device is shown in Fig. 4.42.

4.4.4 Real-Time Monitoring of Laser-Induced Fluorescence Spectrum

The sound wave intensity monitoring method focuses on the changes in the strength of the sound wave signal. The flight time measurement method focuses on the transition between shock waves and acoustic waves. The Laser Induced Fluorescence Spectroscopy (LIFS) and Laser Induced Breakdown Spectroscopy (LIBS) focus on the characteristic spectral information during the laser cleaning process with different emphasis. The former focuses on the visible light band, while the latter focuses on the ultraviolet band. Among these types of information, the TOF that focuses the transition from shock waves to acoustic waves and LIBS that focuses on the characteristic spectra of elements have universality, that is, they can be used as monitoring methods for cleaning judgments for many types of materials during the laser cleaning process.

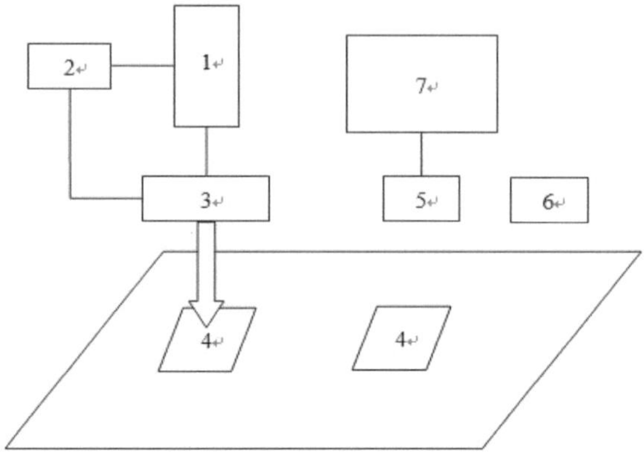

Fig. 4.43 Schematic diagram of hyperspectral imaging technology monitoring device. 1—Laser cleaning equipment, 2—Control system, 3—Optical system, 4—Cleaning object, 5—Hyperspectral acquisition system, 6—Light source, 7—Hyperspectral data processing system

High spectral imaging technology is used to observe the samples during cleaning, which can monitor the reflectivity of the materials being cleaned under different wavelengths in real time. The specific device schematic is shown in Fig. 4.43.

From Fig. 4.43, we can see that the control system communicates with the laser cleaning equipment. The control system can change the average power output of the laser (by adjusting the input current of the laser) and the repetition rate (affecting the pulse width, and thus affecting the peak power). The output laser is shaped by the optical system device (spot size and light intensity distribution). The laser control device can also control the optical system, adjusting the spot size of the laser on the sample by changing the distance of the lens group, thereby adjusting the energy density of the laser pulse. The light source is connected to the high-spectral acquisition system to carry out collection work during laser cleaning, and then the collected data is transmitted to the high-spectral data processing system for processing and analysis, and then feedback to the laser control system, forming a closed-loop feedback. The specific operation steps are as follows:

(1) Place the base material (no pollutants) and the cleaning object (surface pollutants) in the high-spectral monitoring system to obtain their high-spectral images;
(2) Place the sample during cleaning under the same conditions in the high-spectral monitoring system to obtain the high-spectral image of the sample during cleaning;
(3) Process the obtained high-spectral data and images;
(4) Use the following formula to process the above images to obtain the corrected ratio of reflectivity at different wavelengths:

$$CL(\lambda) = \frac{RT(\lambda) - RC(\lambda)}{RG(\lambda) - RC(\lambda)} \tag{4.4.1}$$

where $CL(\lambda)$ is the reflectivity correction ratio corresponding to a certain wavelength, $RT(\lambda)$, $RC(\lambda)$ and $RG(\lambda)$ are the emissivity of the cleaning area, the reflectivity of the surface material to be treated, and the reflectivity of the original matrix, respectively. When $CL(\lambda) \geq 90\%$, which is a judgement for well cleaning in this area, is reached, the corresponding data is output. The specific flowchart is shown in Fig. 4.44.

The processing of high-spectral data and images in the above step (3) includes steps of whiteboard correction, dark current correction, and noise removal. The whiteboard correction uses the flat field correction method, and a completely white high-spectral image is obtained after scanning a white standard correction board with a reflectivity of 99.99% with a high-spectral imager. The reflectivity change intention diagram of high-spectral detection is shown in Fig. 4.45.

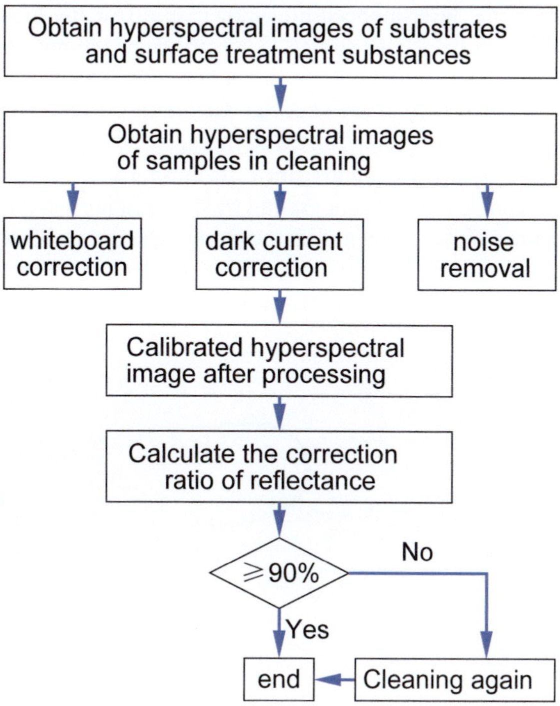

Fig. 4.44 Method flowchart

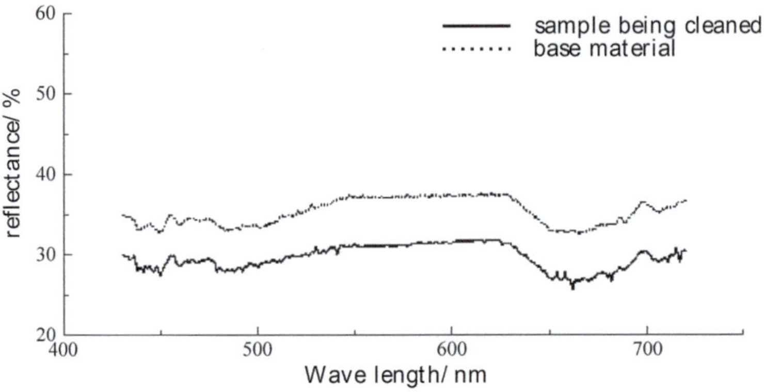

Fig. 4.45 Hyperspectral data schematic

4.4.5 Real-Time Monitoring of Laser Induced Breakdown Spectroscopy (LIBS)

The composition of the laser-induced breakdown spectroscopy monitoring device includes three main parts: a laser source, a spectrometer (composed of a spectroscopic element and CCD), and an elemental characteristic spectrum table, as shown in Fig. 4.46. The laser source is used for the excitation of the plasma, often using ultraviolet lasers; the spectrometer is used to measure the intensity of the plasma's light emission at different wavelengths, draw the plasma spectrum, where the spectroscopic element can be a prism or a grating (mainly grating), and finally compare the plasma spectrum with the elemental characteristic spectrum table to determine the elemental composition of the current layer affected by the laser, thereby determining the cleaning information.

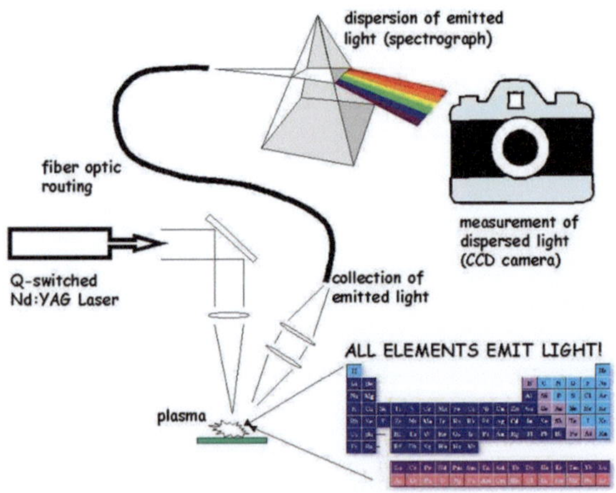

Fig. 4.46 Schematic diagram of LIBS device

The excitation laser light irradiates the cleaning object. When the laser energy density exceeds the vaporization threshold of the material, it generally goes through six stages: absorption, heating, vaporization, plasma formation, light emission, and the formation of ablation pits, as shown in Fig. 4.47. After the material absorbs the laser energy, it undergoes a complex phase change process, such as melting, vaporization, sublimation, fragmentation, plasma formation, etc. The generated plasma emits a continuous spectrum through radiation, which cannot establish a direct connection with the emitting elements due to its higher ionization state, and is useless information; the elemental characteristic spectrum after a delay is useful information, which can be used for spectrum comparison. The characteristic spectral lines of elements are like human fingerprints or irises, capable of performing element "identity verification", serving as a judgment basis, and can accurately judge the degree of laser cleaning based on the changes in the current layer affected by the cleaning. The process of laser-induced breakdown emission is shown in Fig. 4.48.

The LIBS implementation device consists of three main functional units: excitation, delay, and spectroscopy, as shown in Fig. 4.49. The entire device needs to consider various factors, including the reflectivity of the target surface, heat capacity,

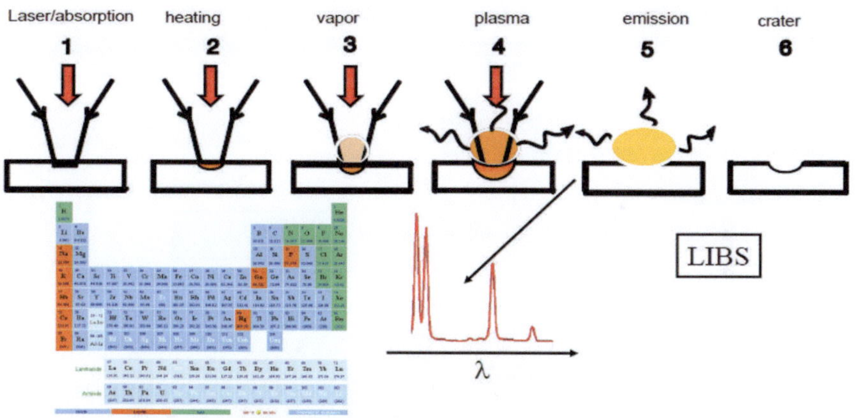

Fig. 4.47 Pulsed laser ablation process

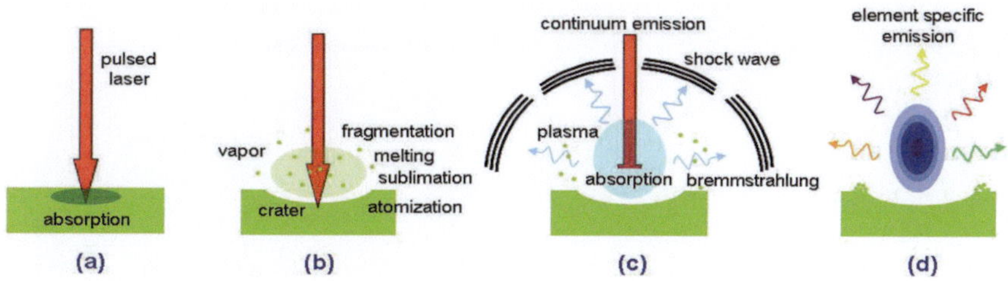

Fig. 4.48 Schematic diagram of laser induced breakdown process

melting point, boiling point, thermal conductivity, the composition of the atmospheric environment, pressure, the pulse energy of the laser, pulse width, spot size, wavelength, delay time, the choice of spectrometer and light detector, etc. Since the characteristic spectrum of the element needs to be measured, there must be a necessary delay device between the laser source and the spectrometer in the specific LIBS device, which is realized through the control unit. The finished delay device on the market is shown in Fig. 4.50, which can accurately adjust the time delay between the laser and the detector. In addition to the single-pulse excitation plasma model, because the plasma generated by two-pulse excitation can obtain a stronger spectral intensity, two types of double-pulse LIBS devices have been proposed in practical applications, among which the vertical double-pulse type has the best effect in practical applications.

The composition and pressure of the atmospheric environment will affect the expansion characteristics of the plasma ablation plume, as shown in Fig. 4.51. Figure 4.52 shows the typical curves of the continuous spectrum and characteristic

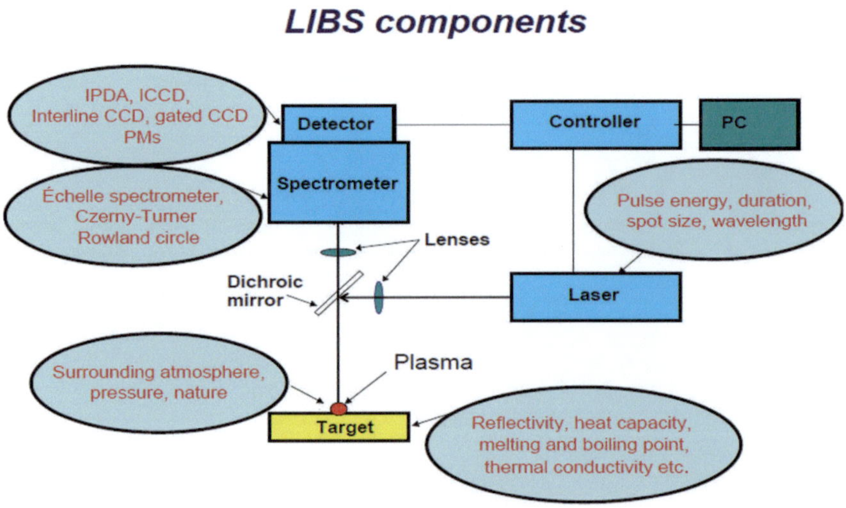

Fig. 4.49 Composition of laser cleaning device monitored by LIBS

Fig. 4.50 Digital delay generator

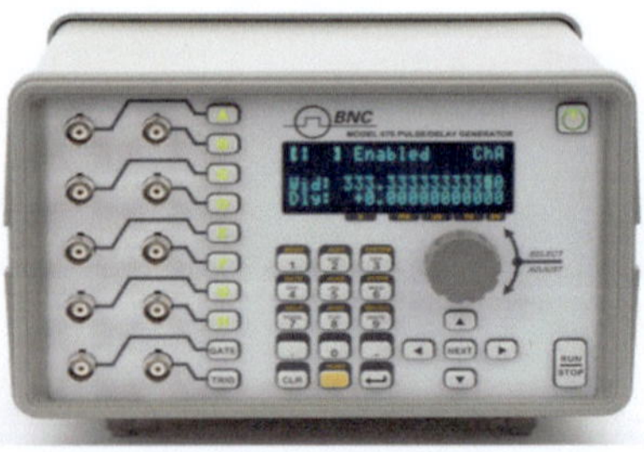

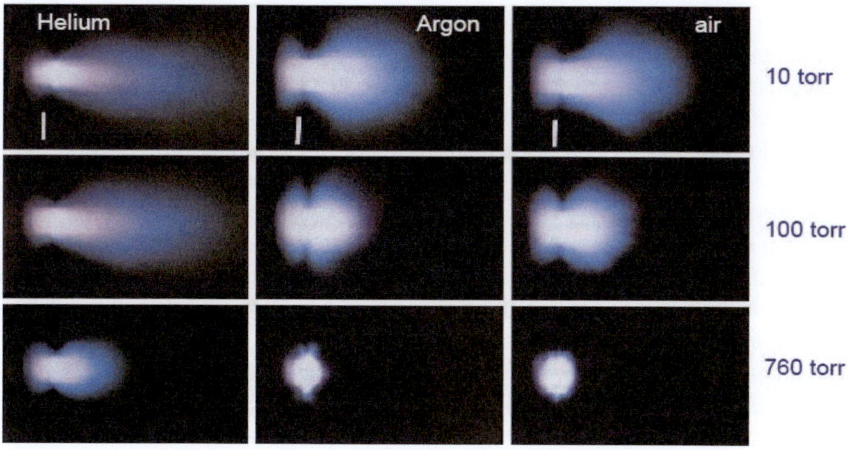

Fig. 4.51 Expansion of ablation plume under different pressures and atmospheric conditions

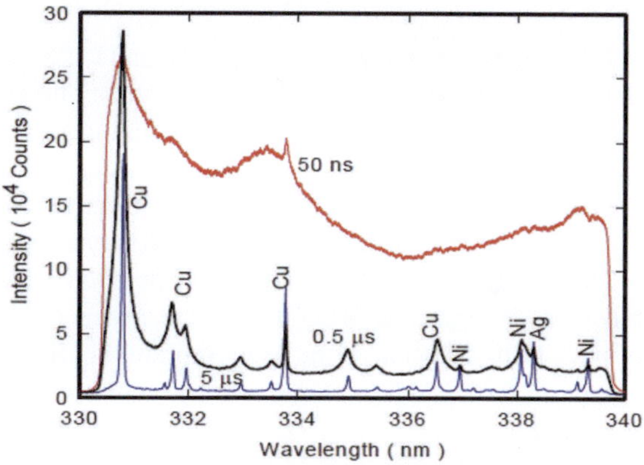

Fig. 4.52 LIBS spectrum of copper alloy

spectrum of copper alloy. The red line is the continuous spectrum under a delay of 50 ns, from which the corresponding element information cannot be obtained. The black and blue lines are the elemental characteristic spectra at delays of 0.5 μs and 5 μs respectively. Although the intensity of the characteristic spectrum varies, the spectrum type is the same, demonstrating the repeatability and effectiveness of the LIBS method.

4.4.6 Real-Time Monitoring of Photoacoustic Signals

Among the aforementioned measurement methods, TOF measures the transition from laser ablation to non-ablation, with the decisive factor being the surface reflectivity of each layer. Therefore, TOF measurement mainly distinguishes surfaces with significant differences in reflectivity. LIBS measures the characteristic spectra of elements, distinguishing materials with different core elements. However, when the object of laser cleaning is complex, especially when it presents as multi-layered material, single TOF and LIBS measurements cannot achieve monitoring functions. For example, in a three-layer structure of steel substrate-rust-coating, TOF measurement cannot distinguish between the removal of the coating and the removal of rust layer, but it can distinguish between the iron layer surface and the rust layer. Conversely, LIBS cannot distinguish the elemental differences between the rust layer and the substrate layer (because the core elements are iron and oxygen), but it can distinguish between the coating and the rust. Similar issues arise with artifacts that have a "metal-corrosion-soil clump" three-layer structure, or a "metal-corrosion-paraffin" three-layer structure. For these more complex cleaning objects, combining TOF and LIBS measurements can achieve layer-by-layer monitoring. Also, when TOF and LIBS reach the substrate layer, although plasma is still produced, it is a constant reaction, causing minimal damage to the substrate. This feature makes the combined TOF and LIBS monitoring technology very suitable for the field of cultural relic protection.

Combining the characteristics of TOF and LIBS measurement techniques, a laser cleaning device with photoacoustic spectrum monitoring can be used. By simultaneously obtaining the vibration information and spectral information generated by laser plasma, the current cleaning stage can be analyzed, and then the cleaning parameters can be controlled so that selective non-destructive cleaning can be achieved. This is especially suitable for more complex multi-layer cleaning objects.

Due to the universality of the laser ablation cleaning mechanism, the photoacoustic spectrum monitoring device has a wide range of applications. The characteristics of the photoacoustic spectrum are: (1) Suitable for multi-layer composite structures. Compared to single spectral monitoring for reflectivity information of different interfaces and spectral monitoring for elemental composition information of different interfaces, photoacoustic spectrum monitoring can measure the laser cleaning process of multi-layer structures with the same components and multiple interfaces. For example, in three or even multi-layer structures such as steel substrate-rust-paint, steel substrate-primer-topcoat, newly unearthed artifacts with metal substrate-corrosion product-soil clumps, etc. Using LIBS spectral qualitative component detection analysis can distinguish different material layers (atomic spectral changes), and TOF detection can distinguish different physical bonding layers (interface reflection, absorption changes), thus fully mastering the cleaning status in the cleaning process of multi-layer materials, especially suitable for online monitoring of laser cleaning of cultural artifacts. (2) Universality. Since laser ablation cleaning is suitable for the removal of various pollutants, the photoacoustic spectrum monitoring method using plasma shock wave characteristics and luminescent characteristics also has a correspondingly wide range of applications, becoming another universal online monitoring technology besides optical imaging.

4.5 Examples of Online Monitoring of Laser Cleaning

A laser cleaning device with a monitoring system is given in this section as an example. By simultaneously collecting and recording the light signals and vibration signals generated during the laser cleaning process, it quantitatively analyzes the process of cleaning, determines the stage of cleaning (no cleaning effect, under-cleaning, complete cleaned, over-cleaned, cleaning damage, etc.), and achieves non-destructive cleaning of the optional area on the surface of the cleaning object. It is a comprehensive application of laser spectroscopy, laser thermoelastic waves, plasma shock waves, etc.

The photoacoustic spectroscopy monitoring device, as a collection, analysis, judgment, and control system for laser cleaning information, adjusts the laser parameters in real time during the cleaning process, achieves observable cleaning efficiency, avoids over-cleaned damage during the laser cleaning process, or controls the rapid rise and fall of the surface temperature of the material to achieve beneficial additional effects such as surface hardening and passivation. Compared with the singularity and limitations of conventional vibration and spectral information collection systems, this cleaning monitoring device can simultaneously obtain photoacoustic spectral information during the laser vaporization-ionization process, and characterize different cleaning states through the combination of spectral and acoustic spectral feature information (binary information methods), making the judgment of the feedback control system more simplified, accurate, and rapid, which is beneficial for the non-destructive cleaning or controllable modification of the laser to pollutants. This section uses this as an example to describe the online monitoring of laser cleaning.

4.5.1 Photoacoustic Spectroscopy Detection System

I. Monitoring System Structure

This laser cleaning equipment with photoacoustic spectroscopy monitoring includes a laser device, a beam transmission device, beam shaping, focusing, control device, photoacoustic spectroscopy monitoring device, recycling device, and mechanical transmission device.

The mechanical and electrical connections of each component device are described as follows: The laser device uses a pulsed laser. The laser beam emitted by the pulsed laser enters the beam transmission device through the coupling device (lens group). The function of the beam transmission device is to achieve low-loss closed-cavity transmission of the laser beam over a long distance through the light guide arm or energy transmission fiber. The beam is reshaped, focused, and controlled by the beam control device to redistribute the energy density of the beam, restrict its shape, focus it, set the overlap rate of multiple beam spots, and set the scanning path of the beam spot. The X–Y–Z three-dimensional scanning system in the beam control system and the x–y two-dimensional planar motion of the mechanical transmission device jointly control the relative motion between the cleaning beam and the cleaning object. The recycling device is located near the area where the laser beam acts on the surface of the cleaning object, and recycle the vaporized and broken dust particles of the pollutants on the surface of the cleaning object, to avoid possible solid waste pollution to the environment caused by the cleaning operation. The photoacoustic spectrum monitoring device consists of a vibration monitoring unit and a spectrum monitoring unit. The vibration monitoring unit consists of a detection laser, an angle meter, a photoelectric detector, and an oscilloscope. The angle meter records the deflection angle of the detection laser after passing over the laser action area, with the deflection angle caused by the acoustic wave as the reference angle. The deflection angle provides cleaning information. The spectrum monitoring unit consists of a photoelectric detector, a spectrometer, a timer, and an oscilloscope. It measures the signal of the characteristic spectrum of the core elements of different layers. Place the photoelectric detector near the laser action area to measure the characteristic emission of the laser plasma. The characteristic emission of the plasma includes two stages. First is the continuous spectrum emission, which is useless information. By setting the time interval between the pulse emission time of the laser and the trigger time of the oscilloscope through the timer, the continuous emission stage is shielded, and only the discrete characteristic spectrum after the continuous spectrum is measured by the oscilloscope.

The photoacoustic spectrum monitoring device consists of an acoustic signal measurement unit and an optical signal measurement unit; the cleaning laser is used as the first light source to excite the plasma, and the laser parallel or perpendicular to the cleaning laser beam, as an optional second light source, whose function is to effectively enhance the luminescence intensity of the laser plasma through dual-pulse technology, making the spectral features more prominent; the detection laser used by the acoustic signal measurement unit is usually a He–Ne laser or a visible semiconductor laser. Multiple photoelectric detectors are distributed at equal angular widths with the laser action point, recording the angular deflection of the detection laser caused by the air phase grating induced by the density changes due to the propagation of shock waves and acoustic waves in the air. The optical signal measurement unit measures the luminescence spectrum of the plasma through the spectrometer structure of the monochromator, photoelectric detector, and oscilloscope, connects the delay device between the pulsed laser and the oscilloscope, filters out the continuous spectrum luminescence at the beginning stage of the plasma, and records the characteristic spectral lines of the material in the current cleaning area, finally determining the elemental composition of the current laser action area by comparing the atomic emission spectra of the elements.

II. Photoacoustic Spectrum Detection Equipment

The photoacoustic spectroscopy monitoring device is shown in Fig. 4.53. Under the laser ablation-ionization mechanism, when the laser causes the surface material to

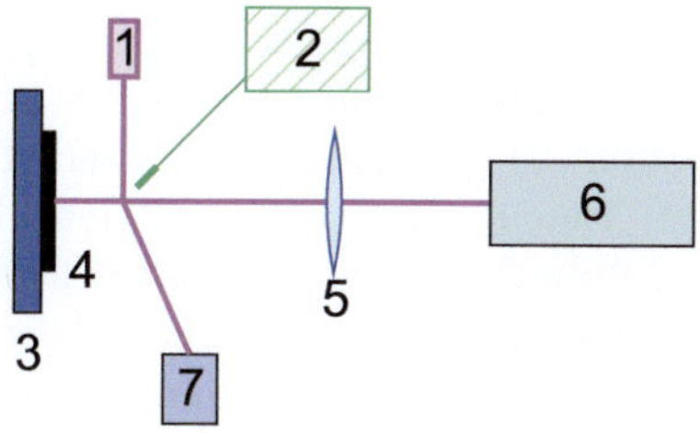

Fig. 4.53 Schematic diagram of photoacoustic spectrum detection device 1—Laser (such as semiconductor laser or helium–neon laser), 2—Photoelectric detector, 3—Mobile platform, 4—Cleaning object, 5—Focusing shaping system, 6—Laser system for cleaning, 7—Pollutant recovery and scattered light barrier device

vaporize and ionize, forming a plasma, it simultaneously generates optical signals of the plasma spectrum and vibration signals of the shock wave. The optical signal measurement unit uses a multi-channel fiber optic spectrometer, and its photodetector receives the spectral signals during the laser plasma process near the laser action area. The characteristic emission spectrum of the plasma is recorded by delaying the opening of the shutter switch of the spectrometer. The acoustic signal measurement unit uses a He–Ne laser as the detection laser, and uses multiple photodetectors distributed at equal angular distances as the receivers of the deflected beam, measuring the deflection of the grazing incident He–Ne laser caused by the propagation of shock waves or acoustic waves. The photodetector group distributed at equal angular distances consists of at least three photodetectors, one in the propagation direction of the detection laser, one in the deflection angle of the acoustic wave, and the other photodetectors together complete the measurement of the deflection angle caused by the shock wave. The photoacoustic spectroscopy monitoring device, as an online detection device for cleaning conditions, feeds back the measurement information to the control system, adjusts the laser parameters (laser energy density, repetition rate, scanning speed, start and stop, etc.) through the control system, and realizes laser non-damaging cleaning.

The digital photoacoustic spectroscopy monitoring device provides cleaning status information for laser cleaning operations. Before the formal cleaning operation, the elemental composition of each layer of the cleaning object is first determined, and if necessary, it can be detected by pre-cleaning. Then, several characteristic peaks of the characteristic elements of each layer of material are selected as the objects to be detected. Each characteristic peak corresponds to a light signal intensity detection. Although the branches of spectral detection increase, the transmission wavelength of the spectroscopic element of each branch is locked, eliminating the need for grating scanning, reducing the detection time. At this time, the acquisition of spectral signals is transformed from a spectrogram to a combination of several binary digital signals. The spectral information in the current plasma can be characterized by a segment of binary numbers, using the discreteness of the formal plasma characteristic spectrum. Similarly, in the acoustic signal measurement unit, two characteristic photodetector angles are selected as the characterization of the acoustic signal, which are the parallel position of the He–Ne light and the deflection angle of the acoustic wave. Four combinations can give various situations of acoustic wave monitoring, which are 00 (shock wave generation), 01 (acoustic wave generation), 10 (no cleaning beam action), 11 (invalid signal). Here 0 indicates that no signal has been received, and 1 indicates that the signal from the detection laser has been received. The binary combinations characterizing the spectrum and the sound spectrum together characterize the current cleaning status. This method is more rapid in obtaining monitoring information, especially suitable for high repetition rate laser cleaning systems.

4.5.2 Example of Acoustic Monitoring in Laser Cleaning

Using sound waves for monitoring in laser cleaning is simple, convenient, and practical. In this section, we introduce the acoustic monitoring effect in laser rust removal.

Figure 4.54 shows the sound wave intensity diagram (a) and sound wave spectrum diagram (b) of a lightly rusted sample when the incident laser energy density is at its cleaning threshold. The time interval for each pulse action is about 1.3 s. The sound wave signal obtained by the sound wave sampling device is first processed by Cool Edit software for noise reduction, and then the time-domain sound wave signal is processed by MATLAB software for Fast Fourier Transform (FFT). Since the sampling frequency is 8 kHz, according to the Nyquist sampling theorem, the frequency that can be truly restored should be 4 kHz. At the same time, since the low-frequency noise has been removed, the frequency band of the sound wave is mainly concentrated in the 2–4 kHz range.

From the sound wave intensity graph 4.54a, it can be seen that there is no obvious pattern in the sound wave intensity changes caused by each pulse during the cleaning process, and the waveforms are roughly similar. Therefore, it is difficult to analyze and judge the cleaning effect only by the changes in the intensity of the sound wave signal. From the sound wave spectrum graph 4.54b, it can be seen that the two peaks near 3178 Hz and 3600 Hz in the spectrum line have the highest relative intensity. For the first two pulses, the peak intensity at 3178 Hz in the spectrum line is greater than that at 3600 Hz. Combined with the threshold experiment results, it can be considered that this is the characteristic spectrum of the initial stage of laser rust removal due to the rust not being completely removed. By the third pulse, the peaks at 3178 Hz and 3600 Hz suddenly decrease, while the intensities of the peaks at other places relatively increase. The reason for this abnormal change in the spectrum line is likely due to the laser breakdown of dust in the air at the focus point, which reduces the incident laser energy and affects the cleaning effect. By the fourth and fifth pulses, the peak intensity at 3600 Hz is greater than that at 3178 Hz. Since the rust is basically cleaned at this time, it can be considered as the characteristic spectrum when the cleaning is completed. In practical applications, the transition of the two characteristic spectra represented by pulses 1, 2 and pulses 4, 5 can be used as a judgment standard for whether the rust removal is complete.

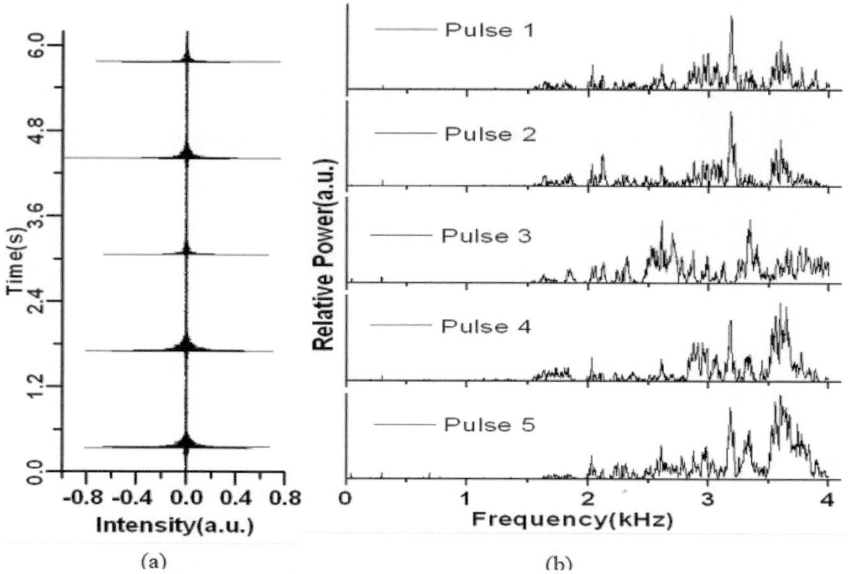

Fig. 4.54 Relationship between sound wave signal strength and frequency under light rust conditions with the number of laser pulse actions

Figure 4.55 is the sound wave spectrum of severely rusted samples under the condition that the incident laser energy density is its cleaning threshold. From the spectrum, it can be found that due to the different physical properties such as shape and mass of the severely rusted samples and the lightly rusted samples, their characteristic spectra of the severely rusted samples are different from the lightly rusted samples: in the case of the first few pulses, the spectrum lines near 2900 Hz, 3240 Hz and 3670 Hz all have strong peaks, and there are many peaks in the sound wave at this time; but as the laser pulses act, the rust gradually decreases, the peak intensities at 2900 Hz and 3670 Hz relatively weaken, while by the fifth pulse, the peak at 3240 Hz becomes the strongest, and the peak at 3000 Hz is enhanced; by the eighth and ninth pulses, the rust is basically cleaned, and the peak intensity at 3000 Hz has exceeded that at 3240 Hz; by the tenth pulse, the peak intensity at 3000 Hz is the largest, and the other peaks are relatively small. From the above spectrum changes, it can be inferred: for the severely rusted samples in the experiment, the peak at 3000 Hz is related to the vibration spectrum of the steel itself, as the rust decreases until it is completely removed, its peak will gradually increase from weak, while the other peaks will relatively weaken. Therefore, the change in the intensity of the 3000 Hz peak relative to other peaks can be used as a simple judgment standard for whether the rust is clean.

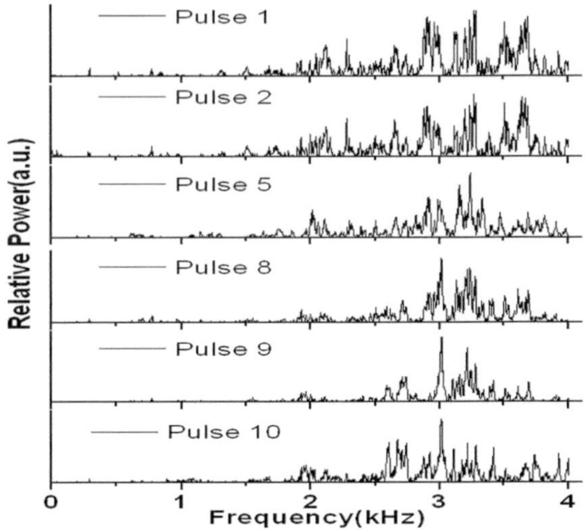

Fig. 4.55 Change in acoustic signal frequency under severely rust conditions with the number of laser pulse actions

In summary, in the experiment, sound wave signals below 4 kHz were collected for laser rust removal, and it was found through analysis that although there is no obvious change pattern in the intensity of the sound wave signals, the change in its spectrum has a specific pattern with the cleaning effect, which can be used for simple judgment of the cleaning process. However, due to limitations, no analysis was conducted on the ultrasonic frequency band, and existing experiments show that dry laser cleaning also has significant signals in the ultrasonic band. At the same time, it can be simply calculated from the double-layer thermoelastic vibration model that when the cleaned sample is a metal substrate, the vibration frequency directly caused by its own elastic vibration during the cleaning process is basically in the range of 0.1–10 MHz. Therefore, if we want to conduct more in-depth and detailed research and accurate judgment on the cleaning process, we need to detect and analyze the frequency band above 4 kHz and even the ultrasonic frequency band.

4.5.3 *Laser-Induced Breakdown Spectroscopy (LIBS)*

LIBS is widely and successfully used in laser cleaning. Through qualitative analysis, the composition of materials can be monitored. Figure 4.56 shows the LIBS spectrum of marble through laser cleaning, where (a) to (h) represent that the ablation depth increase with laser pulse number while keeping the pulse energy density constant. It can characterize the changes of Fe, Si, and Ca elements, that is, Fe and Si decrease while Ca increases, the chemical composition of which is the same as those of the hard layer (soil)–limestone ($CaCO_3$).

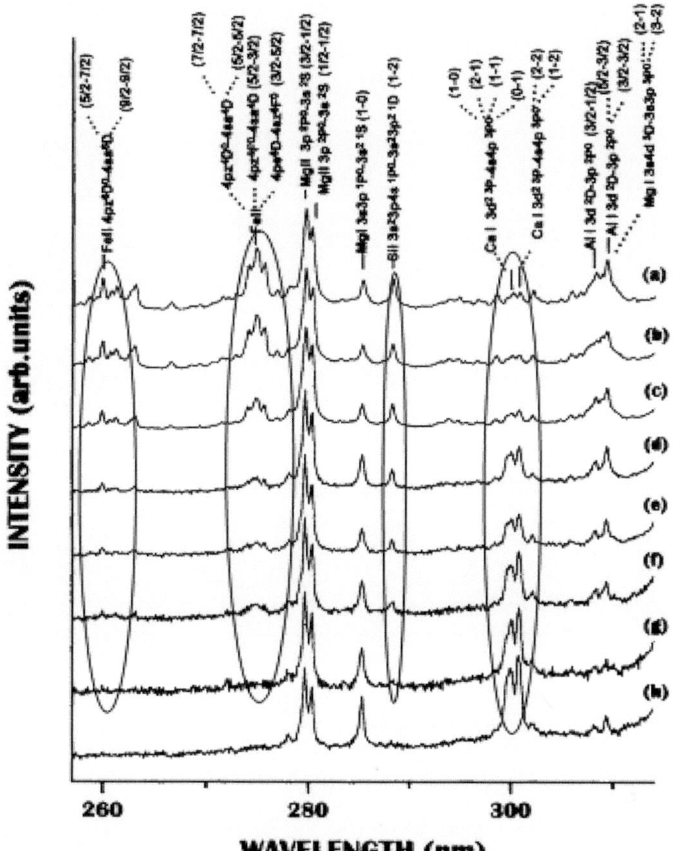

Fig. 4.56 LIBS spectrum of marble through laser cleaning

Through quantitative analysis, we can focus on the intensity change of a certain spectral line of the material, thereby evaluating the cleaning effect. As shown in Fig. 4.57, from top to bottom are the results after action of 3, 15, and 30 laser pulses. By quantitatively monitoring the intensity of Si(I) 288.16/Al(I) 309.27, we can quantitatively assess the current degree of laser cleaning (cleaning depth). Currently, LIBS is becoming more and more widely used and mature in laser cleaning, especially in the cleaning of cultural relics.

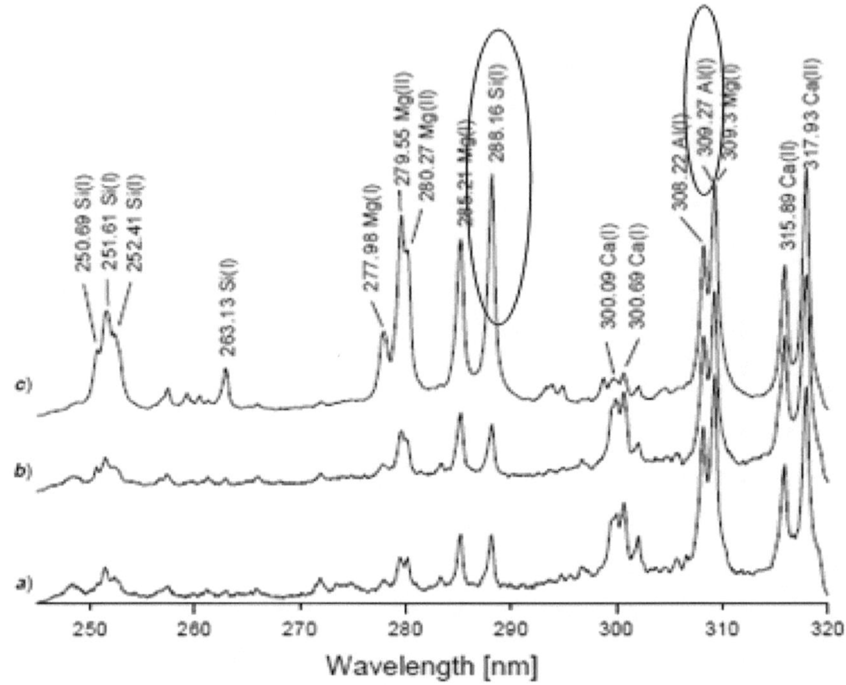

Fig. 4.57 LIBS spectrum of laser cleaning limestone

References

1. Song F, Tian B, Zou W F, Liu S J. An introduction to laser cleaning equipment. Cleaning World, 2006, 22(04): 27–32.
2. Yi X, Guo Q, Zha G S, Wang S Z. Three wave length pulse laser system with high stability and high energy. Chinese Journal of Quantum Electronics, 2017, 34(04): 436–440.
3. Wang H Q, Shi Y S, Huang S H. The control arithmetic of laser galvanometer scanning system with 3D and its application. J. Huazhong Univ. of Sci. & Tech. (Nature Science Edition), 2003, 31(05): 70–71.
4. Song Y Y, Zhu M Z, Wei X C. High-power laser cleaning equipment. China Chemical Trade, 2014, (10): 117–118.
5. Wei X C, Shi L, Zhu M Z, Song Y Y. Hand-holding DPSS laser cleaning machine. Optics & Optoelectronic Technology, 2013,11(03): 20–23.
6. Lu L, Wang F, Zhao Y N, Wang Z Q, Zhang G Y, Che Y, Wang X H. Back-pack dual-wavelength all-solid-state laser cleaner. Aeronautical Manufacturing Technology, 2012,10: 83–85.
7. Li H P, Guo B L. Research on Nd:YAG pulse laser cleaning technology. Electro-Optic Technology Application, 2018, 33(02): 67–71.
8. Zha G S, Meng X G, Jia X D, Guo Q, Fan Y S. Design of optical path control system for a lamp-pumped Nd:YAG four-wavelength laser cleaning machine. Chinese Journal of Quantum Electronics, 2018, 35(02): 165–172.
9. Yi X. Three wavelength pulse laser system with high stability and high energy. Beijing: University of Chinese Academy of Sciences, 2016.
10. Liu P F. Development of high-power all-solid-state laser cleaner. Changchun, Changchun University of Science and Technology, 2013.

11. Lu L. Research on the Key Technology of all solidstate 1064 nm/532 nm dual-wavelength laser cleaner. Changchun: Changchun University of Science and Technology, 2012.
12. Barriga P, Zhao C, Blair D G. Optical design of a high-power mode-cleaner for AIGO. General Relativity and Gravitation, 2005, 37(09): 1609–1619.
13. Ohtake H, Ono S, Liu Z L, Sarukura N, Ohtam, Watanabe K, Matsumoto Y. Enhanced THz radiation from femtosecond laser pulse irradiated InAs clean surface. Japanese Journal of Applied Physics Part 2-Letters, 1999, 38(10B): L1186–L1187.
14. Dubuisson C, Cox A G, Mcleod C W, Whiteside I, Jowitt R, Falk H. Characterisation of inclusions in clean steels via laser ablation-ICP mass spectrometry. ISIJ International, 2004, 44(11): 1859–1866.
15. Araya A, Mio N, Tsubono K, Suehiro K, Telada S, Ohashi M, Fujimoto M K. Optical mode cleaner with suspended mirrors. Applied Optics, 1997, 36(7): 1446–1453.

Chapter 5
Application of Laser in Cleaning Electronic Components

5.1 Origin of Laser Cleaning of Electronic Components

In the manufacturing of electronic circuit boards, tiny particles and other contaminants that adhere to the board during etching, deposition, and plating processes can greatly reduce electronic efficiency and even cause damage to components. These attached particles include tiny particles of silicon, silica, alumina, and polystyrene produced during the manufacturing process. These tiny particles can cause short circuits or significant performance degradation in large-scale integrated circuits and micro high-density storage devices, leading to fatal damage such as scratches or cracks on the surface of micro-mechanical devices. This is an urgent problem to be solved in semiconductor, microelectronics, micro-mechanics, and other high-tech fields. As the size of semiconductors and microelectronic devices becomes smaller, the particles that need to be removed are also getting smaller, reaching the micron and sub-micron level [1, 2]. Such small contaminant particles have an adhesion force on the circuit board far greater than their gravity, which brings great difficulty to cleaning.

In the 1970s and 1980s, IBM in the United States conducted a large amount of research and development on photolithography technology, including Electron Beam Printing (EBP) technology, which has high resolution and overlay capability of the electron beam and fast printing speed, making it very suitable for high-speed mass production of storage chips [3]. This technology first manufactures a mask, and then projects the pattern on the mask onto the photoresist at a 1:1 ratio using an electron beam. The mask is made of silicon wafers and is the key to EBP lithography technology. At that time, the minimum linewidth of the mask pattern was 0.3 μm.

Next is the photolithography exposure process, where the mask is projected onto the substrate. The mask pattern is printed onto the substrate at a 1:1 replication ratio. During the printing process, if the mask is contaminated, such as any defects including contaminant particles larger than 1/4 of the minimum width, it will lead to printing failure. At that time (mid-1980s), the current style of Dynamic Random

F. Song and X. Lin, *Laser Cleaning*, https://doi.org/10.1007/978-981-97-6631-4_5

Access Memory (DRAM) was about 1 μm. If there are particles larger than 250 nm in diameter on the mask, it will affect the printing quality, so it is necessary to remove the contaminant particles on the mask. The adhesion force of a 1 μm particle, including van der Waals force, electrostatic force, or capillary action, is about 10^6 times its gravity [4].

For larger diameter contaminant particles, the traditional simple and practical cleaning method is water washing, but for particles of 1 μm and smaller, water washing is ineffective. Therefore, researchers at IBM tried to use new methods, such as ultrasonic cleaning technology, to remove particles from the mask, but the film would immediately rupture and re-contaminate the mask. They also tried some other cleaning methods, but the results were not good.

In 1982, Zapka from the German IBM German Manufacturing Technology Center (GMTC) discovered that under short-pulse laser irradiation, attached particles can be removed without damaging the pattern on the silicon mask surface. Subsequently, Zapka and Tam at IBM's Almaden Laboratory in the United States tried steam laser cleaning. Compared to "dry" laser cleaning, particle removal is cleaner and more efficient [5, 6]. Further research and application have shown that laser cleaning can easily and efficiently solve the problem of particle removal on the mask plate, making it the most effective cleaning method currently available.

For particles on semiconductor materials, including silicon, silica, metal spheres, and other contaminants on the mask version, and various dust on the electronic circuit board, laser cleaning technology is a relatively suitable technique to ensure the safety and insulation of the substrate during the cleaning process.

5.2 Main Mechanism for Laser Cleaning of Microelectronic Components

5.2.1 Adhesion of Particles

As discussed in Chap. 3: There are three kinds of adhesion forces of particles on the substrate surface: van der Waals force, capillary force, and electrostatic force [7, 8].

When the substrate surface is very dry and the particle size on the substrate surface is less than a few micrometers, the main adhesion force between the particles and the substrate surface is the van der Waals force. The existence of the van der Waals force will further lead to the generation of stress.

When the substrate surface is wet, a very thin layer of liquid will accumulate in the gap between the particles and the substrate, and the cohesion between the particles and the substrate due to the liquid layer becomes capillary force, which is [9]

$$F_c = 4\pi \sigma r$$

where σ is the surface tension coefficient of the liquid, and r is the particle radius.

Due to the transfer of charge between the particles and the substrate, the two carry opposite charges, and therefore there is a mutual attraction of the double-layer electrostatic force, which can be calculated as follows [9]

$$F_e = \pi \varepsilon U^2 r / z$$

where ε is the dielectric constant, U is the contact potential, r is the particle radius, and z is the microscopic gap between the particles and the substrate. From the above formula, it can be seen that the adhesion force is related to the size of the particles, F_e is proportional to r, the particle mass is proportional to r^3, and according to Newton's second law $a = F_e/m$, so a is proportional to r^{-2}. The smaller the particle size, the greater the acceleration required to remove it, making the removal more difficult.

5.2.2 *Near-Field Optical Focusing Phenomenon*

When laser cleaning particles on microelectronic components, because the size of the particles is close to the order of magnitude of the laser wavelength, a special phenomenon occurs—the near-field optical focusing phenomenon [10]. This phenomenon greatly affects the laser cleaning of microelectronic components. As shown in Fig. 5.1, when the laser is incident on the object to be cleaned, the small particles act like a lens, focusing the laser, so the laser intensity at the center is much greater than the incident laser intensity. This is the near-field focusing phenomenon. The near-field focusing phenomenon enhances the thermal expansion of the substrate, and because the local laser intensity is too high, it may cause damage to the substrate.

For particles smaller than 100 nm, the laser intensity can be enhanced several tens of times in the near-field area on the substrate underneath, and the focusing effect is significant. Luk'yanchuk [11] have theoretically analyzed this strong near-field focusing effect of submicron particles. Mosbacher [12] have experimentally verified it.

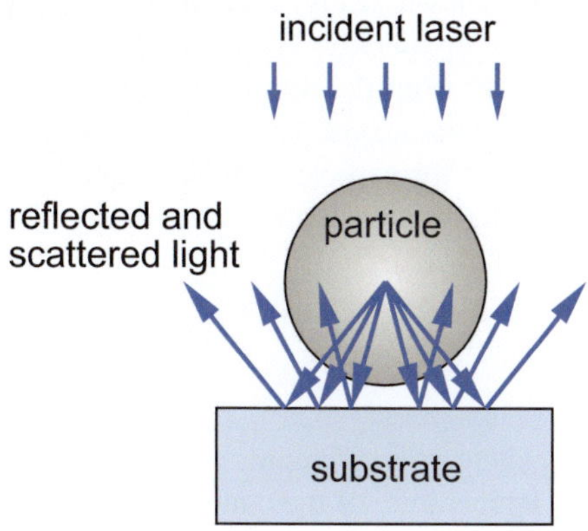

Fig. 5.1 Diagram of near-field optical focusing phenomenon

5.2.3 Mechanism of Dry Laser Cleaning Particles

In dry laser cleaning particles, the laser directly irradiates the surface of the substrate, and the substrate or particles heat up and expand instantaneously, generating a huge rebound force to remove the particles. Good cleaning effects can be achieved for the three cases: the substrate strongly absorbs the laser and the particles weakly absorb it, the substrate weakly absorbs the laser and the particles strongly absorb it, and both the substrate and particles strongly absorb it [13, 14].

1. When the substrate strongly absorbs the laser and the particles are transparent or weakly absorbing (as shown in Fig. 5.2), the substrate rapidly absorbs heat and expands quickly. The particles adsorbed on the surface of the substrate gain speed and acceleration in a short time. After the pulse, the substrate cools and contracts quickly, and the particles leave the surface of the substrate due to inertia. Under the irradiation of a uniform energy density pulsed laser, the temperature rise of the substrate surface can be approximately described by the following formula [15]:

$$\Delta T = (1 - R)F/\rho c\mu \tag{5.2.1}$$

where R is the reflectivity of the laser on the substrate surface, F is the laser energy density, ρ and c are the density and specific heat capacity of the substrate respectively; μ is the thermal diffusion length of the substrate during the laser action. The normal thermal expansion amount caused by the temperature rise of the substrate surface is [15]

$$H = \alpha\mu\Delta T = (1 - R)F\alpha/\rho c \tag{5.2.2}$$

where α is the thermal expansion coefficient of the material. Although the thermal expansion amount is small, the pulse time is very short, the acceleration gained by the particles is large. So the particles can be removed.

2. When the particles strongly absorb the laser and the substrate weakly absorbs it (as shown in Fig. 5.2), under laser irradiation, the particles absorb energy and heat up quickly. The temperature rise inside the particles is uneven, the further away from the irradiated surface, the smaller the temperature rise. The rapid temperature rise causes the particles to expand quickly, causing thermal stress inside the particles, which is equal to the adhesion force per unit area. According to the relationship between stress and strain, the cleaning condition can be derived as [14, 15]

$$\frac{\sigma(d,t)}{E} + \gamma\Delta T(d,t) = \varepsilon(d,t) > 0 \tag{5.2.3}$$

where σ, ε, ΔT represent the thermal stress, relative displacement, and increased temperature of the contact point particle surface layer at time t, respectively; E

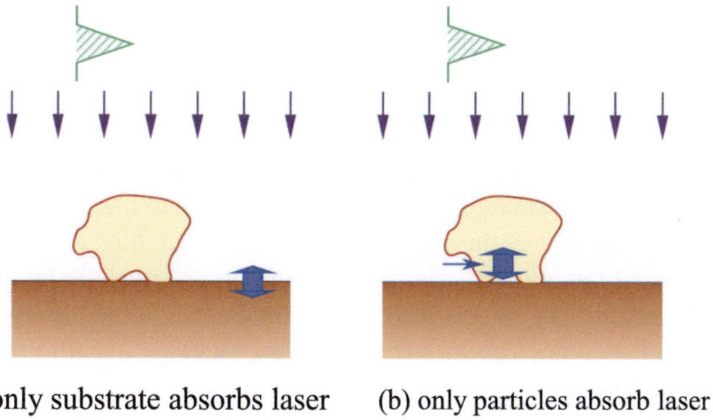

Fig. 5.2 The two special cases for dry laser cleaning methods. **a** Only substrate absorbs laser, **b** only particles absorb laser

is the elastic modulus of the particle; γ is the thermal expansion coefficient of the particle.

5.2.4 Mechanism of Wet Laser Cleaning of Particles

The process of wet laser cleaning mainly involves the liquid absorbing heat, boiling, evaporating, and exploding to remove the particles. Compared to dry laser cleaning, wet laser cleaning adds the resistance needed to remove particles, the resistance is caused by the surface tension of the liquid, thus it has a better cleaning effect. It can remove smaller particles at a lower energy density and is less likely to cause damage to the substrate. For example, dry laser cleaning has difficulty to remove SiO_2 particles below 0.5 μm on a Si substrate [16], while wet laser cleaning can effectively remove SiO_2 particles of 0.1 μm on a Si substrate [17].

For micron and sub-micron particles, the main adhesion forces are van der Waals force and capillary force [18].

$$F_{ad} = \frac{hr}{8\pi z^2} + \frac{h\delta^2}{8\pi z^3} + 4\pi\sigma r \tag{5.2.4}$$

where h, δ and z are the material-dependent Lifshitz–van der Waals constant, the radius of the adhesion interface area and the atomic separation between the bottom surface of the particle and the substrate surface, respectively. σ is the surface tension of the liquid at the temperature T. Wet laser cleaning requires a nozzle to deposit a thin film of liquid onto the particle-contaminated surface. The laser directly heats the liquid or the substrate absorbs the laser energy and then transfers heat to the liquid. The liquid will rapidly heat up and undergo super-rapid boiling. A large number of vapor bubbles nucleate and rapidly expand to squeeze the liquid to generate pressure waves, providing a strong force for the particles to detach. The cleaning force generated for particles with a radius of r is [18]:

$$F = \pi r^2 \times \sqrt{2c\rho f v(P_v - P_\infty)} \tag{5.2.5}$$

where ρ is the liquid density, c is the propagation speed of the pressure wave in the liquid, P_v is the vapor pressure of the vapor inside the bubble, P_∞ is the pressure of the surrounding liquid, f is the volume percentage of bubbles in the liquid, and v is the growth rate of the bubbles. The above formula is derived under the assumption that the liquid is stationary, non-viscous, incompressible, infinitely extended, and the bubble pressure does not change over time. Laser heating of thin liquid films leads to explosive boiling, this process is short and complex, the heat and fluid seriously deviate from the classical thermodynamics and dynamics theory, which needs to be further studied [19–22].

For substrate materials that strongly absorb incident laser and liquid films that are transparent at the wavelength of the incident laser, high cleaning efficiency can be achieved. The process of rapid evaporation of thin liquid films (such as water or alcohol) on opaque solid surfaces (for instance, silicon or chromium) heated by nanosecond pulsed laser radiation involves instantaneous development and phase change of bubbles in the nucleation process, which may cause loss of laser reflectivity due to bubble nucleation and growth at the liquid–solid interface [20]. Through heat conduction calculations, it can be concluded that the solid surface reaches a temperature tens of degrees higher than the boiling point of the liquid before explosive evaporation occurs. Using Maxwell Garnett's [23] effective medium theory to analyze the theoretical transient reflection signal, it can be found that the peak transient pressure can reach 1 MPa.

Under irradiation with sufficiently high laser flux, bubble growth at the liquid or substrate interface is more likely to occur. In this case, the vapor pressure inside the bubble is approximately equal to the saturated vapor pressure at the liquid temperature. The upper limit of bubble growth rate can be described as follows [18]:

$$v(T) = \left(\frac{2}{3}\frac{P_v(T) - P_\infty}{\rho_1(T)}\right)^{1/2} \tag{5.2.6}$$

where $v(T)$ and $\rho_1(T)$ are the bubble growth rate and liquid density at temperature T, respectively. P_∞ is the ambient liquid pressure. In the region near the liquid–solid interface, many growing bubbles act as a vapor layer, with a volume fraction of f, an expansion rate of v, compressing the adjacent liquid and generating stress waves. The energy per unit area of the adjacent liquid obtained by the expansion of this vapor layer is:

$$dE_v = (P_v - P_\infty) f v dt \tag{5.2.7}$$

For the generated stress wave, the energy per unit area is calculated by the following formula:

$$dE_w = \frac{P^2}{2\rho c} dt \tag{5.2.8}$$

where ρ, c, and P are the liquid density, transmission speed, and stress wave pressure, respectively. According to the conservation of energy, the stress wave pressure at the gas/liquid interface is:

$$P = \left[2\rho c(P_v - P_\infty) f v\right]^{1/2} \tag{5.2.9}$$

In order to remove micro and submicro particles, the pressure attenuation can be ignored after this short travel distance, and the cleaning force caused by this pressure on adhered particles with a radius of r can be easily calculated.

A special case in wet laser cleaning is to gradually spray the liquid on the substrate surface, which requires controlling the liquid jet to control the thickness of the liquid film, so that the liquid film immediately vaporizes after laser irradiation, which is the reason why it is also called as steam laser cleaning. In general wet laser cleaning, the substrate is directly immersed in a high humidity liquid, and it is not necessary to control the thickness of the liquid film during cleaning.

Chen [24] simulated the CO_2 pulsed laser wet cleaning of printed circuit boards, using the following assumptions in the simulation: the effect of laser pulse width on the process is ignored; the laser beam is uniform, the water film thickness is 10 μm, and the radius of the laser spot is 5 ~ 10 mm. Since the radius of the laser spot is much larger than the thickness of the water film, a quasi-two-dimensional method can be used for simulation. This can reduce the time of computation, and the accuracy is relatively good. The energy density of the laser pulse is 1 J/cm^2. Because water has a high absorption coefficient for CO_2 lasers, the water quickly heats up to 300 °C, and the instantaneous pressure inside can reach 10 MPa, causing the water film to expand violently and explode. In addition, the simulation results also show that the peak pressure in the water film is about 6 MPa, the pressure inside the liquid is greater than the surface, and it decreases relatively slowly. The pressure near the printed circuit board is greater than 5 MPa within 100 ns, and it decreases relatively quickly after 100 ns. At 100 ns, the deformation of the water film is already large, so the pulse width should be controlled within 100 ns, otherwise the pulse energy will not be effectively utilized. In addition, it was found that the pressure generated by the explosion of the water film due to the action of the laser is much greater than the adsorption pressure of micron-sized particles, and is greater than the adsorption pressure of particles with a radius of 0.1 pm, so under the conditions assumed in this study, micron-sized particles can be easily removed, and particles of 0.1 μm can also be cleaned. Since the damage threshold of the substrate material of the printed circuit board is generally tens of MPa, the pressure generated by the explosion of the water film due to the action of the laser will not cause damage to the substrate of the printed circuit board.

5.3 Dry Laser Cleaning of Electronic Components

Since Zapka and Tam reported dry laser cleaning of particles in 1991 [5, 6], dry laser cleaning particles have been thoroughly studied and applied to the cleaning of electronic mask plates in the following decade. Different wavelengths, pulse widths, energy densities, and incident angles all affect the cleaning effect. The substrates studied are mostly Si, mainly because the removal of particles in the semiconductor industry is the most urgent, and other substrates include PI, PMMA, Ge, NiP, lithium, quartz glass, etc. The particle materials studied include S, Cu, W, SiO_2, Al_2O_3, rubber, etc., with shapes including spherical, flat and irregular shapes, and sizes ranging from tens of nanometers to hundreds of micrometers [15].

I. Laser Energy Density

The laser energy density plays an important role in the cleaning effect. The research of Lu and others [14, 16] shows that laser cleaning has a cleaning threshold and a damage threshold. When the energy density is lower than the cleaning threshold, the cleaning rate is almost 0; when the energy density is higher than the cleaning threshold, the cleaning rate will increase as the laser energy density increases; and when the energy density is higher than the damage threshold, the substrate will be damaged. Therefore, to avoid substrate damage, the energy density needs to be controlled between the cleaning and damage thresholds. Lu used a KrF excimer laser with a wavelength of 248 nm, a pulse width of 23 ns, and a maximum pulse repetition rate of 30 Hz to clean particles on a Si substrate. The experiment found that adjusting the laser energy density from 0 to 350 mJ/cm^2, the laser cleaning efficiency sharply increases with the laser energy density, and the laser energy density exceeds 350 mJ/cm^2. At this time, the base will suffer thermal damage.

Image processing technology can be used to explore the impact of parameters such as laser energy density on cleaning effects. This method is efficient, fast, and accurate. Wang [25] used a KrF excimer laser with a wavelength of 248 nm to conduct cleaning experiments on 1 μm Al_2O_3 particles on polished silicon wafers. Laser parameters: wavelength 248 nm, pulse width 30 ns, frequency 5 Hz, the accuracy of particle count statistics is 97.6%, and the accuracy of cleaning rate is 99.2%. The experimental results show that the cleaning rate increases with the increase of laser energy density. The cleaning effect begins when the laser energy density is 30 mJ/cm^2, and the cleaning rate reaches as high as 89.4% when the laser energy density is 90 mJ/cm^2. As the energy density continues to increase, the cleaning rate is basically stable at around 90%.

II. **Laser Wavelength**

The laser wavelength affects the absorption rate of the base and particles to the laser, thereby affecting the removal effect.

Fourrier [26] used a KrF excimer laser (Lambda Physik LPX 205) to clean PS and SiO_2 particles on a PMMA substrate. They explored the cleaning situation when the wavelength was 248 and 193 nm. Laser parameters: maximum pulse energy 470 mJ, pulse length about 31 ns, repetition rate 1 ~ 30 Hz. For 248 nm, the PMMA base is almost transparent to the laser, and it has almost no removal effect when cleaning transparent SiO_2 particles on PMMA; while PMMA absorbs about 25 mJ/cm^2 of laser energy density at a wavelength of 193 nm, after 20 pulses, the removal rate of 400 nm SiO_2 particles on PMMA reaches 80%.

Song [27] used a KrF excimer laser and YAG laser to clean Cu particles and silicon resin on a quartz base. The experiment used an air propagation particle counter to monitor organic pollutants on the solid surface in real time, and measured that the cleaning threshold of the excimer laser cleaning silicon resin on the silicon base at 248 nm was about 20 mJ/cm^2; and the cleaning threshold observed through optical microscopy was about 16 mJ/cm^2. Then, the influence of wavelength on the cleaning threshold of silicon resin on the Si substrate was studied using 1064, 532, and 355 nm Nd:YAG lasers. When using a 355 nm laser for cleaning, the laser cleaning threshold is about 92 mJ/cm^2, when using a 532 nm laser, it is 122 mJ/cm^2, and when using a 1064 nm laser, it is 540 mJ/cm^2. It can be seen that the cleaning threshold increases with the increase of laser wavelength. The experiment also measured the cleaning threshold of Cu particles of different sizes on a quartz substrate under different wavelengths at 355, 532, and 1064 nm. The cleaning threshold obtained by detecting particles with an airborne particle counter is close to the cleaning threshold obtained from optical microscope photographs. From the experimental results, it can be seen that the cleaning threshold of Cu particles on a quartz substrate increases with the increase of laser wavelength. Therefore, for the laser cleaning of silicon resin and copper particles, the cleaning effect of 355 nm laser is the best, followed by 532 nm, and the effect of 1064 nm is the worst. Generally, the shorter the wavelength, the higher the absorption rate of the material to the laser, and the higher the cleaning rate [28].

III. **Laser Pulse Width**

The laser pulse width affects the thermal diffusion length and action time of the material, thereby affecting the removal effect. Dobler et al. used a frequency-doubled ($\lambda = 532$ nm) Q-switched Nd:YAG laser to clean silicon wafers, with a laser pulse width of 10–20 ns. The study found that the acceleration of particles does not change as strongly with pulse width as expected from a standard Gaussian beam, suggesting that the pulse shape is not an ideal Gaussian beam. However, it can be determined that the shorter the pulse width, the greater the acceleration obtained by the particles, and the easier the particles are removed. The fitting curve shows that the maximum acceleration that particles can obtain is inversely proportional to the square of the pulse width. At the same time, if the pulse width is too short, energy accumulates on the surface, which can easily cause substrate damage [19].

IV. **Pulse Number**

Fourrier and Song et al. [26–30] studied the effect of the number of pulses on the cleaning effect. Fourrier used a KrF excimer laser with a wavelength of 193 nm and an energy density of 25 mJ/cm^2 to clean 400 nm SiO_2 particles on the surface of PMMA. The single pulse cleaning rate was 35%, and no morphological changes were observed in the SEM images. After 5 pulses, the cleaning rate reached 80%, and the cleaning rate did not increase significantly with the increase to 20 pulses. It can be seen that the cleaning rate initially increases with the number of pulses, but the effect of the number of pulses on the cleaning rate saturates. The initial pulse cleaning rate is high, the later pulse cleaning rate is low, and after a certain number of pulses, it no longer has a cleaning effect. Song [27] used a KrF excimer laser or YAG laser to clean Cu particles on a quartz substrate. When the wavelength was 355 nm, the pulse duration was 7 ns, and the laser energy density was 643 mJ/cm^2, the number of particles ejected from the sample was recorded during 5 pulse cleaning periods. It was found that the number of particles detected decreased with the increase of pulses. After 50 pulses, the number of particles detected was close to zero. This means that most of the particles have been removed.

V. **The Impact of the Laser Incident Angle**

Although short-pulse lasers can effectively remove fine particles, the cleaning speed is relatively low due to the small size of the laser spot, and excessive irradiation of the laser on the substrate can cause damage to the substrate. Studies by Vereecke and Lee et al. [31, 32] show that the cleaning efficiency is significantly improved by increasing the angle of incidence of the laser while keeping the area and energy constant. When the laser is incident vertically, if the particles are opaque to the laser, the surface directly below the particles cannot be directly irradiated by the laser. However, when the angle of incidence is increased, the laser beam can directly irradiate the underside of the particles, thus acting at the interface between the particles and the substrate. The cleaning force caused by the laser will also vary at different angles of incidence, and the cleaning efficiency is higher than when the laser is incident vertically. Lee et al. [32] used a Q-switched Nd:YAG laser with a wavelength of 532 nm and a pulse width of 10 ns to clean copper particles with a diameter of about 10 μm on a copper substrate. The cleaning areas at incidence angles of 10° and 90° were 1.35 cm^2 and 0.13 cm^2, respectively. This means that the cleaning efficiency is much higher at grazing incidence. The actual laser energy density at 10° is 0.10 J/cm^2, much less than the energy density of 1.08 J/cm^2 at 90°. The experiment used 10 pulses. At vertical incidence, the cleaning threshold is about 0.15 J/cm^2. As the laser energy density is uniformly increased, the cleaning rate increases, reaching 100% at a laser energy density of 1.0 J/cm^2. However, damage to the substrate occurs at about 0.8 J/cm^2, meaning it is difficult to completely remove copper particles from the surface. At grazing incidence, the cleaning threshold is about 0.01 J/cm^2, an order of magnitude smaller than at vertical incidence. The cleaning rate also increases with the increase in laser energy density, reaching 100% at 0.08 J/cm^2 and above. It can be seen that the laser cleaning effect is better at grazing incidence, and particles can be more easily and effectively removed from the surface without causing any substrate damage.

In addition, for some transparent materials such as quartz, the cleaning results are greatly affected by whether the laser is incident from the front or the back. When the laser is incident from the front, the surface under the particles is covered, resulting in a smaller temperature increase; when the laser is incident from the back, the contact point between the particles and the substrate is not covered, resulting in a larger temperature increase. Research by Lu, Song and others [14, 33] shows that the cleaning rate of backside incidence is higher than that of front side incidence. Lu and others [16] used an excimer pulse laser with a wavelength of 248 nm, a pulse width of 23 ns, and a laser frequency of 10 Hz to conduct experiments on the removal of Al particles from a quartz substrate. They found that at a laser power density of 50 mJ/cm^2, the particle removal rates on both the front and back sides were close to 0. As the laser energy density increased, the cleaning rate rose, and the energy density increased to 100 mJ/cm^2. At times, although the cleaning efficiency of laser irradiation from the front is only about 24%, the cleaning efficiency is 100% when the laser is irradiated from the back. It can be seen that the laser irradiation from the back is more effective in removing Al particles than from the front. Experiments found that for front irradiation, larger particles are harder to remove than smaller ones, while it is easier to remove larger particles through the back irradiation of laser.

VI. The Influence of Materials

The different materials of the substrate and particles have different absorption for lasers, which affects the cleaning effect. Zheng [2] used an excimer laser with a wavelength of 248 nm and a pulse width of 23 ns to successfully remove transparent SiO_2 particles from Si, Ge, NiP substrates. The experiment found that surface roughness has a significant impact in dry laser cleaning, as it leads to a reduction in adhesion. Therefore, as roughness increases, particles can be easily removed. Fourrier [26] used a KrF excimer laser with a wavelength of 248 nm and an energy density of 25 mJ/cm^2 to clean SiO_2 particles and polystyrene (PS) particles on polyimide (PI) and polymethyl methacrylate (PMMA) substrates. It was found that on the PI substrate, the energy density transmitted when the laser is less than the damage threshold is about 19 mJ/cm^2. The laser has a high cleaning rate for particles of a few hundred nm, and the removal efficiency of SP particles is about $(95 \pm 5)\%$. The cleaning rate of transparent SiO_2 particles removed from PI is comparable to that of silicon wafers. However, the PMMA substrate is almost completely transparent to the 248 nm laser, with a transmitted energy density of about 160 mJ/cm^2. The removal efficiency of SP particles is about 80%, and SiO_2 is almost impossible to remove.

In addition, Lu [14] used a laser with wavelength of 248 nm to successfully remove Al particles from a quartz substrate. Kerry [34] used a laser with wavelength of 1.06 μm, pulse width of 20 ns, and energy density of 650 mJ/cm^2 to clean micrometer-sized tungsten particles on a lithium substrate. After 100 pulses, the removal rate reached 95%, while the same laser parameters had a removal rate of less than 5% for rubber particles on a lithium substrate. The removal rate may be proportional to the elastic modulus of the material, as the elastic modulus of rubber is much smaller than that of tungsten. This shows the influence of different materials.

VII. **The Influence of the Particles**

Lasers and particles will affect each other. For transparent particles, when the laser wavelength and particle size are similar, due to laser scattering, optical resonance, near-field effects, and lens effects, the laser at the contact point between the particle and the substrate will be greatly enhanced, which can reach 1–2 orders of magnitude. Although this is beneficial for particles to detach from the substrate, it can also easily damage the substrate. Zheng [2] shows that the enhancement effect of particles on lasers is related to the laser wavelength, particle size, laser incidence angle, substrate surface morphology, etc. Not only spherical particles, irregular Al_2O_3 particles also have a laser enhancement effect.

Lu et al. [14] found that different particle sizes have a significant impact on the cleaning threshold when using a KrF excimer laser with wavelength of 248 nm, pulse width of 23 ns, and maximum pulse repetition rate of 30 Hz to clean particles on a Si substrate. For particles of 0.5 and 1.0 mm in size, the threshold laser energy is approximately 100 mJ/cm^2 and 225 mJ/cm^2, respectively. For larger particles of 2.5 and 5.0 mm, the threshold laser energy is less than 5 mJ/cm^2. In practical applications, the laser energy density cannot exceed 350 mJ/cm^2, otherwise thermal damage may occur in the base part of the irradiation area. Lu et al. found that even using this laser energy density, laser dry cleaning cannot completely remove 0.5 mm particles [14].

Fourrier et al. [26] used a KrF excimer laser (Lambda Physik LPX 205) to clean PS and SiO_2 particles on a PI substrate. They found that 400 and 800 nm particles on PI can be removed by 193 and 248 nm radiation, and the single pulse efficiency is as high as $(95 \pm 5)\%$, which can remove spherical particles with a diameter as low as 110 nm. The single pulse efficiency near the surface damage threshold decreases significantly with the particle size, and for the smallest particles, they are only $(10 \pm 10)\%$.

5.4 Wet Laser Cleaning of Electronic Components

5.4.1 Main Factors Affecting Wet Laser Cleaning of Particles

A large number of experiments have shown that the efficiency of wet cleaning is higher than that of dry cleaning. Wet laser cleaning is based on dry laser cleaning, and a thin layer of liquid is sprayed on the microelectronic components that need to be cleaned before laser cleaning. Therefore, the factors affecting the cleaning efficiency are the same, such as particle size, substrate properties, and pulse width. But there are also different factors, among which the reason for the influence of wavelength on cleaning efficiency is different from dry cleaning. And the composition of the liquid thin film is also a factor affecting the cleaning efficiency. Here we mainly introduce the differences from the dry laser cleaning of particles.

I. Laser Wavelength

The laser wavelength affects the absorption rate of the substrate and the liquid layer to the laser, which will affect the cleaning effect. Regarding wet laser cleaning, researchers from IBM, Zapka et al. [5], deposited a layer of liquid thin film on the Si surface, and then used ultraviolet laser (pulse width of 16 nm, wavelength of 248 nm KrF laser) and infrared laser (pulse width 10 ns, wavelength of 2.94 μm Er:YAG laser) to clean its surface particles of approximately 0.35 μm Al_2O_3 and metal particles. They found that the efficiency was much higher than that of dry laser cleaning, so they started the study of wet laser cleaning particles on silicon wafers.

Through experiments, it was found that both ultraviolet and infrared lasers can effectively clean 0.35 μm Al_2O_3 particles. However, only ultraviolet laser can more effectively and thoroughly clean the metal particles that is more difficult to clean than the non-metal particles. Infrared lasers can only partially clean metal particles, and even increasing the energy density of the infrared laser cannot completely clean them. Moreover, as the energy density increases, it can be observed that the metal particles begin to melt. For infrared laser cleaning, it is the heat absorption and boiling explosion of the liquid film that remove the particles, and the very thin liquid layer is not completely transparent to the infrared laser, which causes the laser to reflect on it, and the metal particles also have a high reflection to the laser. For ultraviolet laser cleaning, it is mainly the Si substrate that absorbs heat, causing the liquid film at the interface to explode and clean all kinds of particles. Therefore, in order to more effectively remove particles, an appropriate laser wavelength is generally chosen so that the substrate mainly absorbs laser energy, rather than the liquid film or particles absorbing heat.

Tam [21] used lasers of different wavelengths to study the removal behavior of particles on Si substrates. The liquid film layer is a transparent water film. After the substrate strongly absorbs the laser energy with a wavelength of 248 nm, it transfers the heat to the liquid. The liquid layer at the interface between the substrate and the liquid overheats and boils explosively, and the particle removal effect is best at this wavelength. For a laser with a wavelength of 10.6 μm, both the liquid and the substrate partially absorb it. The penetration depth of the laser in the liquid is about 20 μm, so if the thickness of the liquid layer is only a few micrometers, part of the laser is absorbed by the liquid, and part of the laser passes through the liquid layer and is absorbed by the substrate. After the substrate absorbs the laser energy, it transfers the heat to the liquid. The liquid layer at the interface between the substrate and the liquid overheats and boils explosively, but because the energy is not concentrated enough, more laser energy is needed to achieve the same removal effect as the 248 nm laser. For a laser with a wavelength of 2.94 μm, only the upper surface of the liquid strongly absorbs the laser, so it only boils explosively on the upper surface of the liquid, not at the interface between the substrate and the liquid, and the particle removal effect is not ideal.

For steam laser cleaning, Mosbacher [35] used a 5 mW He–Ne laser to perform steam laser cleaning experiments on spherical alumina particles on Si wafers. Laser parameters: frequency 1 kHz, irradiation area size 100 μm, and particle diameter 80–600 nm. In the experiment of exploring the cleaning threshold of lasers of different wavelengths, the results show that for a pulsed laser with a pulse width of 2.5 ns, and for lasers with wavelengths of 532 and 583 nm, the cleaning threshold is 50 mJ/cm^2. The cleaning threshold of different laser parameters is determined by when the bubbles nucleate. When the wavelength decreases to ultraviolet, there will be different cleaning thresholds.

Other research institutions have also used KrF excimer lasers, Er:YAG lasers, and TEA CO_2 lasers to clean fine particles on opaque metal substrates, and found that ultraviolet wavelength laser cleaning is more effective. Now most of the lasers used for cleaning microelectronic components are of wavelength 248 nm KrF excimer laser. The laser of this wavelength can penetrate through water, alcohol, or their mixtures, allowing the interface to effectively absorb laser and achieve high cleaning efficiency.

II. **Laser Pulse Width**

The width of the laser pulse affects the heat diffusion length in the substrate and the liquid, thereby affecting the cleaning effect. The shorter the pulse width, the shorter the heat diffusion length, and the more the degree of overheating of the liquid layer at the interface with the substrate, so that the stronger cleaning force can be obtained. However, if the pulse width is too short, it may damage the surface of the substrate. Generally, it is believed that the effect of nanosecond lasers is better than that of microsecond lasers, and millisecond lasers basically have no cleaning effect.

Lu et al. [18] used a 5 mW He–Ne laser to perform steam laser cleaning experiments on spherical alumina particles on Si wafers, and they compared the different effects of ns pulse and ps pulse laser cleaning. When the pulse duration decreases from ns to ps, the heat energy diffused in depth in the Si wafer decreases during the pulse action, and the melting threshold of the bare Si surface will slightly decrease; the shorter pulse length will affect the nucleation of bubbles. When applying ps pulses, due to the shorter pulse length, higher liquid temperatures can be achieved, leading to higher bubble growth rate and therefore higher cleaning efficiency; if the heating of the liquid layer occurs on longer time than the duration of the ps pulse and the heating time of Si, this will result in a slight decrease in the cleaning threshold.

The experimental results show that for the same ns pulse laser, the cleaning thresholds of 2.5 and 8 ns pulses are both 50 mJ/cm^2, but for 2.5 ns pulses, the energy density reaches about 75 mJ/cm^2 and the cleaning rate exceeds 90%; for 8 ns pulses, the energy density needs to reach about 100 mJ/cm^2. As for ns pulses, Lu et al. [18] found a lower cleaning threshold, about 20 mJ/cm^2. The threshold at which the Si wafer begins to melt is about 220 mJ/cm^2. It can be seen that ps pulses can effectively clean at an energy density less than 3 times of the melting threshold. The reduction of the cleaning threshold and the fact that particles can be effectively removed by applying ps pulses indicate that the time scale of the substrate heating is not so long that there is not enough heat to diffuse into the liquid.

III. **Laser Energy Density**

The laser energy density affects the liquid or substrate to absorb laser energy, thereby affecting the cleaning rate. The research of Mosbacher, Lu and others [17, 18, 35] shows that wet and steam laser cleaning also have cleaning thresholds and damage thresholds. When the energy density is lower than the cleaning threshold, there is no cleaning effect; when the energy density is higher than the cleaning threshold, the cleaning rate will increase with the increase of laser energy density; when the energy density is higher than the damage threshold, the substrate will be damaged.

IV. **Liquid Thin Film**

The physical properties of the liquid affect the removal of particles. The liquid films currently used are mainly water, alcohol, methanol, isopropanol or their mixtures. Their particle removal thresholds and removal efficiencies have different specific values, but the influencing rules are similar [18, 21]. The selection of the liquid film hopes to finally leave no residue on the substrate. For different substrates, different types and sizes of particles, there is an optimal choice of liquid film, which needs to be found through experiments to optimize the liquid composition. Tam [21] from IBM Research Institute found through experiments that under the conditions of using KrF excimer laser, larger Al_2O_3 particles on silicon substrates are better cleaned with pure alcohol, and smaller Al_2O_3 particles are better cleaned with water.

Research by Tam [21] shows that water is better than ethanol as a liquid film for laser cleaning, but water has a larger surface tension and is therefore not easy to wet the substrate surface, making it difficult to form a continuous liquid film layer. If a solution of water and 10 ~ 20% methanol, ethanol or isopropanol is used, better cleaning results can be obtained. The purpose of adding methanol, ethanol or isopropanol is mainly to increase the wetness of the substrate surface, so as to form a continuous and uniform liquid film layer on the substrate surface.

Lu [18] found through calculations that when using a 248 nm wavelength and 23 ns laser to clean silicon surface particles, using acetone and ethanol as liquid films, the laser energy density thresholds for cleaning particles of size 1 μm are about 80 mJ/cm^2 and 90 mJ/cm^2 respectively. For particles with a diameter of 0.3 μm, using ethanol as a liquid film during laser irradiation can have a smaller cleaning threshold. Due to some differences in the thermodynamic properties of the two liquid films, the cleaning behavior of these two liquids is different. As the laser energy density increases, the difference between the cleaning force and the adhesion force increases faster on the ethanol film than on the acetone film. This large force difference will lead to high cleaning efficiency.

V. **Particle Material, Size**

Research by Mosbacher, Lu, Lang and others [12, 17, 18, 35] shows that for particles of different materials and sizes, under the same or similar liquid properties and laser parameters, wet (steam) laser cleaning has almost the same cleaning threshold. This value corresponds to the energy density required for the liquid to start boiling, which is related to the start and growth of bubble nucleation, and has little to do with the material and size of the particles. Mosbacher [12] used a laser with a wavelength of 532 nm, pulse widths of 8 and 2.5 ns, and a wavelength of 583 nm, pulse width of 2.5 ns, to clean Al_2O_3 particles with diameters of 60 ~ 800 nm on a crystal Si substrate. They found that the cleaning threshold was basically all at 50 mJ/cm^2, indicating the same removal threshold. Lang [35] used a laser with a wavelength of

532 nm and a pulse width of 8 ns. The ns laser conducted a laser cleaning experiment on PS particles on the Si substrate. The experimental particles had diameters ranging from 140 to 1300 nm. The liquid films used were water and isopropanol, and it was found that both types of liquid film layers had the same cleaning threshold.

Luo [36] studied the laser cleaning of Al particles on the Si substrate, using a Nd:YAG pulsed electro-optic Q-switched laser with a pulse width of 12.8 ns, a repetition rate of 5 Hz, and a single pulse energy of 430 mJ. Under the action of multiple pulses, it was found that large particles would break and turn into small particles, and some particles would undergo phase changes to form smooth spheres after reaching the melting point. They believe this is the result of the thermodynamic effects of the plasma generated after laser irradiation.

From the previous analysis, it can be seen that in the laser cleaning of microelectronic components, the actual cleaning effect is affected by various factors, which requires that the cleaning mode and parameters should be reasonably selected for the actual situation and cleaning requirements in the specific application to achieve higher laser cleaning efficiency.

5.5 Application of Laser Cleaning in Electronic Components

Through a large number of experimental studies, laser cleaning of microelectronic components has entered the stage of practical application. The following is a brief introduction to the actual application situation.

I. **Laser Cleaning of Polyimide Films, Electronic Circuit Boards, and Silicon Wafers**

In the past decade, the technology of using excimer lasers to strip organic polymers has developed rapidly and is now used throughout the process of electronic component packaging. Polyimide film is a dielectric material for the internal connection structure of multilayer packaging films for high-speed, high-density electronic components. Excimer lasers can remove Ti, Cr, W, Ni, and Pb microparticles that contaminate polyimide films. Excimer lasers can also clean Cu_2O passivation films on the surfaces of microelectronic systems. The cleaning effect of Al_2O_3, SiO_2, and PSL (polystyrene latex) particles on the surface of silicon wafers with multimode pulsed CO_2 lasers is good [37–39]. Femtosecond laser ablation of the silicon material surface to obtain an anti-reflection structure, not only can it effectively eliminate the laser surface oxide layer, but it can also create small-scale microstructures. Using a focused elliptical laser spot, large-area and energy-continuous attenuation multi-laser cleaning is realized, solving the problem of new oxide layer formation during the cleaning process [40]. Using a Q-switched Nd:YAG laser to remove uranium dioxide and thorium dioxide particles on a zirconium substrate, this process minimizes the production of secondary waste, and has become the most attractive technology to eliminate radioactive surface contamination. Using a Nd:YAG laser to act on the back of a wafer immersed in water, the laser energy causes a shock wave in the wafer, which is transmitted to the water, creating a flowing bubble stream. The shock wave and bubble stream removed 0.5 μm alumina particles. Nd:YAG laser with pulse energy of 2 mJ and pulse duration of 100 ns is used for carbon cleaning experiments [41, 42]. Nanosecond pulsed Nd:YAG laser is also used to clean the gold layer on the molten quartz substrate. The surface roughness under the gold film does not change after laser cleaning. The effects of pulse width, beam incidence angle, spot overlap, laser intensity, and number of channels on cleaning efficiency were studied. The results show that within 3 min, the laser can effectively clean the deposited gold layer with a thickness of 48 nm. Although the laser shock wave cleaning process provides a promising alternative to traditional dry cleaning processes for the removal of nanoscale particles, it has always been unable to explain the problem of organic particle removal [43–45]. This work clarifies the physical basis for the ineffective removal of organic particles by laser shock cleaning using polystyrene latex particles on a silicon substrate. Contamination on the surface of optical components can degrade the quality of the laser beam and cause damage to the optical components. Particles and grease contamination are two common types of contamination on optical surfaces. In this work, a 1064 nm laser-induced plasma shock wave cleaning technique is used to remove SiO_2 contamination particles from the surface of K9 glass, with a removal rate of over 95%. The laser cleaning technology using a KrF excimer laser to remove photoresist from the silicon wafer surface has a cleaning efficiency of nearly 90% as shown by contact angle measurements [46].

II. **Deflashing and Demarking of Integrated Circuit Components**

As the integration of ICs increases, more and more pins and the holes are getting smaller. Traditional cleaning methods are difficult to remove mold flash, i.e., fine adhesion, from small holes. Deflashing with an excimer laser has a clear advantage and has become the most suitable deflashing technology. When a Nd:YAG laser with wavelength of 532 nm and pulse width of 7 ns is used to clean a 7 μm mold flash, as shown in Fig. 5.3a, 4 pulses of laser with energy density of 300 mJ/cm^2 can completely clean the mold flash, as shown in Fig. 5.3b. While using laser demarking, the dust, grease, and oxides on the marked surface are also cleaned, and the durability of the re-marking is better [47].

III. **Laser Cleaning of Inkjet Printed Flexible Circuits**

With the development of printing technology, in order to achieve better printing quality, the inkjet holes are getting smaller. This makes the removal of contaminants increasingly important. Figure 5.4 shows the flexible circuit on the polyimide film before and after cleaning. The black line in Fig. 5.4a is the contaminated area, and Fig. 5.4b is the magnified contaminated area, where many tiny inkjet holes can be seen. With laser cleaning, just position the cleaning area around the tiny inkjet holes, so the laser will not interact with the conductive circuit. Laser cleaning can clean the inkjet printed flexible circuit without damaging the polyimide film and the conductive circuit, which is efficient, high-yield, and low-cost [48, 49]. The laser with wavelength of 1.06 μm (Nd:YAG) and 10.6 μm (CO_2) was adopted for selective removal with

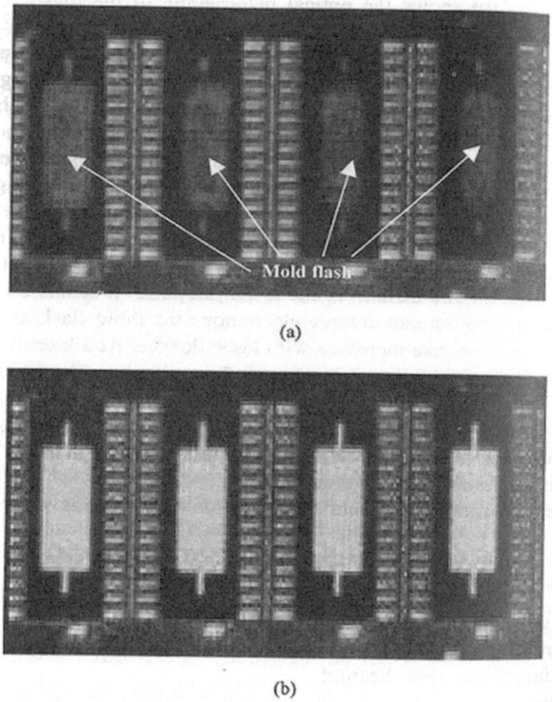

Fig. 5.3 Diagram of Nd:YAG laser cleaning flash

direct and indirect burning of the paper-based surface conductive aluminum film (25 nm thickness). Laser pulses were used to treat samples of natural and artificially contaminated paper, and the damage thresholds and cleaning thresholds of pure paper and contaminated paper under different laser treatment conditions were measured. The experimental results show that femtosecond laser radiation has a higher cleaning efficiency than that with the nanosecond laser [50, 51].

IV. **Laser Cleaning Optoelectronic Devices**

Promising nanoelectronic and optoelectronic materials have been widely studied. Because these optoelectronic materials have adsorbed a large amount of oxygen and water molecules on the surface, reducing the performance of the device and thus hindering precise applications. By irradiating optoelectronic material devices with lasers and exploring the impact on transmission and laser response, it has been proven that this laser cleaning is a direct method to remove physically adsorbed pollutants [52].

Through the above introduction, it is known that laser cleaning provides an effective and reliable technology for the situation where the device size in the microelectronics industry is small and fine particles are not easy to remove. At the same time, laser cleaning does not wear and corrode the substrate material and has good environmental performance, and can also perform cleaning operations in narrow spaces. It can be foreseen that with the development of laser cleaning technology,

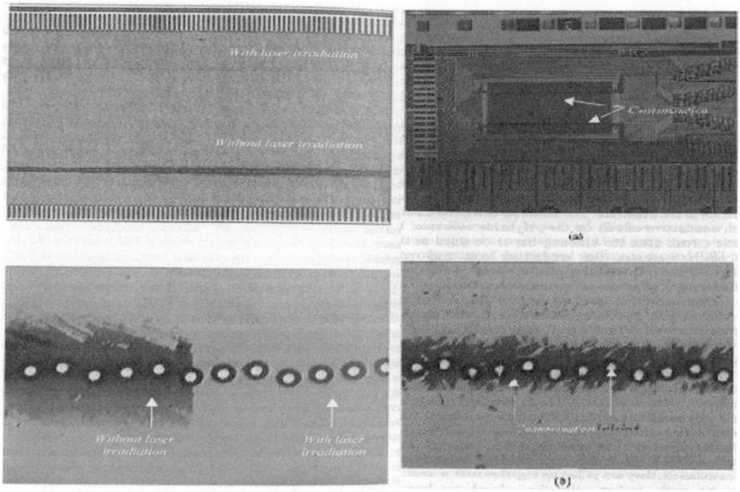

(a) (b)

Fig. 5.4 Laser cleaning of flexible circuits on polyimide film

the application in the field of microelectronic component cleaning will become more widespread and deepened.

References

1. Zhao Z M. The application of laser cleaning technology in microelectronics in industry field [J]. Cleaning Technology, 2004, 2(8): 29–34.
2. Zheng Y W, Lukyanchuk B S, Lu Y F, Song W D, Mai Z H. Dry laser cleaning of particles from solid substrates experiments and theory [J]. Journal of Applied Physics, 2001, 90(5): 2137–2142.
3. Mauer J L, Pfeiffer H C, Stickel W. Electron optics of an electron-beam lithographic system [J]. IBM Journal of Research & Development, 1977, 21(6): 514–521.
4. Mittal K L. Particles on Surfaces, Detection, Adhesion and Removal [M]. New York: Plenum Press, 1988.
5. Zapka W, Ziemlich W, Tam A C. Efficient pulsed laser removal of 0.2 μm sized particles from a solid surface [J]. Applied Physics Letters, 1991, 58(20): 2217–2219.
6. Tam A C, Do N, Klees L, Leung W P. Explosion of a liquid film in contact with a pulse-heated solid surface detected by the probe-beam deflection method [J]. Optics Letters, 1992, 17(24): 1809–1811.
7. Kelley J D, Hovis F E. A thermal detachment mechanism for particle removal from surfaces by pulsed laser irradiation [J]. Microelectronic Engineering, 1993, 20(1–2): 159–170.
8. Mittal K L. Particles on Surfaces [M]. New York: Marcel Dekker, 1995: 1–33.
9. Kane D M. Laser Cleaning II [M]. Singapore: World Scientific, 2006: 9–11.
10. Born M, Wolf E, Principles of Optics, 7-th Edition [M]. Cambridge University Press, 1999: 491–555.
11. Luk'yanchuk B S, Huang S M, Hong M H. 3D effects in dry laser cleaning [J]. SPIE Proceedings, 2002, 4760: 204–210.

12. Mosbacher M, Chaoui N, Siegel J, Dobler V, Solis J, Boneberg J, Afonso C N, Leiderer P. A comparison of ns and ps steam laser cleaning of Si surfaces [J]. Applied Physics A, 1999, 69(1): S331-S334.
13. Gu H Y. Study on the mechanism of adsorbed particles on the surface of ULSI silicon single crystal substrate sheet and removal technology [D]. Tianjin: Hebei University of Technology, 2001.
14. Lu Y F, Song W D, Low T S. Laser cleaning of micro-particles from a solid surface-theory and applications [J]. Materials Chemistry and Physics, 1998, 54(1): 181–185.
15. Chen J F, Zhang Y K, Kong D J, Ye X. Research progress of cleaning tiny particles by short-pulsed laser [J]. Laser Technology: 2007, 31(3):301–305.
16. Lu Y F, Zheng Y W, Song W D. Laser induced removal of spherical particles from silicon wafers [J]. Journal of Applied Physics, 2000, 87(3): 1534–1539.
17. Mosbacher M, Dobler V, Boneberg J, Leiderer P. Universal threshold for the steam laser cleaning of sub-micron spherical particles from silicon [J]. Applied Physics A, 2000, 70(6): 669–672.
18. Lu Y F, Zhang Y, Wan Y H, Song W D. Laser cleaning of silicon surface with deposition of different liquid films [J]. Applied Surface Science, 1999, 138(1): 140–144.
19. Tam A C, Park H K, Grigoropoulos C P. Laser cleaning of surface contaminants [J]. Applied Surface Science, 1998, 127–129: 721–725.
20. Yavas O, Leiderer P, Park H K, Grigoropoulos C P, Poop C C, Leung W P, Do N, Tam A C. Optical reflectance and scattering studies of nucleation and growth of bubbles at a liquid-solid interface induced by pulsed laser heating [J]. Physical Review Letters, 1993, 70(12): 1830.
21. Tam A C, Leung W P, Zapka W, Ziemlich W. Laser-cleaning techniques for removal of surface particulates [J]. Journal of Applied Physics, 1992, 71(7): 3515–3523.
22. Jin R X, Huai X L, Liu D Y. The applicability of classical theory for rapid transient explosive boiling [J]. Journal of Engineering Thermo Physics, 2003, 24(6): 1013–1015.
23. Markel V A. Introduction to Maxwell Garnett approximation: tutorial [J]. Journal of the Optical Society of America A, 2016, 33(7): 1244–1256.
24. Chen H, Zhu H H, Cheng Z H. Numerical simulation of propulsive force in steam laser cleaning of printed circuit board [J]. Chinese Journal of Lasers: 2006, 33(suppl):432–434.
25. Wang X Y, Xu W X, Sima Y, Wu D J, Kang R K, Guo D M. Evaluation of efficiency for silicon wafer cleaning by image processing [J]. Optics and Precision Engineering: 2007, 15(8): 1263–1268.
26. Fourrier T, Schrems G, Muhlberger T, Heitz J. Laser Cleaning of Polymer Surfaces [J]. Appl Phys A, 2001, 72(1): 1–6.
27. Song W D, Hong M H, Lee S H, Lu Y F, Chong T C. Real-time monitoring of laser cleaning by an airborne particle counter [J]. Applied Surface Science, 2003, 208(2): 306–310.
28. Li J C. Calculation of Laser Diffraction and Heat Interaction [M]. Beijing: Science Press, 2001. 341–432.
29. Dobler V, Oltra R, Boouillon J P, Mosbacher M, Boneberg J, Leiderer P. Surface acceleration during dry laser cleaning of silicon [J]. Applied Physics A - Materials Science & Processing, 1999, 69(7): 335–337.
30. Mosbacher M, Munzer H J, Zimmermann J. Optical field enhancement effects in laser-assisted particle removal [J]. Applied Physics A - Materials Science & Processing, 2001, 72(1): 41–44.
31. Vereecke G, Rohr E, Heyns M M. Influence of beam incidence angle on dry laser cleaning of surface particles [J]. Applied Surface Science, 2000, 157(1): 67–73.
32. Lee J M, Watkins K G, Steen W M. Angular laser cleaning for effective removal of particles from a solid surface [J]. Applied Physics A - Materials Science & Processing, 2000, 71(6): 671–674.
33. Song W D, Hong M H, Koh H L, Wang W J, Zheng Y W, Lu Y F, Chong T C. Laser-induced removal of plate-like particles from solid surface [J]. Applied Surface Science, 2002, 186(1): 69–74.
34. Kerry J D, Stuff M I, Hovus F E, Linford G J. Removal of small particles from surfaces by pulsed laser irradiation [J]. Proc SPIE, 1991, 1415: 211–219.

35. Lang F, Mosbacher M, Leiderer P. Near field induced defects and influence of the liquid layer thickness in steam laser cleaning of silicon wafers [J]. Applied Physics A, 2003, 77(1): 117–123.
36. Luo J, Feng G Y, Han J H, Shen X, Zhang L J, Ding K Y. Thermodynamics of laser plasma removal of micro and nano-particles [J]. Acta Physica Sinica: 2020, 69(8):084201.
37. Xu S Z, Dou H Q, Han F M, Zu X T. Laser cleaning of particulate contaminants on K9 glass surface [J]. Research and Exploration in Laboratory: 2017, 36(6): 5–8.
38. Zhang K W. Laser Cleaning Technical Review [J]. Applied Laser. 2002, 22(2):264–268.
39. Chen D S. Interaction between polymer and laser (III) manufacture of fine circuit on polyimide film by selective electroless plating using silver as seeding [D]. Shanghai: Shanghai Jiao Tong University, 2007.
40. Chen T, Wang W J, Tao T, Pan A F, Mei X S. Multi-scale micro-nano structures prepared by laser cleaning assisted laser ablation for broadband ultralow reflectivity silicon surfaces in ambient air [J]. Applied Surface Science, 2020, 509: 145182.
41. Kumar A, Biswas D J. Particulate size and shape effects in laser cleaning of heavy metal oxide loose contamination off clad surface [J]. Optics and Laser Technology, 2018, 106: 286–293.
42. Tsai C H, Peng W S. Laser cleaning technique using laser-induced acoustic streaming for silicon wafers [J]. Journal of Laser Micro Nanoengineering, 2017, 12(1): 1–5.
43. Singh A, Choubey A, Modi M H, Upadhyaya B N, Oak S M, Lodha G S, Deb S K. Cleaning of carbon layer from the gold films using a pulsed Nd:YAG laser [J]. Applied Surface Science, 2013, 283: 612–616.
44. Choubey A, Singh A, Modi M H, Upadhyaya B N, Lodha G S, Oak S M. Study on effective cleaning of gold layer from fused silica mirrors using nanosecond-pulsed Nd:YAG laser [J]. Applied Optics, 2013, 52(31): 7540–7548.
45. Ye Y Y, Yuan X D, Xiang X, Cheng X F, Miao X X. Laser cleaning of particle and grease contaminations on the surface of optics [J]. Optik, 2012, 123(12): 1056–1060.
46. Baek J Y, Jeong H, Lee M H, Song J D, Kim S B, Lee K W, Kim G S, Kim S H. Contact angle evaluation for laser cleaning efficiency [J]. Electronics Letters, 2009, 45(11): 553–554.
47. Wang X, Wu Z, Zeng K, Ma H, Li M. Vision orientation and laser scanning optical coaxial device of laser deflash machine, has telecentric field lens positioned on lower part of scanning head, at same vertical axis as central line of scanning head: China, CN201632772-U [P]. Jiang yin ace electronic equip co LTD. 2010, 11, 17.
48. Xu X L. Laser print document inspection analysis [J]. Science and Informatization (Kexue yu Xinxihua):2019, 1:190–191.
49. Tang Y J. Current situation of laser cutting application in industry [J]. Laser & Optoelectronics Progress: 2002, 39(1):53–56.
50. Rahimi R, Ochoa M, Ziaie B. A comparison of direct and indirect laser ablation of metallized paper for inexpensive paper-based sensors [J]. ACS Applied Materials & Interfaces, 2018, 10(42): 36332–36341.
51. Pentzien S, Conradi A, Koter R, Kruger J. Cleaning of artificially soiled paper using nanosecond, picosecond and femtosecond laser pulses [J]. Applied Physics A - Materials Science & Processing, 2010, 101(2): 441–446.
52. Zhang S N, Li R J, Yao Z X, Liao P C, Li Y F, Tian H F, Wang J H, Liu P Z, Guo J J, Liu K H. Laser annealing towards high-performance monolayer MoS_2 and WSe_2 field effect transistors [J]. Nanotechnology, 2020, 31(30): 30LT02.

Chapter 6
Laser Cleaning Paint Layer

6.1 Basic Introduction to Laser Paint Removal

Coating serves as a crucial protective measure for substrate materials across various industries and everyday applications. Whether it is the surfaces of ships, airplanes, cars, bridges, furniture, or artworks, paint layers are applied to safeguard the underlying substrate from damage. Be it metal materials like steel and aluminum alloy or non-metal materials such as plastics and rubber, coating is indispensable. However, over time, instruments, equipment, architectural structures, and transportation vehicles with paint layers inevitably suffer from environmental wear and tear. This deterioration compromises the integrity of the coating, leading to exposure of the substrate and subsequent damage. To mitigate corrosion risks and uphold structural integrity, periodic recoating becomes imperative, with paint removal emerging as a critical preparatory step. Figure 6.1 illustrates a schematic depiction of the paint removal process on a ship.

Traditional industrial paint removal methods mainly encompass shot blasting cleaning, chemical cleaning, high-pressure water jet cleaning. For smaller areas or irregular components, manual grinding, namely the mechanical friction method, is often employed to remove paint.

(1) Mechanical friction method

This method involves the use of specialized equipment or manual tools, including wire brushes, grinding wheels, or abrasive papers, to eliminate paint layers from equipment surfaces. It represents one of the earliest paint removal techniques. However, its efficiency is limited, often yielding unsatisfactory cleaning results, and necessitating a significant labor investment. Presently, it is primarily employed for areas with intricate structures in equipment and functions as a supplementary method to shot blasting.

F. Song and X. Lin, *Laser Cleaning*, https://doi.org/10.1007/978-981-97-6631-4_6

Fig. 6.1 Ship paint removal

(2) Shot blasting paint removal method

This method utilizes a high-pressure air pump to propel particles like sand grains, plastic beads, and metal pellets onto the paint layer, imparting a forceful impact that fractures and dislodges the paint. Shot blasting stands as the most efficient and widely adopted technique for paint removal, which is often deployed in scenarios necessitating extensive paint removal, such as the surfaces of ships and other large-scale equipment. Figure 6.2 provides a schematic representation of the shot blasting paint removal process. However, this method generates substantial dust and emits high levels of noise during paint removal, posing health and safety risks to operators.

(3) High-pressure water jet method

This method entails pressurizing ordinary water to tens or even hundreds of megapascals using a high-pressure pump, resulting in a concentrated, high-velocity water jet traveling at speeds of 200–500 m/s. Directed at the surface of equipment, this jet effectively removes the paint layer, as illustrated in Fig. 6.3. However, challenges arise with

Fig. 6.2 Shot blasting for paint removal

Fig. 6.3 Paint removal by high-pressure water jet

this technique, including the substantial water consumption required as a medium and the rapid rusting tendency observed on cleaned equipment surfaces thereafter. Furthermore, the disposal of wastewater containing paint fragments presents environmental concerns. Direct discharge into natural water bodies risks pollution, while secondary treatment methods increase the overall cost. To enhance cleaning efficiency, small sand grains, plastic beads, or metal pellets are introduced into the water, combined with shot peening method, which compounds the environmental pollution problem.

(4) Chemical Reagent Method

Chemical cleaning involves decomposing the paint layer on a metal surface using specialized cleaning agents through methods such as dipping, spraying, or brushing. This approach aims to remove the paint effectively. However, the process typically requires multiple treatments for complete surface cleaning, resulting in extended cleaning cycles and complex procedures. Moreover, there is a risk of corrosion on the metal surface due to prolonged exposure to chemical agents. Additionally, chemical paint removal is non-recoverable and can lead to secondary pollution of soil and rivers. Furthermore, these agents may contribute to atmospheric pollution and pose health risks to humans if not managed properly.

Among the various paint removal methods, shot blasting and chemical reagents are mainly employed in China, with shot blasting being the most widely utilized. Presently, the primary demand for paint removal in the industrial field arises from the shipping and aviation industries. For instance, in the shipbuilding sector, propelled by China's rapid economic growth and burgeoning foreign trade, the country has witnessed a substantial rise in the number of Chinese ships. With nearly a thousand ship maintenance enterprises nationwide, approximately 90% employ shot blasting for paint removal. Similarly, in the aviation domain, stringent safety standards mandate regular paint removal and repainting for aircraft and other aviation equipment. However, traditional paint removal methods pose significant environmental pollution risks, impacting both economic development and human health. Consequently, there is an urgent need for a new paint removal method that is efficient, cost-effective, and environmentally friendly. Laser paint removal is a promising technology that could supplant traditional methods like shot blasting, chemical reagents, and high-pressure water jets. Laser paint removal offers numerous advantages, including minimal damage to the ship's hull, elimination of raw material requirements, zero emissions, and low pollution levels.

Laser paint removal garnered significant interest from researchers as early as the 1980s, with pioneering work undertaken by Woodroffe and collaborators [1] in the United States. Their experiments involved the use of a CO_2 laser with a pulse energy of 1 kJ, a pulse width of 20 μs, and a repetition frequency of 10 Hz, confirming the feasibility of laser paint removal. Over the subsequent decades, many research teams dedicated considerable efforts to investigating various aspects of laser paint removal, including the selection of laser types, wavelengths, Q-switching methods, and the regulation of peak power density. These collective endeavors significantly propelled the field of laser paint removal technology, paving the way for its practical implementation across a wide range of industrial applications.

6.2 Theoretical Research on Laser Paint Removal

Laser cleaning techniques encompass both dry laser cleaning and wet laser cleaning methods. The mechanisms underlying laser cleaning include ablation, vibration, shockwave (acoustic shock), plasma effect, and photo pressure effect. Dry laser cleaning is considered more effective than wet laser cleaning and is therefore more commonly utilized. However, theoretical research on dry laser cleaning has primarily focused on the removal of particulates, while studies on the removal of thin films and coatings from sample surfaces are scarce. Currently, the acknowledged effects in laser paint removal include ablation, vibration, and shockwave effects. At sufficiently high laser energy densities, ablation occurs concurrently with vibration and shockwave effects, leading to physicochemical alterations. Conversely, at lower laser energy densities, the process is driven by vibration and other physical changes. Regardless of the specific mechanism involved, laser-matter interactions in the cleaning process adhere to the principles of the conservation of energy and Newton's laws of motion.

The ablative effect in laser paint removal refers to the process wherein laser irradiation heats the surface of the paint, causing its temperature to rise to the point of melting, boiling, or ignition. This elevated temperature leads to the melting, vaporization, burning, or decomposition of the paint layer, ultimately resulting in its removal [2]. In continuous lasers, heat accumulates on the paint's surface, gradually increasing its temperature, whereas pulsed lasers rapidly heat the paint, causing it to vaporize and ablate upon absorbing the laser energy. Another mechanism involves the laser heating the substrate, leading to deformation and detachment of the paint layer, which can be classified as a vibrational effect [3]. Many studies on the mechanism of laser paint removal have utilized heat conduction equations to elucidate these processes. Abdullah et al. [4] have conducted substrate damage analysis, revealing that low-power laser applications primarily induce chemical bond breaking in the molecular components of coatings. This includes the exchange and rearrangement of double bonds of carbon and oxygen, and even single carbon bonds with three oxygen bonds, resulting in photochemical interactions that toughen and fracture the coating surface. Conversely, in medium to high-power applications, photothermal interactions dominate, initiating the thermal decomposition process of the coating.

Chen et al. [2] discussed the mechanism of paint removal by using CO_2 laser. For CO_2 lasers, laser irradiation does not induce photolytic decomposition of paint molecules. Therefore, the primary mechanism of paint removal by CO_2 laser is the instantaneous combustion and vaporization of the paint layer, which leads to the rapid detachment of the layer. Additionally, thermal effects also cause swift thermal expansion of the paint layer. In experiments, a minor amount of paint residue was observed, which could be removed by directing auxiliary nitrogen gas onto the substrate surface. During the laser paint removal process, the temperature rising T_a of the surface at the beam center can be approximately calculated by [3]:

$$T_a = \frac{2(1-R)F}{\kappa}\sqrt{\frac{at}{\pi}} \tag{6.2.1}$$

where $(1 - R)$ represents the absorptivity of the surface to the laser, F is the average laser power density, κ is the thermal conductivity, a is the thermal diffusion rate, and t is the laser action time. T_v represents the vaporization temperature of the paint layer, while T_m indicates the melting point of the substrate; considering $T_v < T_m$ to ensure the removal of the paint layer without damaging the substrate, the temperature of the paint layer (T_s) must satisfy: $T_v \le T_s \le T_m$. Corresponding to this requirement, the average laser power density must fulfill $F_v \le F_s \le F_m$. When F_s equals the lower bound, the initiation threshold for laser paint removal is reached; when F_s equals the upper bound, it is the threshold for substrate damage during laser cleaning. Therefore, when the laser power satisfies the above conditions, the paint layer can be removed without substrate damage. Zou et al. [5] attribute the removal of paint by nanosecond pulsed lasers primarily to thermal stress. They calculated the theoretical damage threshold by solving the one-dimensional heat conduction equation and quantitatively analyzed the thermal stress. They determined the threshold conditions for laser cleaning by comparing thermal stress and adhesive force and experimentally validated the theoretical model using a Q-switched Nd:YAG laser.

Zhang et al. [6] believed that the short laser pulses could cause some solid paints to immediately liquefy, with some potentially overheating above their boiling points for nanosecond pulsed lasers with high repetition frequency such as pulsed Nd: YAG lasers. Due to the short duration and high energy of laser pulses in the cleaning process, paints that cannot immediately evaporate from the heated surface may be further heated by subsequent laser pulses [7–9].

6.2.1 One-Dimensional Thermal Stress Model

I. One-dimensional heat conduction equation

When the laser irradiates the surface of the paint, it undergoes reflection, transmission, and absorption. Upon absorption, the paint layer's temperature rises, followed by irradiation of the substrate by the laser passing through the paint. This process results in an increase in the substrate's temperature, which conforms to the heat conduction equation. Cartesian coordinate systems are established for the paint (x, y, z) and the substrate (x′, y′, z′), respectively, as illustrated in Fig. 6.4 [5].

In industrial cleaning, the surface area of the object being cleaned vastly exceeds its thickness. Hence, studying the temperature rise during laser cleaning can be modeled as the laser heating of an infinitely large thin plate. Heat diffusion in the tangential direction of the thin plate is negligible, allowing for the use of a one-dimensional heat conduction equation. The following assumptions can be made: (1) The laser intensity exhibits a flat-top shape on the x–y plane; (2) The laser intensity variation in the z direction conforms to Beer-Lambert's law of absorption; (3) The x–y plane is treated as an infinite plane.

The one-dimensional heat conduction equation is given by [9]:

$$\begin{aligned} &\rho c\frac{\partial T(z,t)}{\partial t} - k\frac{\partial^2 T(z,t)}{\partial z^2} = Q(z,t) \\ &Q(z,t) = (1-A)\beta I_0 e^{-\beta z} g(t) \\ &g(t) = e^{-\left(\frac{t}{\tau}\right)^2} \end{aligned} \tag{6.2.2}$$

where $T(z, t)$ represents the temperature of the coating, depending on depth and time. Q is the laser heat source, related to depth and time. $g(t)$ is the time-dependent laser pulse shape, which has a Gaussian profile. k is the thermal conductivity. α is the material's absorption coefficient. R is the reflectivity of the object. I_0 is the laser intensity. ρ is the mass density of the paint. c is the specific heat capacity of the paint. τ is the pulse width of the laser. The boundary and initial conditions of the equation are given by Eq. (6.2.3). And Eq. (6.2.4) implies that the room temperature is chosen to be 300 K.

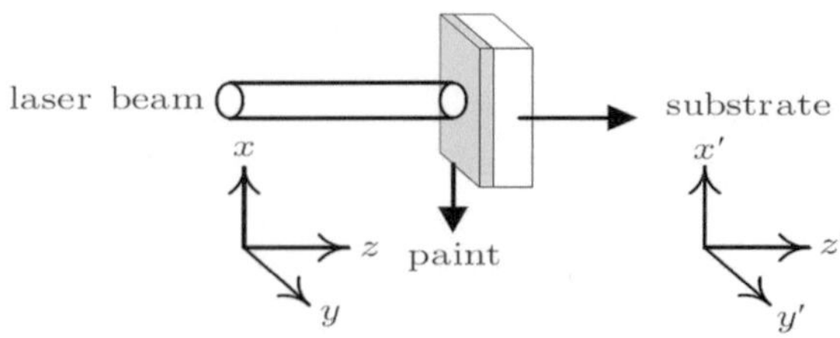

Fig. 6.4 Simplified schematic of laser paint removal

$$\left\{\lambda\frac{\partial T}{\partial z}\right|_{z=0}=0\left\{\lambda\frac{\partial T}{\partial z}\right|_{z=l}=0 \tag{6.2.3}$$

$$T(z,0)=300\ (\mathrm{K}) \tag{6.2.4}$$

II. Temperature of paint and substrate

Both the paint and the substrate material meet the aforementioned boundary and initial conditions. By solving Eq. (6.2.2), one can obtain the temperature at time t and position z to be:

$$\begin{aligned} T(z,t)=\sum_{n=1}^{\infty}\frac{C_nI_0}{a_t}\left(\frac{l}{n\pi}\right)^2\cos\left(\frac{n\pi}{l}z\right)\left[1-\mathrm{e}^{-a_t\left(\frac{n\pi}{l}\right)^2t}\right] \\ +C_0I_0t+300\ (\mathrm{K}),\quad n=1,2,3,\ldots, \\ 0\le t<\tau,\quad 0\le z\le l. \end{aligned} \tag{6.2.5}$$

where ΔT represent the increment of temperature, then we can get:

$$\begin{aligned} \Delta T(z,t) &= T(z,t)-T(z,0) \\ &=\sum_{n=1}^{\infty}\frac{C_nI_0}{a_t}\left(\frac{l}{n\pi}\right)^2\cos\left(\frac{n\pi}{l}z\right)\left[1-\mathrm{e}^{-a_t\left(\frac{n\pi}{l}\right)^2t}\right] \\ &\quad +C_0I_0t\ (\mathrm{K}),\quad n=1,2,3,\ldots, \\ &\quad 0\le t<\tau,\quad 0\le z\le l. \end{aligned} \tag{6.2.6}$$

where

$$C_n=\frac{2\alpha}{\rho cl}\frac{1-\cos(n\pi)e^{-al}}{1+\left(\frac{n\pi}{\alpha l}\right)^2}\quad n=1,2,3,\cdots \tag{6.2.7}$$

$$C_0=\frac{\alpha}{\rho cl}\left(1-e^{-al}\right) \tag{6.2.8}$$

If the temperature increases to the melting or gasifying point of the paint, the paint will detach from the substrate through ablative effects. If the temperature is insufficient to cause phase transitions or chemical reactions in the paint, the detachment occurs due to thermal stress caused by the temperature gradient, surpassing the adhesive force of the paint to the substrate. This model solely addresses the temperature increase from laser heating and does not incorporate heat loss due to the vaporization of the paint layer, which would diminish heat accumulation. Furthermore, the decomposition of organic materials also releases energy.

III. **Thermal stress**

The thermal expansion of an object under laser irradiation leads to the generation of thermal stress within the material. And the thermal expansion length in the z-direction can be given by the following formula [10]:

$$\Delta l = l\gamma \Delta T \tag{6.2.9}$$

where γ is the linear coefficient of thermal expansion. The parameters (ρ, c, α, and γ) typically vary with temperature. During the paint removal process, the temperature rise is small, so for simplicity, they are assumed to be constant here. When the laser energy density value approaches the cleaning threshold, the temperature change of the paint and the substrate is minimal. Regarding the paint, it can be approximated as a uniform elastic body; therefore, at $z = l$, the thermal stress per unit area of the paint is given by:

$$\sigma_P = Y_P \varepsilon_P = Y_P \frac{\Delta l_P}{l_P} = Y_P \gamma_P \Delta T_P(l, T) \tag{6.2.10}$$

where ε_p is the strain in the paint caused by temperature change, Y_p is the modulus of elasticity, l_p is the thermal expansion length of the paint, and γ_p is the coefficient of linear thermal expansion of the paint.

Only the light transmitted through the paint layer can strike the substrate. And t_p is the transmissivity of the paint, I_0 is the intensity of the laser incident on the paint. The intensity of the light after passing through the paint layer can be expressed by $I_0' = t_p I_0$.

At $z = 0$, the thermal stress on the surface of the substrate material can be expressed as:

$$\sigma_S = Y_S \varepsilon_S = Y_S \frac{\Delta l_S}{l_S} = Y_S \gamma_S \Delta T_S(0, T) \tag{6.2.11}$$

Considering that only the transmitted laser can heat the substrate material when using Eq. (6.2.11) to calculate the thermal stress on the substrate, I_0 needs to be replaced with t_p I_0. The condition for laser paint removal is $\sigma \geq f$, where σ is the sum of the thermal stresses at the interface between the bottom layer of the paint and the surface layer of the substrate (taking direction into account), and f is the adhesive force between the paint and the substrate.

6.2.2 Cleaning Threshold in Laser Paint Removal

The cleaning threshold refers to the laser energy density (peak power density) value that is just sufficient to remove contaminants in dry laser cleaning. When the laser energy density is below this critical value, no matter how many laser pulses are applied or how long the sample is exposed, contaminants cannot be removed from the sample surface. However, when the laser energy density exceeds this threshold, applying a certain number of laser pulses or a certain duration of laser exposure will have an effective removal impact on the contaminants.

From Eq. (6.2.6), we know that ΔT reaches its maximum at the end of the laser pulse. By using the relationship expressions given in Eqs. (6.2.10) and (6.2.11) between thermal stress and temperature rise, it becomes apparent that the maximum thermal stress varies with laser energy density. When the maximum thermal stress equals the adhesion force, the corresponding laser energy density represents the cleaning threshold for laser paint removal.

Zou et al. [5] simulated the variation of maximum thermal stress with energy density. The resulting curves for maximum thermal stress on paint and iron substrates at different energy densities are shown in Figs. 6.5 and 6.6, respectively. It is evident that as energy density increases, the maximum thermal stress grows linearly. Notably, the thermal stress exerted by the iron substrate significantly surpasses that of the paint. Therefore, it is primarily the thermal stress of the iron substrate that triggers paint detachment. From the simulation results, the maximum thermal stress of the iron substrate ranges from $(1.28–1.66) \times 10^8$ N/m^2, with corresponding to laser energy density ranging from 0.43 to 0.56 J/cm^2. Therefore, the theoretical cleaning threshold for laser paint removal is between 0.43 and 0.56 J/cm^2, with paint removal achievable when the laser energy density equals or exceeds this threshold.

The simulation also indicates that the temperature of the paint increases progressively with prolonged laser irradiation until paint removal occurs. Moreover, a higher laser energy density corresponds to a swifter rate of paint removal. Simulations at $z = 0$ show that the temperature of the paint remains well below the melting point of 500 K for the paint. Hence, no phase transition occurs in the paint, which is

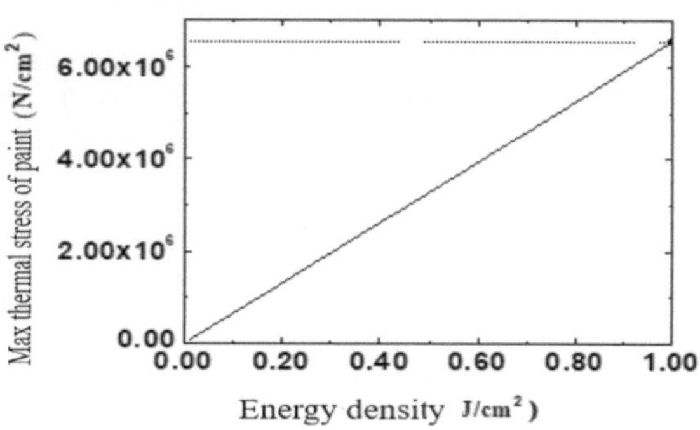

Fig. 6.5 The relationship between the maximum thermal stress of paint and the laser energy density

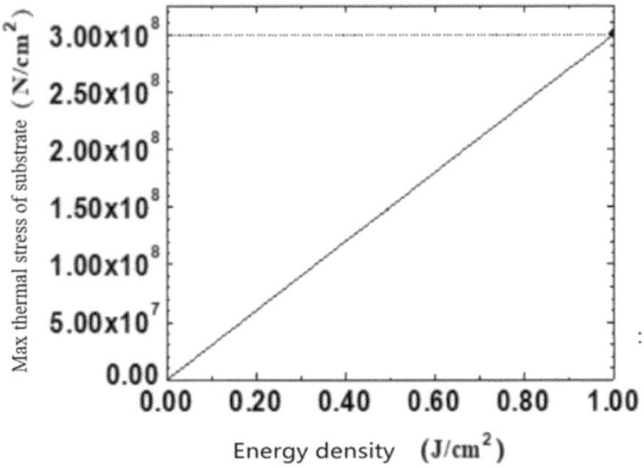

Fig. 6.6 The relationship between the maximum thermal stress of iron substrate and the laser energy density

consistent with experimental results. Once the paint separates from the substrate, the temperature of the paint no longer increases.

6.2.3 *Damage Threshold of Laser Paint Removal*

During the laser cleaning process, careful adjustment of the laser energy density (peak power density) is crucial, tailored to the specific properties of the target object. Failure to maintain the energy density within an optimal range can risk substrate damage. The damage threshold represents the laser intensity at which the substrate material begins to incur damage. Below this threshold, the substrate surface remains unaffected, irrespective of the laser pulse count or exposure duration. However, surpassing this threshold can lead to substrate damage. Unlike the paint layer, the substrate's temperature evolution follows a distinct pattern. Upon reaching the cleaning threshold, the substrate temperature escalates swiftly, peaking at the conclusion of the laser pulse. For example, in the case of an iron (Fe) substrate, oxidation initiates rapidly at 900 K before approaching its melting point of 1808 K. Discrepancies between theoretical and experimental damage thresholds arise due to alterations in substrate parameters (ρ, c, α, γ) with increasing laser energy density. Therefore, it is imperative to exercise caution in adjusting the laser energy density. Optimal selection ensures a balance between paint layer removal efficiency and substrate safety.

Zou et al. [5] plotted the temperature of a certain iron substrate sample at the end of the pulse ($t = \tau = 10$ ns) as a function of energy density. This visualization, presented in Fig. 6.7, delineates the temperature variations for different energy densities, considering adhesion forces of the paint and substrate at 1.28×10^8 N/m^2 and 1.66×10^8 N/m^2, respectively.

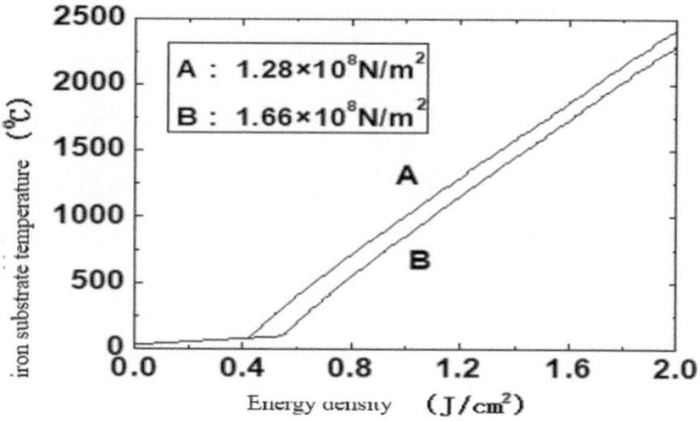

Fig. 6.7 The temperature of the iron substrate varies with energy density

As depicted in Fig. 6.7, the temperature of the iron substrate escalates with the energy density, displaying a pivotal juncture where the substrate temperature experiences a sudden surge. This critical juncture aligns with the energy density recognized as the cleaning threshold. Beyond this threshold, the paint layer can be stripped away, leading to direct exposure of the iron substrate to the laser, hence the abrupt elevation in temperature. Given that the melting point of iron is 1535 °C, analysis of the chart reveals that when the adhesion force is 1.28×10^8 N/m^2, the damage threshold stands at 1.36 J/cm^2; whereas with an adhesion force of 1.66×108 N/m^2, the damage threshold corresponds to 1.46 J/cm^2. Consequently, the damage threshold range for the sample spans from 1.36 to 1.46 J/cm^2. It is worth noting that due to the substantial temperature changes post-paint removal, the values for ρ, c, λ, α_t, and γ can no longer be considered constants, potentially resulting in deviations between the calculated damage threshold and practical outcomes. Therefore, Fig. 6.7 should be utilized cautiously as a guideline rather than a definitive metric.

6.2.4 *Three-Dimensional Cleaning Model*

If we account for the spatial distribution of the laser in the x and y directions, a three-dimensional model becomes imperative. Gao et al. [11] employed COMSOL Multiphysics to construct a three-dimensional finite element model for removing a polyurethane paint layer from the surface of aluminum alloy using nanosecond pulsed laser. Drawing upon Fourier's law of heat conduction and the principle of energy conservation, the transient three-dimensional heat conduction equation [12] under the rectangular coordinate system is formulated as follows:

$$\kappa\left(\frac{\partial^2 T_S}{\partial^2 x}+\frac{\partial^2 T_S}{\partial^2 y}+\frac{\partial^2 T_S}{\partial^2 z}\right)=\rho c\frac{\partial T_S}{\partial t} \tag{6.2.12}$$

where κ represents the thermal conductivity of the material, T_s is the instantaneous temperature of the material, t denotes the time of heat conduction, and ρ and c represent the material's density and specific heat capacity, respectively.

By numerically solving and analyzing the impact of various parameters on the temperature field and cleaning depth during laser cleaning, and corroborating these findings through experiments, it has been revealed that the scanning speed influences cleaning efficiency in the guise of overlap ratio. A slower scanning speed correlates with reduced cleaning rate, and an optimal cleaning efficiency is attained when the overlap ratio stands at 50%. With the escalation of laser energy density, the peak temperatures of both the paint layer surface and the substrate surface ascend linearly. Upon reaching a laser energy density of 25 J/cm^2, the material of the paint layer within the laser irradiation region is entirely removed, while the ablation depth of the aluminum alloy substrate measures 50 μm. Employing experimental parameters of a laser energy density at 25 J/cm^2 and an overlap ratio of 50%, the peak-to-valley height of the substrate surface groove amounts to 50.234 μm. This parameter combination yields a surface that aligns with the prerequisites of the coating process.

6.3 Laser Paint Removal Experimental Research Method

6.3.1 Laser Paint Removal Equipment

Laser paint removal equipment, like general laser cleaning machines, includes components such as the laser, beam shaping and transmission, movement, recovery, and monitoring systems. As shown in Fig. 6.8, some companies have already developed commercial laser paint removal machines. For instance, the American company HDE has designed a paint removal workstation, which is towed by a locomotive and employed for paint removal from bridge piers and vehicles. This setup utilizes an 1800 W Nd:YAG laser to deliver laser energy to the workpiece through a 0.6 mm diameter fiber, spanning distances of several tens of meters. Meanwhile, a company in New Jersey, USA, leverages a specialized pulsed CO_2 laser for ablating paint layers. Remarkably, this laser system can vaporize a 5 μm thick layer of paint with a single pulse while ensuring the substrate remains unscathed. By expanding the beam area, a 1 mm thick, 36 m^2 area of paint can be entirely stripped within one hour, with waste slag managed through a vacuum system and filters. Notably, employing multiple laser systems enables the complete removal of the exterior paint of a Boeing 737 aircraft within a mere 32 h [13].

In the laboratory, the most essential components are the laser, the beam shaping and transmission devices, and the mobile positioning device.

Integrating a cleaning head onto a 3D scanning system facilitates three-dimensional scanning along the x, y, and z axes. To ensure efficient paint removal, it is imperative to recover the dislodged paint from the substrate. This is commonly

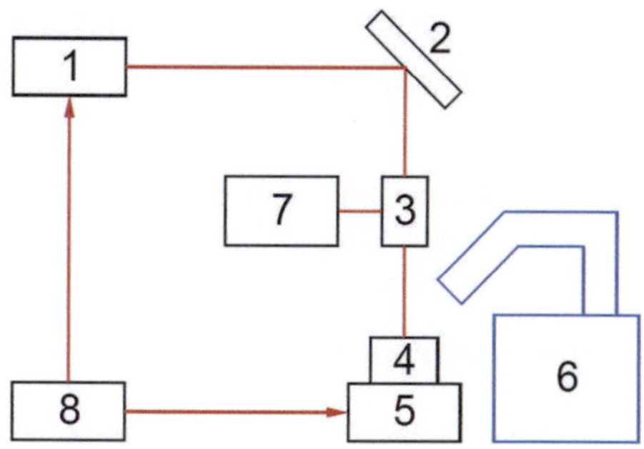

Fig. 6.8 Laser paint removal equipment schematic. 1—laser, 2—reflection mirror, 3—laser beam shaper, 4—workpiece, 5—mobile platform, 6—dust collector, 7—auxiliary gas, 8—computer

achieved using vacuum devices, which swiftly suction away the removed paint particles. Additionally, advanced laser paint removal machines often incorporate real-time monitoring devices to verify whether the paint removal process is complete. In contrast to lasers employed for removing microparticles from semiconductor circuit boards, laser paint removal predominantly relies on two types of lasers: continuous wave and pulsed CO_2 lasers, along with Q-switched Nd:YAG lasers. However, in recent years, there has been a growing adoption of fiber lasers for this purpose. For example, Tian et al. [14] conducted paint removal experiments utilizing a pulsed Nd:YAG laser, while Luo et al. [15] investigated aircraft paint removal employing a high-power continuous wave CO_2 laser. Similarly, Zhu et al. [16] conducted experiments utilizing a 1064 nm pulsed fiber laser.

6.3.2 Research Parameters in Laser Paint Removal

During laser paint removal, various factors such as scanning speeds, defocus amounts, average power, overlap in scanning paths, and paint layer thickness significantly influence the efficacy of the removal process. Moreover, understanding the impact of laser paint removal on the physical properties of the substrate is crucial for practical applications. Instruments used to characterize the laser paint removal effects include metallographic microscopes, Scanning Electron Microscopes (SEM), Stereoscopic Microscopes (SM), Confocal Microscopes (CM), spectrophotometers for observing the cleaning effects, and devices like tensile testing machines, micro-Vickers hardness testers (HV), conductivity meters for characterizing other physical properties of the material after cleaning.

Indeed, for more precise analysis, spectroscopic techniques like Laser-Induced Breakdown Spectroscopy (LIBS) can be employed. LIBS technology involves measuring the emission spectrum of the plasma generated during the paint removal process. By analyzing this spectrum, key parameters such as electron density and temperature of the plasma can be calculated before and after paint removal. This allows the study of how the intensity of spectral feature peaks corresponding to characteristic elements in the paint changes over time [17, 18].

6.3.3 Laser Cleaning of Different Types of Substrates and Paints

For the surface being cleaned, the most critical aspects are the type of substrate material and paint. Regarding the different substrate for laser cleaning, the following summary can be made.

I. **Different types of substrates**

(1) **Research on laser paint removal from aluminum alloy substrates**

The lasers employed in laser paint removal encompass various types such as CO_2 lasers, 1064 nm solid-state lasers, and fiber lasers, among others, featuring continuous wave or pulsed outputs. Notably, aluminum alloy substrates exhibit a lower damage threshold compared to steel during laser paint removal processes. For CO_2 continuous wave laser, the damage threshold typically falls within the range of 1–10 kW/cm^2. In contrast, Nd:YAG laser generally operates at a damage threshold level around 1 J/cm^2. When it comes to scanning speed, it typically ranges between 100 and 1000 mm/s, allowing for efficient and precise paint removal operations.

(2) **Research on laser paint removal from other metal substrates**

In laser paint removal processes involving metallic substrates other than aluminum alloys, a research approach akin to that utilized for aluminum alloys is commonly adopted. Nd:YAG lasers are frequently employed, predominantly operating in pulsed mode. In the case of steel substrates, the damage threshold for laser paint removal generally exceeds that of aluminum alloys. For Nd:YAG lasers, the damage threshold typically hovers around 1 J/cm^2. However, the scanning speeds utilized for laser paint removal on steel substrates are typically slower, ranging from 1 to 10 mm/s, ensuring precise and controlled removal. Silver substrates present an even lower damage threshold, with excimer lasers typically operating at a few hundred millijoules per square centimeter, making them suitable for delicate paint removal operations [19].

(3) **Research on laser paint removal from stone substrates**

In practical laser cleaning applications involving stone substrates, the presence of paint often accompanies additional organic contaminants. Pulsed lasers constitute the primary choice for such operations. However, due to the more intricate surface characteristics and composition of stone compared to metals, the explored damage thresholds in various experiments exhibit significant variation, spanning from 0.01 to 1 J/cm^2. The choice of atmosphere during laser paint removal from stonework can also influence the process outcomes. Furthermore, the presence of cracks in some stones can impact the depth of paint penetration and subsequently affect the efficiency of the paint removal process. A notable phenomenon observed during the laser removal of paint from stonework is discoloration. For instance, the utilization of infrared (IR) lasers for cleaning paint from stone sculptures often results in darkening [20].

(4) **Research on laser paint removal from glass substrates**

Glassware, distinct from other materials, generally has a lower absorption rate and higher transparency to light. When it comes to certain pollutants, laser cleaning is more effective when the laser is directed from the back of the object rather than from the front. Bilmes et al. [21] conducted a study on laser cleaning applied to standard transparent glass and frosted glass coated with black paint, employing a 1064 nm laser with a pulse width of 7 ns. Laser cleaning was performed on black paint layers with varying thicknesses by irradiating them from both the front and back. By using Laser Ablation Induced Photoacoustic Spectroscopy (LAIP), they determined the ablation threshold of the paint, finding it to be consistent regardless of paint thickness or substrate properties. Interestingly, this threshold was also unrelated to the microstructure of the substrate, whether frosted or transparent glass. To assess the relationship between the number of laser irradiations and cleaning efficiency, the researchers measured the transmittance of the glass in the ablated area and the amplitude of the acoustic signals generated during the ablation process [21]. Their experimental findings revealed that laser cleaning achieved greater effectiveness when the glass sample was irradiated from the back. To elucidate this phenomenon, they proposed a phenomenological model that could interpret the cleaning efficiency obtained in the experiments with backside irradiation. This model's analysis of the processes occurring with backside irradiation was different from those with frontside incidence. This model suggested that backside irradiation induced the formation of an overheated layer, which bends upward due to the impossibility of lateral expansion. The high pressure accumulated in the heated area caused the paint to detach from the substrate and be ejected. Additionally, the model predicted the existence of a critical thickness beyond which backside cleaning became less effective. This critical thickness parameter provided valuable insights into optimizing laser cleaning processes on glass substrates.

(4) **Research on laser paint removal from resin and other organic composite material substrates**

When confronted with substrate materials composed of organic composites such as resin or plastic, akin in properties to the coatings themselves, the efficacy of laser cleaning is significantly challenged. This challenge arises from the poor thermal conductivity inherent in composite materials, necessitating meticulous regulation of the laser's power, pulse width, and spot size to mitigate adverse effects. Notably, the thermal rise induced by laser irradiation poses a notable risk of thermal decomposition, precipitating material damage. Jia et al. [10] embarked on an investigative endeavor targeting the removal of epoxy protective paint layers from the surface of aramid fiber-reinforced resin-based composite materials. Employing infrared pulsed lasers as their tool of inquiry, the researchers dissected the myriad mechanisms underpinning laser removal processes, guided by both theoretical conjecture and experimental exploration. Through theoretical calculations and experimental tests, the process parameters for the laser to remove the coating were obtained. Combined with ethanol wiping, the effective removal of the epoxy protective paint layer on the surface of the resin-based composite material parts was achieved. The results indicate that the aramid fiber-reinforced resin-based composite material was undamaged after laser treatment and can meet the requirements of industrial applications.

II. **Paint layer thickness, type, and color**

The absorption spectral characteristics of different types of paint are distinct. Zuo and Jin [22] undertook an exploration into the absorption spectra exhibited by various types of paint. Their investigations unveiled distinctive absorption spectral features spanning the wavelength range of 2.5–15 μm, characterized by pronounced absorption bands, particularly prevalent within the 7 to 9 μm region. Intriguingly, analyses further discerned those paints of identical types, albeit differing in coloration, showcased strikingly similar absorption spectral profiles, thereby dissociating absorption characteristics from pigment hue. Subsequent investigations by Xu et al. [23] delved into the infrared spectral attributes of diverse paint varieties, yielding invaluable insights encapsulated within Table 6.1.

(1) **Influence of paint thickness and type**

In a study conducted by Zhang et al. [6], an examination of paint removal from aluminum alloy substrates was undertaken, employing a quasi-continuous wave Nd:YAG laser operating at a repetition frequency of 20 kHz and a pulse duration of 140 ns. The investigation employed three distinct paint types: an 80 μm thick acrylic ester top layer, a 50 μm thick middle layer composed of polyurethane paint, and an undercoat measuring 50 μm in thickness comprising epoxy ester. Employing the ANSYS finite element analysis method, the research team simulated the temperature distribution and thermal penetration depth in proximity to the laser spot on the paint layer. The experiments found that when the paint layer was thick, the high thermal conductivity of the aluminum alloy substrate had little effect on the temperature rise of the paint, and the simulation results were in good agreement with the

Table 6.1 Infrared characteristic peaks of different automotive paints (unit: cm^{-1})

Type of paint	Mainly film forming agent	Characteristic absorption peak
Alkyd resin paint	Alkyd resin	3460, 2920, 2860, 1730, 1600, 1580, 1450, 1390, 1260, 1131, 1071
Acrylic resin paint	Methacrylate and acrylate copolymer	1727, 1460, 1382, 1250, 1190–1150, 840, 720
Amino resin paint	Amino resin alkyd resin	3300, 1731, 1640, 1546, 1120, 1080–1064, 813
Polyurethane paint	Polyurethane	3260, 1725, 1690, 1540, 1220, 1070
Nitro-lacquer	Nitrocellulose alkyd resin	1730, 1640, 1280, 1150–1110, 834
Vinyl perchloride paint	Perchloroethylene resin alkyd resin	2930, 2870, 1720, 1285, 1130, 1074, 690
Epoxy resin paint	Epoxy resin	3450, 1510, 1235, 1175, 915, 826

experimental results. However, for thinner paint layers, the aluminum alloy substrate absorbed a portion of the laser energy, thereby impacting the temperature rise within the paint layer. Consequently, a notable disparity between the simulation and experimental results emerged under such conditions. Thus, in instances where the paint thickness falls below the maximum thermal penetration depth, a more comprehensive modeling approach incorporating the thermal conduction effects of the substrate material becomes imperative.

(2) **The influence of paint color**

Gomes et al. [24] undertook a comparative analysis to assess the efficacy of mechanical and laser cleaning techniques for paint removal on rock substrates amidst an SO_2 atmosphere, focusing on paints of diverse hues. The disparate absorption rates exhibited by these paints were contingent upon their constituent compositions, chiefly binders, and pigments. Notably, paints in four distinct colors (red, blue, black, and silver) displayed variable absorption properties. Of these, black, red, and blue paints formulated with alkyd resins and polyester polymers. Black paint exhibited commendable absorptive qualities across the entire spectral range, while red and blue paints manifested absorption bands falling within the 500–700 nm range, bordering on the near-ultraviolet spectrum. Conversely, silver paint, comprising polyethylene-type polymers, displayed a lower absorption rate throughout the spectrum. Experimental results showed the red, black, and blue paints can be cleaned effective. However, the removal of silver paint yielded suboptimal outcomes, with an opaque residual film persisting on the granite surface post-cleaning. Additionally, silver paint particles were sparsely dispersed across the surface.

III. **Paint removal environment**

In the realm of paint removal, the prevailing milieu typically comprises air and paint molecules, constituting the micro-environment where the laser interacts with the cleaning object. However, there are some complex cases, such as the laser cleaning of on-site stone paint. Goms et al. [24] compared the effects of mechanical and laser cleaning methods on different types of paint on a rock substrate in an environment rich in SO_2. The experimental regimen centered on granite specimens subjected to prolonged exposure within an artificially enriched SO_2 atmosphere, mimicking the aging process encountered in real-world scenarios. The experiments compared the cleaning effects of four types of graffiti paint sprays. Distinct removal techniques were employed: (1) a mechanical approach utilizing fine abrasive with varying compositions, which included three permutations, (a) a composite of air, water, and silicon dioxide as the abrasive medium, (b) silicon dioxide combined with air, and (c) calcium carbonate in conjunction with air; (2) laser-based cleaning utilizing an $Nd{:}YVO_4$ laser operating at a wavelength of 355 nm and a pulse duration of 25 ns. The painted granite surfaces, featuring an array of hues including red, blue, black, and silver, were prepared for examination. Subsequently, these surfaces underwent a SO_2-rich environment for a duration of two months.

The following cleaning experiments were conducted by employing silicon dioxide with aluminum silicate abrasives, silicon dioxide with calcium carbonate abrasives, as well as a nanosecond Nd:YVO_4 laser operating at 355 nm. Notably, the presence of SO_2 within the ambient milieu posed a notable challenge, necessitating extended cleaning durations for aged specimens. Unlike their unaged samples, the delineation between distinct paint layers on aged surfaces appeared less pronounced, rendering the conventional layer-by-layer cleaning approach less tenable. Therefore, aged surfaces exhibited more conspicuous global color alterations and a higher percentage of residue. Additionally, the aged cleaning samples manifested enhanced water resistance relative to the original granite substrate, albeit with a reduced quantity of paint. The experimental results indicated the profound impact of SO_2 exposure on both mechanical and laser-based cleaning methods. Notably, samples subjected to SO_2-induced aging presented heightened recalcitrance to cleaning. For mechanical methods, the cleaning performance was unaffected by the paint composition. During mechanical cleaning with silicon dioxide combined with air as a micro-abrasive, the stone surface experienced significant morphological alterations. Conversely, employing calcium carbonate in conjunction with air as micro-abrasives, the samples underwent less deformation, but a higher percentage of paint residue was left. Noteworthy, the micro-abrasive formulation comprising air, water, and silicon dioxide yielded the least post-cleaning color deviation and occupied an intermediary position in terms of morphological alterations. In practical applications, this micro-abrasive amalgam emerged as the optimal choice for both unaged and aged paint, minimizing damage to the granite's morphology. Furthermore, the Nd:YVO_4 laser paint removal leaves a mix color of blue and black paint on the surface, and a more visible transparent film can be detected by the naked eye on the surface where silver paint has been cleaned. Conversely, mechanical paint removal elicited more conspicuous overall color change on the surface of the sample. Especially in the case of silver paint, it was found that the residual percentage and reflectivity change were the highest [24].

6.3.4 The Impact of Laser Cleaning on the Performance of the Base Material

Selecting appropriate parameters for laser cleaning can result in the formation of pits on the cleaned surface, which can be beneficial for subsequent painting processes. The relevant studies will be presented in Sect. 6.4 on damage threshold. Some teams have conducted specialized research [25] by using a continuous-wave fiber laser with a maximum power of 2 kW and a wavelength of 1070 nm as the light source for laser cleaning. The size of the laser spot is 5.0 mm, and it was used to clean the paint off an aluminum alloy substrate. The study measured the width and depth of the laser-ablated pits and examined the paint removal efficiency of laser de-coating. Morphological studies have shown that a good cleaning effect can be achieved on the aluminum alloy surface with appropriate parameters. For example, laser cleaning with an intensity of 11.9 W/cm^2 results in a surface with better corrosion resistance and roughness.

Wu [26] investigated the impact of infrared continuous wave laser cleaning on polyurethane black paint coatings within a low-temperature vacuum setting. The study concluded that the infrared laser predominantly influenced the polyurethane black paint thermally. On a macroscopic scale, damage manifested as a decrease in coating color, alterations in color, and ablation. Microscopic examination revealed the appearance of microcracks and particle agglomeration. Laser irradiation and subsequent temperature elevation altered the coating's constituents and chemical composition. Specifically, the urethane groups within the binder polyurethane, including C–N bonds and O–C=O double bonds, were susceptible to breakdown due to laser irradiation and heating, resulting in material degradation and carbonization. Ablation processes generated thermal decomposition gases and volatiles.

Laser cleaning can also enhance the performance of the substrate material. Abeens [27] employed a Nd:YAG Q-switched laser with varied pulse energies (200 mJ, 300 mJ, 400 mJ) to investigate alterations in the mechanical properties of AA-7075-T651 aluminum alloy (commonly utilized in aircraft) after laser paint removal. Upon paint removal using a pulse energy of 400 mJ, the surface hardness of the aluminum alloy reached a peak of 236 HV, accompanied by a deformation layer approximately 500 μm thick and a maximum compressive residual stress of 317 MPa. The wear rate was measured at 1.18×10^{-6} g/m. High-resolution transmission electron microscopy (HRTEM) was utilized to analyze the material's microstructure, revealing that the enhancement in material performance primarily stemmed from the grain refinement and dislocation strengthening effects of the LSP technique.

Hu et al. [28] compared laser paint removal and solvent paint removal methods on aircraft paint-skinned components, assessing parameters such as oxide film layer thickness, film layer integrity, mechanical properties, electrical conductivity, and paint layer adhesion strength after paint removal and repainting. Laser paint removal was found to preserve the mechanical properties of the substrate material. Li et al. [29] conducted experiments revealing that after three times of laser scanning for paint removal, the original surface of steel was exposed, exhibiting an average surface roughness of approximately 2 μm and an adhesion strength of 20 MPa, significantly surpassing the standard requirement of 3 MPa for ship hulls.

6.4 Laser Paint Removal Cleaning Threshold and Damage Threshold

The cleaning threshold is the critical value at which the paint layer just can be removed, while the damage threshold is the critical point when the substrate is damaged. The thresholds here refer to the laser power density when other laser parameters are determined. Given the multitude of paint and substrate types with distinct chemical compositions, the laser cleaning process involves variations in parameters such as average power, peak power, pulse energy, beam size, and focal depth. Therefore, this causes differences in both the laser cleaning threshold and the damage threshold. On one hand, it is essential to calculate an approximate numerical range through theoretical models; on the other hand, experimental validation is required based on sample testing with a laser cleaning machine. The following will illustrate with our experiment as an example [30].

The substrate material selected for cleaning is 45# steel, with dimensions of 200 mm × 50 mm × 1 mm. The paint used is red alkyd paint, and the samples are prepared in a professional spray-painting workshop. Generally, to ensure the protective effect of the paint on the substrate, the adhesive strength between the paint layer and the substrate is required to be 10–100 MPa. The commonly used crosshatch method is employed to test the adhesion of the paint layer. Within a 50 mm × 100 mm area on the painted surface, small squares of 15 mm × 15 mm are marked using a cutting knife. Subsequently, tape is applied to the painted surface with the marked squares, and a pressure of 5 N is exerted on the tape. Holding one end of the tape, it is quickly pulled in a direction perpendicular to the coating surface. The paint coating is then inspected to determine if it adheres to the tape and peels off. If there is no paint peeling, the adhesion is considered satisfactory. If peeling occurs within the coating, it is still acceptable. However, if peeling occurs between the metal and the coating, it is considered unqualified.

As illustrated in Fig. 6.9, the Nd:YAG acousto-optic Q-switched laser is employed as the cleaning light, with a repetition frequency ranging from 3 to 6 kHz. The pulse width spans from 150 to 300 ns, and the maximum average power is 100 W. The focused and shaped spot diameter is approximately 0.5 mm, and the scanning speed of the galvanometer (the moving speed of the sample) is about 0.55 mm/s. With these parameters, the laser pulses can be sequentially irradiated onto the sample surface without repeating the same spot, thus ensuring the accuracy of the cleaning threshold and damage threshold. Experimentally, the paint layer on the sample surface begins to peel off with the growing peak power density. To effectively determine the cleaning and damage threshold, the onset of peeling and a tendency for detachment on the sample surface indicate reaching the cleaning threshold. Complete cleaning is achieved when the paint layer has entirely splattered and detached from the metal surface without significant indentations. The presence of noticeable mesh-patterned indentations on the metal surface indicates reaching the damage threshold. The post-cleaning substrate surface is observed through metallographic microscopy.

In the experiment, the repetition frequency of the laser source is 3 kHz, and the spot area remained constant at 0.19 cm^2. By varying the laser average power, corresponding changes in pulse width can be observed. During the cleaning process, the evolution of the cleaning is monitored; and after cleaning, for samples where the paint is completely stripped, the substrate surface is observed with a metallographic microscopy. The experimental data are presented in Table 6.2.

Experimental results reveal that when the laser's average power is below 11.5 W, there is no observable alteration in the paint layer. However, at an average power of 11.5 W, the paint layer in the scanned region begins to peel off without fracturing after

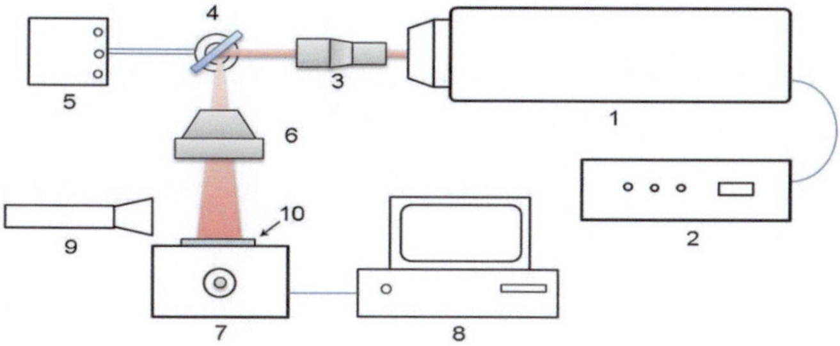

Fig. 6.9 Structure of laser paint removal experimental system. 1—laser head; 2—laser power supply; 3—plastic beam expanding mirror; 4—scanning galvanometer; 5—galvanometer driver; 6—field lens; 7—one-dimensional sliding platform; 8—control system of sliding platform; 9—negative pressure recovery device for pollutants; 10—sample

laser scanning, as depicted in Fig. 6.10a. After tearing the patent leather manually, the metallic substrate was observed to be smooth. Therefore, this laser power is designated as the cleaning threshold, with the peak cleaning threshold power per unit area expressed by:

$$I = \frac{4P}{f \cdot \pi D^2 \cdot \tau} = \frac{4 \times 11.5}{3000 \times 3.14 \times (0.5 \times 10^{-1})^2 \times 450 \times 10^{-9}}$$
$$= 4.48 \times 10^6 \text{ W/cm}^2$$

where f is the repetition frequency, D denotes the beam spot diameter, P is the average power, and τ represents the pulse width.

Figure 6.10b shows the cleaning results at an average power of 13.8 W, and the paint layer begins to detach, even resulting in the fragmentation. One edge of the paint layer has detached from the substrate, while the other remains connected. Besides, the surface of the metallic substrate remains intact. When the average power is increased to 13.8–29.6 W, the paint layer has completely fragmented and begun to eject from the metallic substrate. The metallic substrate retains its smooth surface, and microscopic examination reveals no visible damage. As the power continues to increase, a mesh pattern appears on the metal substrate, exhibiting noticeable roughness. Figure 6.10c shows the substrate image at an average power of 35.6 W, revealing visible damage to the metal matrix after cleaning. It is believed that the paint removal damage threshold has been reached at this point. And the calculated peak power density of the laser is 34.43×10^6 W/cm^2. Through metallographic microscopy, the microstructure of the metal surface after cleaning can be observed more clearly. Figure 6.11a depicts the photo taken after cleaning at a laser peak power density of 4.48×10^6 W/cm^2, demonstrating excellent cleaning results without damage to the metal substrate surface. In Fig. 6.11b, captured after cleaning at a laser peak power density of 6.04×10^6 W/cm^2, the paint layer on the metal surface is completely removed, and there is

Table 6.2 Experimental data of the sample

Number	Average power (W)	Pulse width (ns)	Peak power density (MW/cm^2)	Experimental phenomenon
1	11.5	450	4.48	One side of the paint layer is detached from the substrate, while the other side remains adhered to it
2	13.8	400	6.04	The paint layer begins to fracture, dispersing into a powder and spraying away from the substrate surface, while the substrate remains intact
3	18.9	350	9.48	The paint layer has been completely removed, and the substrate remains undamaged
4	24.6	300	16.68	The paint layer has been completely removed, and the substrate remains undamaged
5	29.6	250	24.76	The paint layer has been completely removed, and the substrate remains undamaged
6	35.6	200	34.43	The paint layer has been entirely removed, but the substrate exhibits slight mesh-patterned markings, indicating minor 150damage to the substrate

(continued)

Table 6.2 (continued)

Number	Average power (W)	Pulse width (ns)	Peak power density (MW/cm^2)	Experimental phenomenon
7	44.0	200	38.25	The mesh-patterned markings on the substrate have become more pronounced, indicating an increased depth of damage
8	49.9	160	54.22	The substrate shows noticeable unevenness, indicating more severe damage

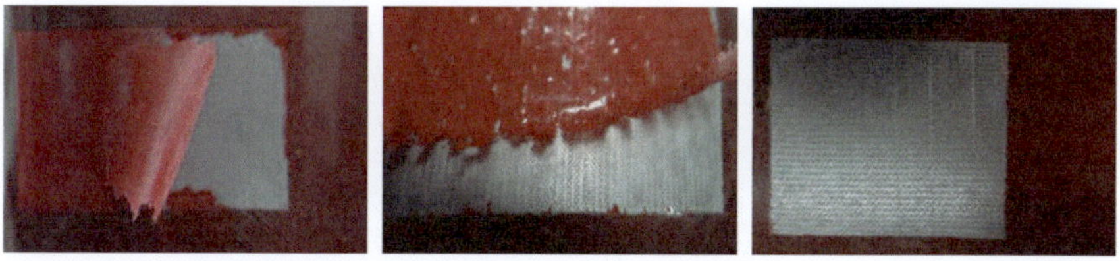

Fig. 6.10 Samples after cleaning at different peak power densities

no damage to the substrate either. Figure 6.11c illustrates the morphology at a laser peak power density of 34.43×10^6 W/cm^2. Although the paint layer is completely removed, the metal surface has been destroyed.

I. The influence of different substrates on the laser paint removal cleaning threshold and damage threshold

Du [30] applied the same paint on iron and aluminum alloy substrates through identical procedures. The parameters of the samples are listed in Table 6.3.

Fig. 6.11 The microstructure of the samples after cleaning at different peak laser power densities

Table 6.3 Relevant physical parameters of the samples

	Sample I		Sample II		Sample III	
	Substrate	Paint	Substrate	Paint	Substrate	Paint
Material	45# Steel	Alkyd	45# Steel	Alkyd	Aluminum alloy	Alkyd
Melting point (°C)	1350		1350		660	
Specific heat capacity c_p (J/(kg °C))	0.49	2.51	0.49	2.51	0.96	2.51
Density ρ $(\times 10^3\ \text{kg/m}^3)$	7.85	1.30	7.85	1.30	2.71	1.30
Thermal conductivity λ (W/(m °C))	51.9	0.3	51.9	0.3	218	0.3
Thermal diffusivity α_t $(\times 10^{-8}\ \text{m}^2/\text{s})$	1450	9.19	1.45	9.19	8.8	9.19
Linear expansion coefficient γ $(\times 10^{-6}/\text{K})$	11.7	1.00	11.7	1.00	24	1.00
Absorptivity A	0.360	0.470	0.358	0.530	0.642	0.500
Reflectance R	0.640	0.410	0.642	0.410	0.358	0.410
Transmittance T	–	0.120	–	0.060	–	0.090
Elastic modulus Y $(\times 10^{10}\ \text{N/m}^2)$	20.5	1.0	20.5	1.0	7.2	1.0

Table 6.4 presents the theoretical and experimental cleaning thresholds for three types of samples (Samples I and II on steel substrates, and Sample III on aluminum alloy substrate). It is evident that the experimental values of the cleaning threshold are consistent with the theoretical values in trend, with the theoretical values slightly lower. One of the reasons for this difference lies in the assumption within the theoretical model that the laser spot is flat-topped, whereas in reality, it follows a Gaussian distribution.

For the steel substrate, the cleaning threshold of Sample II is higher than that of Sample I. This is primarily due to the thicker paint layer and lower laser transmittance of Sample II. Consequently, a higher power density of incident laser is required to make the thermal stress of the substrate exceed the adhesive force of the paint layer.

The damage of the substrate is caused by the instantaneous temperature of the substrate surface exceeding the melting point of the metal. Table 6.5 shows the damage thresholds for the three types of samples. The theoretical calculation value is considered by taking the maximum paint-substrate adhesion force of 1.66 × 108 N/m^2.

Table 6.4 Theoretical and experimental cleaning thresholds of samples (unit: W/m^2)

	Sample I	Sample II	Sample III
Theoretical cleaning threshold	3.56×10^6	4.80×10^6	4.96×10^6
Experimental cleaning threshold	4.48×10^6	6.04×10^6	6.04×10^6

Table 6.5 Theoretical and experimental damage thresholds of samples (unit: W/m^2)

	Sample I	Sample II	Sample III
Theoretical damage threshold	5.42×10^7	8.12×10^7	2.20×10^7
Experimental damage threshold	3.44×10^7	5.42×10^7	1.67×10^7

It can be observed that the experimental damage thresholds are consistent with the theoretical ones in trend. And the theoretical values are higher than the experimental values. This is because the actual cleaning mechanism involves multiple factors such as ablation, vibration, and shock waves. Aluminum alloy has a lower melting point, making it more susceptible to damage compared to 45# steel substrate. Regarding research on aluminum alloy substrates, Chen et al. [2] conducted paint removal experiments using CO_2 continuous laser with a wavelength of 10.6 μm. The paint was a common milky white coating with a thickness of 50 μm. For investigating the effect of laser power density on cleaning efficiency, the scanning speed was kept constant at 3.0 m/min while adjusting the laser power density from 0 to 3.00 kW/m^2. It was found that the paint layer underwent a progression from no significant change to shallow traces of removal on the surface, then to increasingly deepening traces with gradual thinning of the paint layer until complete removal. Further increasing the laser power density would result in slight bending deformation of the substrate. Thus, three power density thresholds for laser paint removal were determined. The initial cleaning threshold was 0.15 kW/cm^2, the complete cleaning threshold was 1.78 kW/cm^2, and the substrate damage threshold was 2.80 kW/cm^2.

In the realm of iron-based substrate research, Shudong Shi et al. [31] from Nankai University conducted experimental and mechanistic studies on the cleaning of paint layers on steel substrates. They utilized a Q-switched Nd:YAG continuous wave laser with a wavelength of 1064 nm, a beam spot diameter of approximately 0.6 mm, and a tunable repetition frequency ranging from 0.5 to 50 kHz. For the steel substrate coated with a 50 μm alkyd paint layer, the cleaning threshold power density stood at 5.31×10^6 W/cm^2 when the average power exceeded 20 W and the overlap rate exceeded 80%. And excessive power density posed risks of substrate damage. Additionally, Shi et al. [31] delved into the mechanisms underlying the cleaning process. Both simulated and experimental results indicated that pulsed laser ablation of paint layers on steel substrates operated primarily under the vibrational effect at the effective cleaning threshold, whereas both vibrational and ablative effects occurred simultaneously when substrate damage was evident.

In the study of stainless-steel substrates, Liu [32] employed finite element method to simulate the temperature field on the surface of stainless-steel samples under the moving pulsed laser. A Nd:YAG laser was utilized, with a pulse width of 7 ns and a wavelength of 1064 nm, exhibiting a Gaussian spatial distribution of laser intensity. Stainless-steel plates coated with red paint of 0.05 mm thickness were selected as

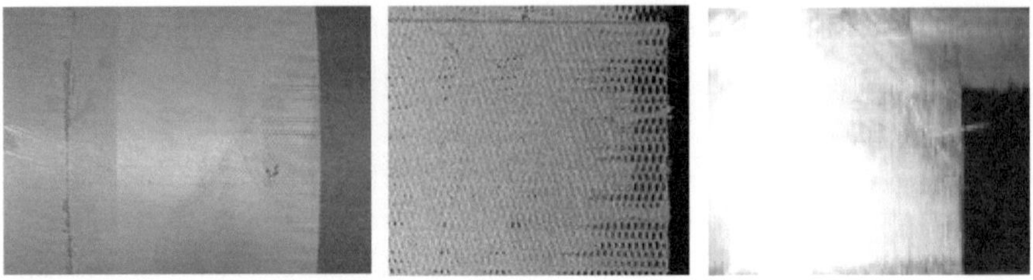

Fig. 6.12 Cleaning of paints with different colors on the same substrate (aluminum alloy)

samples. The paint began to melt at a laser energy density of 0.8 J/cm^2, while the substrate started to incur damage at 1.2 J/cm^2.

II. **Influence of paint color and type on cleaning threshold**

When examining the threshold for removing paint from metal surfaces under various parameters, it is essential to consider not only the substrate type but also the characteristics of the paint itself. As depicted in Fig. 6.12, the paints with various colors (red, green, yellow) and equal thickness on the aluminum alloy substrate are cleaned with Nd:YAG Q-switched laser. The results show that the red and yellow paints are effectively removed, while residue remained from the green paint. This indicates that the cleaning thresholds for paints with different colors are distinct, even on the same substrate.

Guo et al. [33] conducted dry laser paint removal experiments on low carbon steel using a high-power pulsed TEA CO_2 laser with a wavelength of 10.6 μm, a maximum repetition frequency of 20 Hz, and a pulse energy of 5.5 J. They investigated the laser cleaning power density threshold for paints with different compositions and colors. The thickness of the paint layers in the experiments was controlled to be around 0.1 mm. The three types of paints used were red alkyd paint, red metal spray paint, and yellow metal spray paint. The experimental results showed that at a repetition frequency of 1 Hz, the complete cleaning thresholds for all three paints ranged from 9.29 to 11.07 J/cm^2. Among them, the lowest threshold was for red metal spray paint, followed by red alkyd paint, and the highest was for yellow metal spray paint. It can be concluded that for paints with the same composition, lighter-colored paint layers are more prone to reflect the incident laser beam, while darker-colored paint layers are more likely to absorb the laser beam. Consequently, the thresholds for complete cleaning and damage are smaller for darker-colored paint layers, and larger for lighter-colored paint layers. When the paints have identical colors and different compositions, the removal threshold for metal spray paint is slightly lower than that for brushed paint. This is because the spray paint has a smoother surface, whereas the brushed samples result in less uniform surface thickness. Therefore, there are differences in adsorption between the two types of paints with different compositions, leading to variations in cleaning effectiveness even with the same laser power density.

Xian et al. [34] investigated the influence of Nd:YAG Q-switched laser on the cleaning threshold for removing different types of paint on stainless steel surfaces. They utilized commonly used Santana automatic spray paints, including red and black paints. The laser parameters were set as follows: pulse width of 10 ns, repetition frequency of 1 Hz, energy density ranging from 0.05 to 1 J/cm^2, output beam radius of 4 mm, and paint coating thickness of approximately 1.0 mm. The removal threshold was defined as the minimum pulse energy density of the laser when the ratio of the residual paint area within the beam range to the beam area was less than 0.01. The experiments revealed that removal thresholds for Santana red paint and black paint were 0.47 J/cm^2 and 0.63 J/cm^2, respectively.

In laser cleaning, the protection of the substrate is also crucial, especially for substrates like stone artifacts. Therefore, investigating the damage threshold is also a key focus in research. Qi et al. [35] utilized a laser with a wavelength of 1064 nm and a pulse width of 10 ns to conduct laser cleaning on the surface rocks of the Yungang Grottoes with various organic pollutants. Concerning paint removal by laser, surface paint samples began to peel off when the laser energy reached the cleaning threshold, manifesting as the paint layer gradually detaching layer by layer under laser irradiation.

Research indicates that the surface structure of stone materials is often more intricate than that of metals, hence it is important to discuss the influence of substrate structure on paint removal effectiveness. Granite is a commonly used substrate material. Rivas et al. [36] employed a Nd:YVO_4 laser with a wavelength of 355 nm to study the laser cleaning effects on two types of granite and four different colors of graffiti paint. The colors of the four paints were cyan blue, devil red, graphite black, and silver chrome. The optical parameters included a pulse duration of 25 ns, wavelength of 355 nm, beam diameter of approximately 2.2 mm, repetition frequency of 100 kHz, and pulse energy of about 0.1 mJ. Characterization was carried out using X-ray fluorescence, X-ray diffraction, Fourier-transform infrared spectroscopy (FTIR), and scanning electron microscopy. The results revealed the surface cleaning effectiveness of silver paint is different with blue, red, and black paints. The red, blue, and black paints could be effectively removed, while the silver paint remained in the form of a translucent film. This is primarily attributed to the chemical composition of silver paint. The composition of silver paint contains aluminum-rich components, which has a high reflectivity and significantly reduces the energy available for laser ablation, thereby affecting laser cleaning effectiveness. There was not much difference in laser cleaning effectiveness for granite with different compositions. However, the cleaning effectiveness was truly influenced by the types and distribution of cracks in the granite. Polarized light microscopy observation and color measurements indicated that the intensity and distribution of cracks affected the depth of paint penetration, resulting in hue changes within the stone, ultimately impacting the cleaning efficiency of the two types of granite. These color changes were inconsistent with paint traces on the surface. Therefore, because the color is largely influenced by the amount of residual paint beneath the surface, it is necessary to be cautious in using colorimetric change measurements as an evaluation method for laser paint removal effectiveness, especially for rocks with cracks and pores like granite.

6.5 The Influence of Laser Cleaning Parameters on Cleaning Effectiveness

In laser cleaning, the essence lies in the delicate balance between reaching the cleaning threshold and avoiding the damage threshold. However, a measured degree of substrate impairment is permissible, strategically utilized to attain a desired level of surface roughness. The judicious selection of laser parameters is imperative for the expeditious attainment of thorough cleaning, as variations in these parameters exert significant influence on the cleaning outcome. Herein, we delve into the impact of laser power density and energy density, beam scanning speed, overlap amount, and other cleaning parameters on the efficacy of laser cleaning.

I. **Power density and energy density**

Laser power density is an important parameter in the study of laser paint removal. For the aluminum alloy substrate, Zhu et al. [16] used a 1064 nm pulsed fiber laser to remove thermoplastic acrylic aerosol paint on the aluminum alloy plate. Operating with a pulse width of 10 ns and a repetition frequency of 40 kHz, the laser irradiated a fixed light spot onto the cleaning material measuring approximately 5×10^{-5} m in diameter. Remarkably, adjusting the laser power from 9 to 13 W, the researchers observed that the change of power density exerted a negligible influence on the efficacy of the cleaning process. Chen et al. [37] employed a Q-switched Nd:YAG pulsed laser with electro-optical modulation to investigate the effect of laser energy density on the efficacy of paint removal. The laser operated at a wavelength of 1064 nm, with a pulse width of 7 ns, a repetition frequency of 1 Hz, an output energy range of 3 to 40 mJ, and a beam spot radius of 200 μm. By increasing the laser energy from 6.7 to 30 mJ, they observed that the electron density of the plasma gradually increased from 2.23×10^{18} cm^{-3} to 3.228×10^{18} cm^{-3}, exhibiting an approximate linear relationship with energy. This indicates that higher laser energy during the paint-stripping process leads to more chemical bond ruptures, resulting in increased electron density of the plasma. Concurrently, the electron temperature of the plasma also increased with the rise in incident energy, from 1.925×10^3 to 2.059×10^3 K. As the incident laser energy increased, the area of paint ablation gradually expanded, consistent with the variations in plasma electron density and temperature. Finally, elemental analysis unveiled a conspicuous reduction in carbon (C) content with the escalation of incident laser energy, indicative of intensified combustion processes precipitated by the augmented energy input. Li et al. [38] conducted meticulous experiments focused on paint removal, employing a Q-switched laser with a pulse width of 15 ns, a wavelength of 1064 nm, a beam radius of 3.8 mm, and an energy instability of less than 3%. The paint samples utilized in the study were comprised of a blend of epoxy-polyester material, adhered to an aluminum substrate, with a paint thickness of 75 μm. The experimental investigation into the impact of different laser energy densities on paint removal efficacy is illustrated in Fig. 6.13.

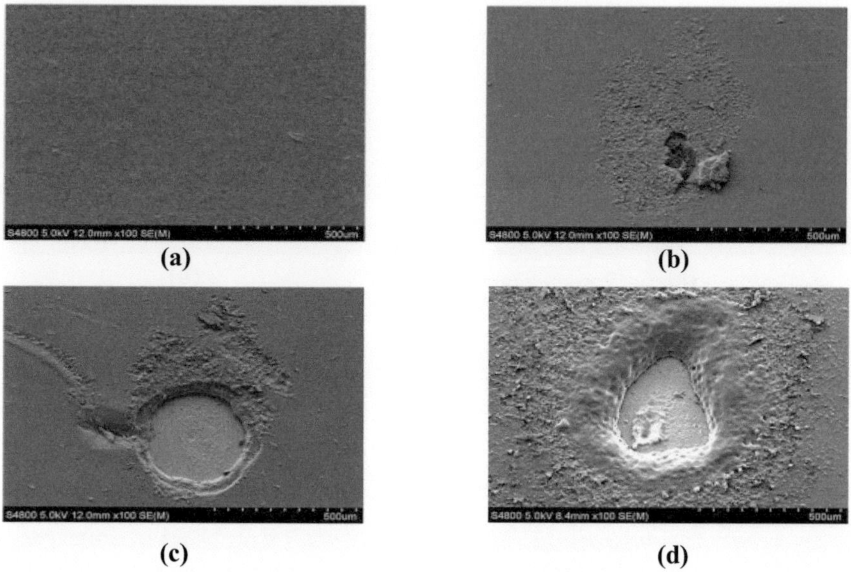

Fig. 6.13 Influence of different laser energy densities on paint removal. **a** The surface of the paint; **b** partial cleaning (0.53 J/cm^2); **c** complete cleaning (0.84 J/cm^2); **d** the damage of the substrate (1.58 J/cm^2)

As delineated in Fig. 6.13b, under irradiation with low laser pulse energy density, the paint is only partially removed, and the removed areas exhibit significant irregularity. The paint exhibits only partial removal, with the eliminated areas displaying pronounced irregularities. The complete paint removal becomes achievable with an increment in energy density, leaving the metal substrate devoid of discernible damage traces, as portrayed in Fig. 6.13c. Nevertheless, at elevated laser energy densities, the area of paint removal widens, accompanied by severe ablation, resulting in damage to the substrate itself, as evidenced in Fig. 6.13d. Analysis of the morphological characteristics of ablated paint allows for the examination of the process and mechanism of laser ablation. The microstructure of ablated paint is depicted in Fig. 6.14.

Figure 6.14a illustrates the case of laser irradiation at an energy density of 0.53 J/cm^2. In the laser-affected area, one can observe ablative cooling traces on the paint surface and significant paint detachment. This suggests that the paint detachment can be attributed to fracture induced by thermal stress. Figure 6.14b displays the microscopic morphology of completely peeled paint under laser pulses with an energy density of 0.84 J/cm^2. It can be observed that there are no apparent signs of ablation on the edges of the paint fracture and the substrate. The ablative removal of paint is not evident, and the heat stress between the paint and the substrate leads to the detachment of the paint from the surface of the substrate. The edge morphology of laser irradiation with high energy density (1.58 J/cm^2) is shown in Fig. 6.14c and d. According to Fig. 6.14, it can be observed that there are evident ablative pits at the

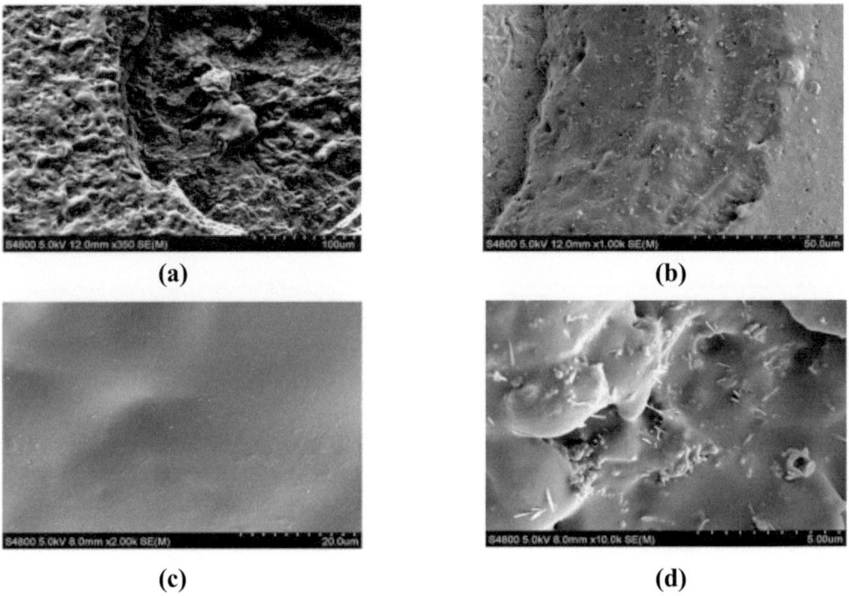

Fig. 6.14 Microstructure of ablative paint under different laser energy densities. **a** partial cleaning (0.53 J/cm^2); **b** complete cleaning (0.84 J/cm^2); **c** the damage on the edges of substrate (1.58 J/cm^2); **d** the damage of substrate (1.58 J/cm^2)

center of laser beam action, evidently caused by solidification cooling. Surrounding these ablation pits are myriad micrometer/nanometer particles [39].

In addition to investigating the threshold for laser paint removal on stainless steel, Shi et al. [31] also explored the impact of other power densities on the cleaning efficacy. As depicted in Fig. 6.15, it is evident that damage to the substrate becomes apparent with the increasing peak power density.

For substrates composed of rock-like materials, Qi et al. [35] utilized a laser with a wavelength of 1064 nm and a pulse width of 10 ns to clean the surfaces of rock at the Yungang Grottoes with different organic pollutants. Figure 6.16 illustrates the paint removal effects under different laser conditions, with the required number of pulses ranging from 5 to 6, a repetition frequency of 1 Hz, and a spot area of approximately

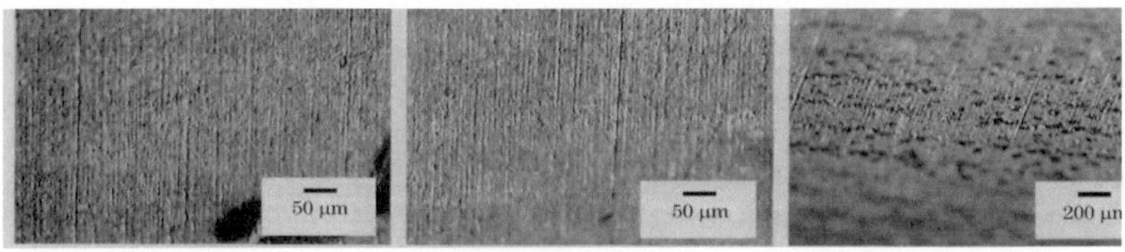

Fig. 6.15 Microscopic surface of the sample under different laser peak power densities. **a** 5.31 × 10^6 W/cm^2; **b** 7.43 × 10^6 W/cm^2; **c** 11.64 × 10^6 W/cm^2

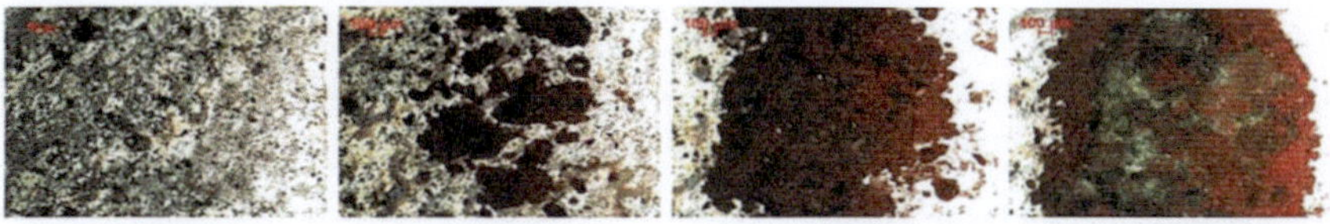

Fig. 6.16 Microscopic picture of paint sample after laser radiation with different powers. **a** The paint; **b** 32.5 mJ; **c** 48 mJ; **d** 64.3 mJ

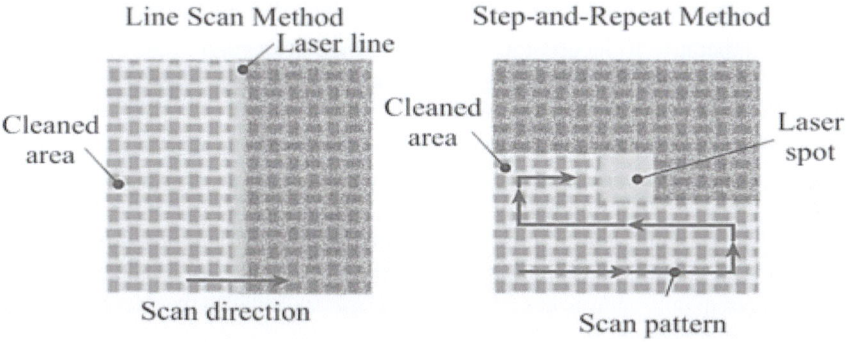

Fig. 6.17 Scanning mode of light spot in laser cleaning

7 mm^2. Additionally, experiments were conducted on the same paint sample using a wet laser cleaning method. The cleaning efficacy of the dry and wet laser cleaning methods showed no significant difference with the same laser parameters.

II. **Scanning method and scanning speed**

Nepomnyashchii [40] presented two scanning methods for laser cleaning: line scanning and point scanning, as depicted in Fig. 6.17.

The scanning speed plays a crucial role in determining the efficacy of laser paint removal. Excessively high scanning speeds may result in incomplete removal of the paint layer. Conversely, excessively slow scanning speeds can indeed ensure the complete removal of the paint layer. However, this approach may inadvertently lead to substrate deformation or surface damage. Hence, it is imperative to regulate the scanning speed within a range that guarantees thorough paint removal while mitigating potential substrate damage.

Zhu et al. [16] investigated paint removal on aluminum alloy plates using a 1064 nm pulsed fiber laser. They found that different scanning speeds did not significantly affect the cleaning results, and the cleaning effect was most stable at a speed of 1500 mm/s. Chen et al. [2] used a CO_2 laser operating at a wavelength of 10.6 μm to investigate the influence of scanning speed on cleaning effectiveness. Furthermore, they delved into the correlation between cleaning power and scanning speed.

Notably, their observations revealed that at relatively low laser power levels, approximately 300 W, the maximum scanning speed for achieving complete paint removal was close to the minimum scanning speed of substrate damage.

III. **Overlap of the laser beam**

The overlap of the laser beam during scanning is another pivotal factor influencing the efficacy of paint removal, which often intertwines with scanning speed to dictate the outcome. Laser beams typically exhibit a Gaussian distribution, leading to an uneven energy spread wherein the central region possesses higher energy while the periphery registers lower energy levels. Consequently, to achieve more uniform cleaning, it becomes imperative to incorporate a certain degree of overlap at the edges of the laser beam. Termed as overlap of the laser beam, this parameter assumes a critical role in facilitating consistent paint removal.

Zhang et al. [41] employed the spot overlap of pulsed laser illustrated in Fig. 6.18. Here, R denotes the diameter of the spot, L_1 represents the overlap between spots, and L_2 indicates the overlap between scanning lines. They defined $\gamma_1 = L_1/2\text{R}$ and $\gamma_2 = L_2/2\text{R}$ as the spot overlap ratio and scanning line overlap ratio, respectively. Experimental findings revealed that complete paint removal can be achieved when the spot overlap ratio approaches 50% between laser spots with a fixed laser scanning speed of $F = 150$ mm/min.

Zhang et al. [6] investigated the spot overlap ratio during the removal of paint layers from T-shaped structures utilizing a quasi-continuous wave Nd:YAG laser operating at 20 kHz with a pulse duration of 140 ns. They initially defined the overlap rate to elucidate the overlap between adjacent laser spots, which was calculated by the number of pulses to remove each paint layer. Experimentally, the laser spots were circular, and a two-dimensional laser cleaning system program was utilized to control the laser scanning along both the x and y axes. Besides, the distribution of the laser spots in the x and y directions were scrutinized, while cleanliness and substrate damage were evaluated through electron microscope (NanoSEM650). The results revealed that the ablation depth at the periphery of the laser spot was inferior to that

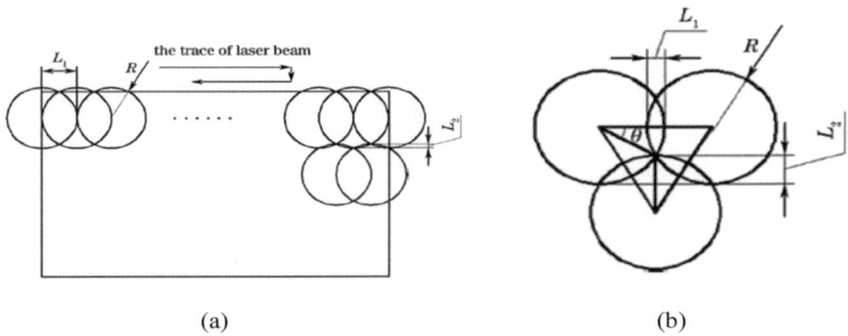

Fig. 6.18 **a** Schematic diagram of laser spot overlap during scanning and **b** the relationship between the overlap ratios

at the center when the laser did not overlap. To mitigate this issue, the translation speed in the y-direction was reduced to augment the spot overlap ratio. Concurrently, the scanning speed in the x-direction was increased to reduce the ablation depth induced by the reduction of translation speed in the y-direction. By continual adjusting these parameters, the uneven energy distribution of laser spots caused by overlap ratio and the residual paint caused by gaps between adjacent circular laser spots were ultimately addressed. Zhang et al. [42] also utilized a pulsed laser with an average power of 30 W, a pulse width ranging from 80 to 200 ns, and a repetition frequency from 30 to 300 kHz for cleaning acrylic paint with a thickness of 0.18 mm on aluminum alloy (6061). By controlling the overlap of laser spots in both the x and y directions, they derived a method to calculate the same overlap ratio in both directions under the "bow" scanning mode. Both theoretical analysis and experimental results indicated that the pulse interval had minimal impact on the material temperature variation, and the primary removal mechanism was found to be ablation.

Shi et al. [31] undertook paint removal experiments by employing a quasi-continuous wave Nd:YAG laser with a wavelength of 1064 nm, a spot diameter of approximately 0.6 mm, and an adjustable repetition frequency ranging from 0.5 to 50 kHz. They defined the overlap ratio as $l/2r$, where r represented the radius of the irradiated spot and l denoted the overlap length of each point. Notably, from the perspective of the scanning line's movement direction, it was discerned that a slower scanning speed correlated with a higher overlap ratio. Experimentally, the specimens subjected to cleaning comprised steel substrates coated with red alkyd paint, featuring a paint layer thickness of 50 μm. The laser power was set to a level for achieving complete cleaning, while the repetition frequency remained fixed at 3 kHz. By manipulating the translation speed of the workbench, the scanning speed was modulated to 1.88 cm/s. It was observed that the paint layer could be entirely removed within the pulse operating range of the laser scanning area. However, residual paint streaks were evident along the scanning direction, which could be attributed to the low overlap ratio caused by excessive scanning speed. Upon reducing the scanning speed to 1.39 cm/s, the residual paint streaks between the scanning lines became less pronounced. Ultimately, at a scanning speed of 0.76 cm/s, the paint layer was completely removed without any discernible damage to the metal substrate. Therefore, it was imperative for the overlap ratio between laser spots to exceed 80% for achieving an optimal paint removal.

In the laser cleaning experiments conducted by Yang [43] on acrylic resin paint adhered to 304 stainless steel substrates using a nanosecond pulsed laser, it was revealed that the overlap ratio of laser spots exerted the most significant influence on surface composition, while laser power predominantly affected surface roughness. The study identified that the optimal laser paint removal effect was attained when the parameters were set at 19.18 W for laser power, 46% for the overlap ratio, and three times for the scanning process. However, it was noted that excessively high overlap ratio could elevate the substrate surface temperature beyond its melting point. Then the surface paint layer became ensnared by molten slag before vaporized, augmenting the carbon content of the recast layer. Hence, excessively high overlap ratio was unfavorable for achieving superior surface cleaning quality.

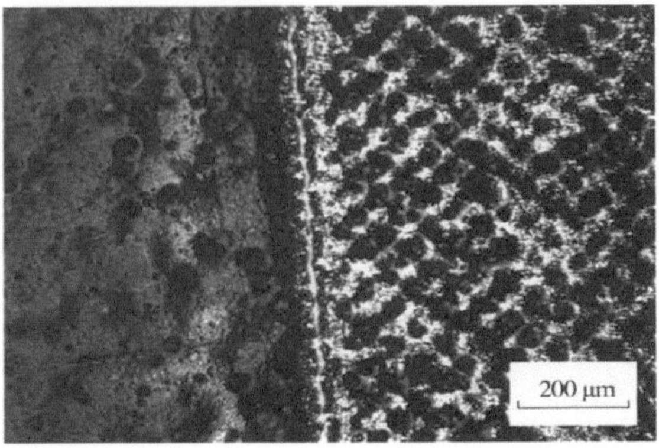

Fig. 6.19 Microstructure of aluminum alloy after cleaning, with the original morphology of paint layer on the left and the surface morphology after laser cleaning on the right

IV. **Defocusing distance**

The focus of the laser beam is minimal, and the cleaning target does not need to be placed at the focal point of the laser in high-power cases. And the distance from the cleaning object to the focus is termed the defocusing distance. In practical applications, the laser cleaning equipment is often fixed (encompassing beam quality, focusing lens, and even laser power), while the cleaning threshold of the contaminants is different. Thus, cleaning is facilitated by adjusting the defocusing distance. In fact, different defocusing distances imply different power densities of laser cleaning, which significantly affect the laser paint removal. Zhu et al. [16] investigated the influence of various parameters such as scanning speed, defocusing distance, and average power on laser paint removal by utilizing a 1064 nm pulsed fiber laser. Notably, the results revealed that the defocusing distance has the most significant impact on cleaning effectiveness, as illustrated in Fig. 6.19.

Zhu et al. [44] devised a high-speed camera optical filter amplification system to document the dynamic process of paint removal under varying defocusing distances. By employing a scanning electron microscope, they captured the surface morphology after laser paint removal at different defocusing distances. By integrating the dynamic processes through high-speed imaging and the analysis of microscopic morphology via scanning electron microscopy, they studied the discrepancies in paint removal mechanisms under diverse defocusing conditions. Their findings unveiled that the primary mechanism for paint layer removal was thermal stress, melting, and gas impact when the defocusing distance was 0 mm. However, the predominant mechanisms shifted towards thermal stress and melting evaporation for paint layer removal at a defocusing distances of $\pm$ 4 mm.

V. **Laser wavelength**

The laser wavelength distinctly influences laser paint removal, primarily due to variations in the absorption rates of the paint layer and the substrate at different wavelengths. To exemplify this phenomenon, let's consider giltwood—a significant and original art form in Portuguese art, commonly employed for embellishing altars, churches, and various religious or secular structures. Giltwood typically features a surface adorned with gold leaf, often overlaid with brass-based paint. Gaspar et al. [45] investigated the surface paint removal of giltwood by employing a Q-switched Nd:YAG laser with three distinct wavelengths: 1064 nm (near-infrared), 532 nm (visible light, green), and 266 nm (ultraviolet).

The experimental results revealed that distinct interactions between the three laser wavelengths and the surface of giltwood samples. The 266 nm radiation demonstrated efficient removal of the brass-based paint layer at low energy levels. However, the incident radiation caused some degree of damage to the gold leaf due to gold's high absorbance in ultraviolet light. Regarding the 532 nm radiation, while capable of removing the brass-based paint layer, its efficacy in cleaning giltwood samples was inferior to that of the 266 nm wavelength. Additionally, the gold leaf still sustained damage, leading to delamination and alterations to the substrate. In contrast, the 1064 nm radiation proved ineffective in cleaning the giltwood surface, as it could not remove the brass-based paint without compromising the underlying gold leaf, merely altering the paint texture. Therefore, it could be deduced that laser radiation exhibited the capacity to remove the brass-based paint from giltwood samples, with ultraviolet radiation emerging as the most effective option compared to visible light and infrared radiation. An optimal outcome could be attained by utilizing 100 pulses with an energy density of 0.025 J cm^{-2}, as depicted in Fig. 6.20. Indeed, caution must be exercised with the gold leaf, and any direct exposure to light must be avoided to prevent potential surface damage.

VI. **Beam quality**

In exploring the impact of various laser parameters on paint removal from metal substrates, researchers have not only utilized Gaussian beams but have also investigated the efficacy of flat-top beams. Among them, Shi et al. [46] employed flat-top lasers to investigate laser cleaning for eliminating paint coatings from metal substrates. They conducted comparative analyses on the effectiveness of Gaussian

Fig. 6.20 Cleaning effects of 1064 nm (near infrared), 532 nm (visible light, green) and 266 nm (ultraviolet) under 100 pulses with low energy

and flat-top lasers, along with evaluating cleaning performance under different laser beam diameters. Experimentally, they utilized a Q-switched Nd:YAG laser with a wavelength of 1064 nm, pulse energy of 25 mJ, pulse duration ranging from 200 to 400 ns, and a repetition frequency of 3 kHz. To achieve a flat-top laser beam, a lens system with negative spherical aberration lens was integrated into the optical path. Each lens was configured with spherical aberration, and the aperture was adjusted to enable the passage of the beam's center. By selecting lenses with varying coefficients, the diameter of the laser beam could be tailored accordingly. Surface paint layers, approximately 50 μm thick and uniformly sprayed, were then removed from iron substrates. The effectiveness of laser paint removal was assessed by examining changes in surface texture. The findings revealed that an approximately flat laser beam could yield a more optimized cleaning effect compared to Gaussian laser beams. Furthermore, owing to the uniformity of nearly flat laser intensity, minimal thermal effects on the substrate could be achieved, thereby reducing substrate damage and enhancing cleaning efficiency. The paint removal results are visually depicted in Fig. 6.21.

VII. **Continuous-wave and pulsed laser**

Both continuous-wave and pulsed lasers have been utilized for paint removal applications. Initially, CO_2 continuous-wave lasers were commonly used. Subsequently, research expanded to include acousto-optic and electro-optic Q-switched lasers, such

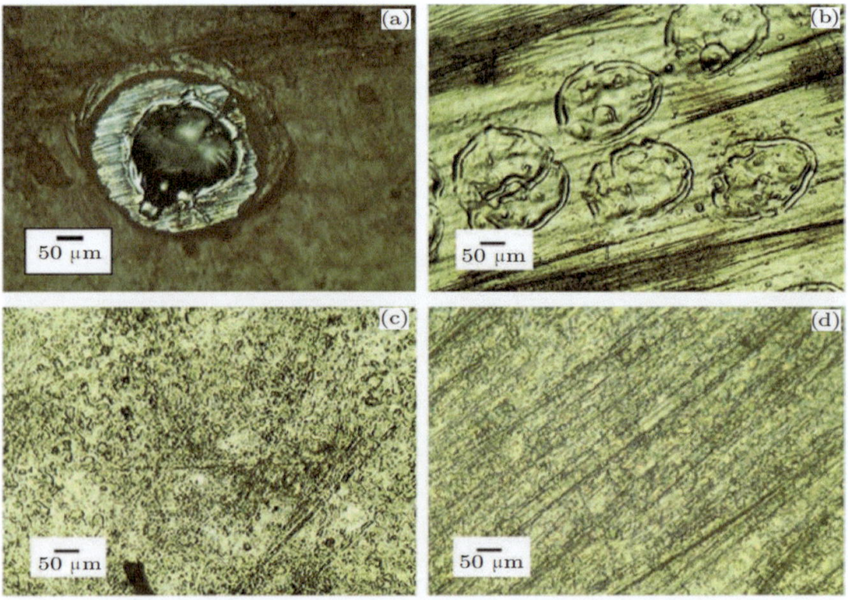

Fig. 6.21 Surface morphologies after laser cleaning with different beam quality. **a** The result of Gaussian laser irradiation; and the results of flat-top laser irradiation with a spot diameter of **b** 0.4 mm, **c** 0.7 mm, and **d** 1.0 mm

as YAG solid-state lasers and fiber lasers. Additionally, mode-locked lasers have also been explored for paint removal purposes. In the realm of pulsed laser cleaning, Zhao [47] employed a 1064 nm high-repetition-frequency fiber laser to eliminate a 50 mm thick layer of polyacrylic resin primer from the outer shell (LY12 aluminum alloy plate) of an aircraft. They delved into the effects of parameters such as scanning speed, repetition frequency, scanning spacing, and laser power on cleaning efficiency. Through comprehensive analysis of the cleaned surface and particles collected during the cleaning process, they proposed three possible cleaning mechanisms: ablation, thermal stress vibration, and plasma impact. On the other hand, in continuous-wave laser cleaning, Lu et al. [48] utilized a continuous-wave fiber laser with a maximum average power of 2 kW, operating at a wavelength of 1070 nm, and a laser spot size of 5.0 mm for laser paint removal. During the cleaning process, argon gas was blown over the cleaning surface to mitigate combustion effects. The experimental samples comprised 25A06 aluminum alloy substrates coated with epoxy polyester (EP) paint, with a thickness of approximately 200 μm. The study revealed a nearly linear relationship between the intensity of the laser and the depth of paint removal, as higher laser intensity facilitated easier evaporation of the paint surface. Furthermore, due to the ablative effect of continuous-wave lasers on the coating layer, photodissociation processes showed significantly advantageous. The paint removal rate demonstrated a linear variation with the output power of the continuous-wave laser, as illustrated in Fig. 6.22.

For paint removal with continuous-wave laser, the process at low incident laser intensities primarily involves evaporation and phase explosion. However, the impact of multiple reflections of the incident laser on the coating surface limits the depth of ablation of the coating. Besides, the ablation rate is effectively inhibited due to heat

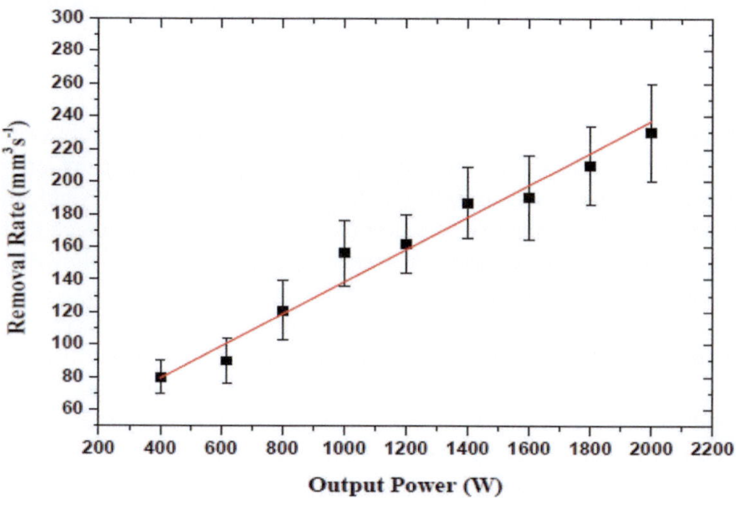

Fig. 6.22 Relationship between removal rate and laser output power

penetration [1, 15]. Conversely, in the case of high incident laser intensities, laser-induced breakdown of air can readily lead to the formation of atmospheric plasma. This phenomenon creates significant pressure gradients that facilitate paint removal.

6.6 Practical Application Research of Laser Paint Removal

In laser paint removal research, a notable portion is dedicated to the practical application of the technique in various industries. This includes stripping paint from aircraft skins, paint removal in ammunition repair, and restoring the surfaces of cultural relics. These studies prioritize practical engineering applications and are often directly aimed at real-world engineering challenges. This section provides an overview of these endeavors.

6.6.1 Laser Paint Removal of Aircraft Skin

Aircraft skins are primarily constructed from two main materials: carbon fiber-reinforced resin-based composites and aluminum alloys. The selection of material depends on the specific location on the aircraft. Surface painting processes for aircraft skins typically involves surface preparation, application of primer layers, and application of topcoat layers. Various types of paints are utilized, such as polyurethane topcoats, epoxy polyamide or polyurethane primers, and alodine coatings. In some cases, an oxide film must be generated on the surface of aluminum [49].

In the laser paint removal for aircraft skins, it is customary to utilize actual aircraft skin materials as experimental specimens. The parameters typically resemble those utilized in studies on laser paint removal of aluminum alloys. A primary focus of investigation revolves around whether lasers induce damage to the substrate surface. Examination of alterations in the surface oxide layer of aluminum alloys serves to assess substrate damage. Furthermore, evaluating the impact of laser paint removal on the mechanical properties of aircraft skin, including tensile strength, yield strength, and Vickers hardness, constitutes another pivotal aspect of the research endeavor. Jiang et al. [50] investigated laser paint removal on LY12 aluminum alloy substrates by using a high-repetition-frequency CO_2 laser with a wavelength of 10.6 μm. These substrates were 1.01 mm thick and coated with a dual-layer composite paint system comprising a gray-black TS96-71 topcoat and a yellow TB06-9 primer. They studied the effects of average power density, peak power density, scanning spacing (overlap ratio), and scanning repetitions on paint removal efficiency. Notably, they found that when the scanning spacing was 0.1 mm and the average power density was 13.1 kW/cm^2, the paint layer was completely removed and the original lettering on the substrate surface became clearly visible after two times of scanning. This specific average power density value served as the threshold for achieving complete paint removal. Remarkably, the researchers astutely utilized the oxidation layer on the material surface as an indicator of substrate damage; the appearance of a pronounced whitening phenomenon in the oxidation layer signified potential substrate damage. Regarding the assessment of substrate damage, Hu et al. [28] adopted a more precise approach. They utilized a CO_2 laser with an output power of 150 W and evaluated the integrity of the oxidation layer by measuring the material conductivity. Additionally, the samples were subjected to continuous salt spray corrosion tests to further evaluate their corrosion resistance. The integrity of the oxidation layer on the aluminum alloy surface was pivotal, as it remained non-conductive when intact, providing excellent corrosion resistance to salt spray. After laser paint removal, the conductivity of the aluminum alloy decreased, and no significant corrosion spots were observed in the salt spray environment, indicating that the oxidation film of the aluminum alloy remained effectively preserved.

Composite materials, such as carbon fiber reinforced composite materials (CFRP), are crucial components of aircraft skin. Niebojewski [38] conducted a study focusing on the paint removal of CFRP. Regarding the impact of laser paint removal on the mechanical properties of aircraft skin, Jiang et al. [50] investigated comprehensive investigations on changes in roughness, tensile strength, yield strength, and Vickers hardness of aircraft skin after paint removal and its original anodized substrate. Their analysis revealed that laser paint removal had no discernible effect on the mechanical properties of aircraft skin. Furthermore, it did not alter the substrate's surface roughness. In a similar vein, Hu et al. [28] explored the effects of laser paint removal on the mechanical properties of aircraft skin, comparing it with solvent paint removal. Their mechanical properties revealed minimal differences in tensile strength, yield strength, elongation, and other properties between parts subjected to solvent and laser paint removal. Importantly, both methods met the requirements of the original material standards. Furthermore, they also studied the adhesion of paint layers after laser and solvent cleaning followed by repainting. The results indicated superior adhesion of paint layers to surfaces treated with laser paint removal compared to solvent paint removal. Specifically, samples subjected to laser paint removal and repainting exhibited excellent paint layer adhesion, while those treated with solvent paint removal and repainting showed signs of paint layer peeling. These findings collectively suggest that laser paint removal not only preserves the mechanical properties of aircraft skin but also offers enhanced adhesion of subsequent paint layers, making it a superior choice compared to solvent-based methods.

Additionally, Zheng et al. [51] delved into the influence of TEA CO2 laser paint removal on the mechanical properties of aircraft skin. Their experiments unveiled that the mechanical properties of the aircraft metal skin remained unchanged both before and after laser paint removal, thereby posing no safety hazards.

6.6.2 Laser Paint Removal in Ammunition Repairment

Ammunition, often stored for extended periods before being utilized, possesses inherent explosive properties. Over time, particularly during prolonged storage in damp environments, the paint applied to the surface of ammunition shells may begin to peel. This deterioration exposes the underlying metal substrate to corrosion, potentially compromising the integrity of the ammunition. In severe cases, corrosion-induced damage may even lead to the exposure of the ammunition contents, posing significant safety risks. To mitigate such hazards and ensure utmost safety, various types of ammunition, including shells and bullets, require periodic refurbishment. This process entails removing the deteriorated paint from the ammunition shell's surface and applying a fresh coat. The efficacy of paint removal directly impacts the quality and reliability of ammunition repairment. Traditionally, sandblasting was a common method used for paint removal. However, this approach necessitated the emptying of ammunition shells before cleaning to prevent any inadvertent safety incidents.

Since the advent of laser cleaning technology, its potential feasibility in ammunition repairment has garnered attention. Song et al. [52] undertook laser paint removal on copper artillery shells utilizing a pulsed fiber laser with an output power of 20 W, a wavelength of 1064 nm, and a pulse width of 80 ns. The experimental results revealed that intense laser irradiation scanning facilitated the removal of the old paint layer from the sample surface, while the old paint layer remained loosely attached and did not immediately detach. Subsequently, a weaker laser was employed for further cleaning to accomplish complete paint removal. Remarkably, areas subjected to laser irradiation exhibited a uniform metallic luster, showing consistent color and texture across the substrate material. Analysis of surface temperature variations indicated minimal changes post-laser paint removal. Consequently, concerns regarding detonation due to thermal influences during surface treatment with a 20 W laser can be alleviated. However, in terms of operational efficiency, if the process entails both rust and paint removal, the efficiency stands at approximately 0.083 s/mm^2, indicating scope for overall improvement.

Kono et al. [53] conducted a study utilizing ultrafast laser on military-grade gold woven fabrics. Their findings showed the ultrafast laser cleaning could efficiently remove contaminants from such fabrics. Moreover, the cleaning process was safe and controllable. The advent of robust and compact femtosecond laser systems further bolsters the potential of ultrafast laser ablation to emerge as a standard tool for safeguarding armories.

6.6.3 Laser Paint Removal in Cultural Relics and Artworks

The application of laser paint removal on stone materials primarily focuses on cleaning stone artifacts. Although laser cleaning for stone artifacts began in Europe in the early 1970s, its widespread adoption only occurred in the past decade. Laser cleaning has emerged as a highly significant method for physically cleaning stone artifacts in many European countries, such as France, Italy, Germany, and Greece. In China, extensive theoretical and practical research has demonstrated that laser cleaning is a safe and effective method for cleaning stone artifacts. For instance,

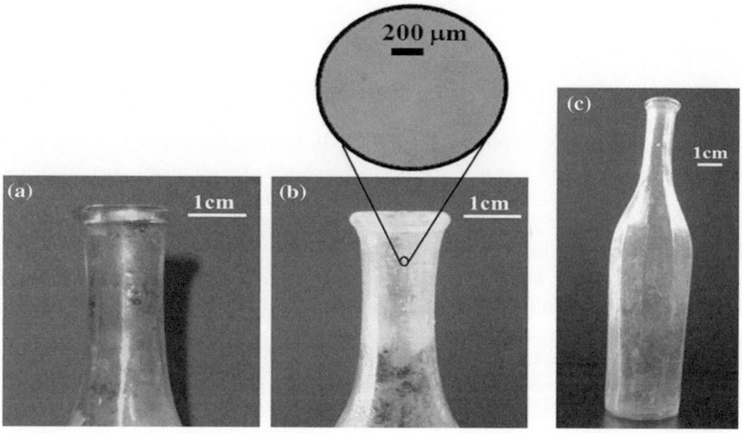

Fig. 6.23 The images before and after laser cleaning. **a** The original state of the upper part of the bottle; **b** bottles partially cleaned by a laser; **c** the final state of the whole bottle after laser cleaning from the outside, with a flux of 1 J/cm^2. The upper circle is a microscope image of the clean part, showing that the glass is not damaged

lasers have been successfully utilized to remove typical paint layers from stone artifacts in sites like the Yungang Grottoes, showing their safety and reliability for stone artifact conservation.

The utilization of laser paint removal on glass materials predominantly manifests in backside irradiation techniques. This method has proven effective in the restoration of two artifacts unearthed from archaeological sites. Specifically, it has been applied to rectify a piece of glass covered with a layer of black paint and an antique glass bottle with internal black blemishes, which were recovered during excavations in Buenos Aires. The implementation of backside irradiation laser cleaning has yielded remarkable results. Figure 6.23 are images capturing the state of these artifacts before and after laser cleaning [53].

After prolonged use or weathering, surfaces of many industrial components, equipment, historical artifacts, and other items often accumulate layers of contamination, including a mixture of paint, oil residue, rust, scale, and other organic coatings. In such cases, laser technology emerges as a viable solution for concurrent paint and rust removal. In specific applications, the selected laser type and laser wavelength should be determined based on the properties, microstructure, and morphological defects of the substrate material, and then the energy threshold of laser cleaning should be determined [54].

References

1. Woodroffe JA. Laser removal of poor thermally-conductive materials-delivers laser beam pulses having wavelength at which material to be removed is opaque and fluence sufficient to decompose material: US4756765-A US33777-E. 1991-12-24.
2. Chen JF, Zhang YK, Xu RJ, et al. Experimental research of paint removement with a fast axis flow CO_2 laser. Laser Technology, 2008, 1: 64-66. (in Chinese)
3. Watkins KG, Curran C, Lee JM. Two new mechanisms for laser cleaning using Nd:YAG sources. Journal of Cultural Heritage, 2003 ,4(SI): 59–64.
4. Jaafar MS, Suhaimi FM, Masri MN, et al. Study on performance of CO2 laser in paint removal over selected national car model. Journal Technology, 2016 , 78 (3): 203–209.
5. Zou WF, Xie YM, Xiao X, et al. Application of thermal stress model to paint removal by Q-switched Nd:YAG laser. Chin. Phys. B, 2014, 23(7): 074205.
6. Zhang ZY, Zhang JY, Wang YB, et al. Removal of paint layer by layer using a 20 kHz 140 ns quasi-continuous wave laser. Optik, 2018, 174: 46–55.
7. Tan XY, Zhang DM, Yu BM, et al. Vaporization effect studying on high-power nanosecond pulsed laser deposition. Physica B, 2005, 358: 86–92.
8. Lu QM, Mao SS, Mao XL, et al. Delayed phase explosion during high-power nanosecond laser ablation of silicon. Appl. Phys. Lett, 2002, 80(17): 3072–3074.
9. Singh N. Pulsed-laser evaporation technique for deposition of thin films: physics and theoretical model. Phys. Rev. B, 1990, 41(13): 8843–8859.
10. Jia BS, Tang HP, Su CZ, et al. Removal of surface coating of resin matrix composites by pulsed laser. Chinese Journal of Lasers, 2019, 46(12): 133–140. (in Chinese)
11. Gao LY, Zhou JZ, Sun Q, et al. Numerical simulation and surface morphology of laser-cleaned aluminum alloy paint layer. Chinese Journal of Lasers, 2019, 46(5): 327–335. (in Chinese)
12. Zou WF, Luo Y, Fan MM. Theoretical analysis of laser cleaning paint on stone relics. Journal of Gannan Normal University, 2018, 39(3): 46–49. (in Chinese)
13. Watkins KG. Mechanisms of laser cleaning. Pro .SPIE, 2000, 3888: 165–174.
14. Tian B, Zou WA, He Z, et al. Paint removal experiment with pulsed Nd:YAG laser. Cleaning World, 2007 ,(10): 1–5.
15. Luo HX, Cheng ZG. High power CW CO2 laser using in aircraft laser paint removing. Laser Journal, 2002, 23(6): 52–53.
16. Zhu W, Meng XW, Dai ZC, et al. Laser paint removal process for Aluminum Alloy flat Surface. Electric Welding Machine, 2015, 45(11): 126–128.
17. Chen L, Debg GL, Feng GY, et al. Mechanism of laser paint removal based on LIBS and time-resolved characteristic peaks. Spectroscopy and Spectral Analysis, 2018, 38(2): 367–371. (in Chinese)
18. Tserevelakis GJ, Pozo-Antonio JS, Siozos P, et al. On-line photoacoustic monitoring of laser cleaning on stone: evaluation of cleaning effectiveness and detection of potential damage to the substrate . Journal of Cultural Heritage, 2019, 35: 108–115.
19. Raza MSS, Tudu P, Saha P. Excimer laser cleaning of black sulphur encrustation from silver surface. Optics and Laser Technology, 2019, 113: 95–103.
20. Pouli P, Oujja M, Castillejo M. Practical issues in laser cleaning of stone and painted artefacts: optimization procedures and side effects. Applied Physics A-Materials Science & Processing, 2012, 106 : 447–464.
21. Bilmes GM, Vallejo J, Vera CC, et al. High efficiencies for laser cleaning of glassware irradiated from the back: application to glassware historical objects. Applied Physics A-Materials Science & Processing, 2018, 124(4): 347.
22. Zuo MG, Jin HZ. Study on infrared absorption spectrum and baking temperature of paint. Infrared Technology, 1983, (6): 23–26. (in Chinese)
23. Xu JL. Application of infrared absorption spectroscopy in comparison and identification of paint evidence. Legality Vision, 2018, (17): 125–126. (in Chinese)

24. Gomes V, Dionisio A, Pozo-Antonio JS, et al. Mechanical and laser cleaning of spray graffiti paints on a granite subjected to a SO_2-rich atmosphere. Construction and Building Materials, 2018, 188: 621–632.
25. Lu Y, Yang LJ, Wang Y, et al. Paint removal on the 5A06 aluminum alloy using a continuous wave fiber laser. Coatings, 2019, 9(8): 488.
26. Wu LX, Ye XS, Liu ZJ. Study on damage mechanism of polyurethane black paint by infrared laser. Chinese Journal of Lasers, 2011, 38(3): 14–18. (in Chinese)
27. Abeens M, Muruganandhan R, Thirumavalavan K, et al. Surface modification of AA7075 T651 by laser shock peening to improve the wear characteristics. Materials Research Express, 2019, 6(6): 066519.
28. Hu J, Ma L, Wang WQ, et al. Effect of laser and solvent paint removal on Performance of Aircraft Skin Parts. Metal World, 2017, (6): 6–9. (in Chinese)
29. Li X, Huang T, Chong AW, et al. Laser cleaning of steel structure surface for paint removal and repaint adhesion. Opto-Electronic Engineering, 2017, 44(03): 340–344.
30. Du P. Experimental research of pulsed laser paint removal and preparation of laser paint removal test. Tianjin: Nankai University, 2012. (in Chinese)
31. Shi SD, Du P, Li W, et al. Study on paint removal by 1064 nm quasi-continuous laser. Chinese Journal of Lasers, 2012, 39(9): 58–64. (in Chinese)
32. Liu CF, Feng GY, Deng GL, et al. Temperature field analysis and experimental study of moving laser paint removal by finite element Method. Laser Technology, 2016, 40(2): 274–279. (in Chinese)
33. Guo WX, Hu QW, Wang ZM, et al. Study on paint removal by high power pulsed TEA CO_2 laser. Optics & Optoelectronics Technology, 2006, 4(3): 32–35. (in Chinese)
34. Xian H, Feng GY, Wang SP. Theoretical analysis and related experiments of laser transmission through paint layer. Journal of Sichuan University, 2012, 49(5): 1036–1042. (in Chinese)
35. Qi Y, Zhou WQ, Chen J, et al. Experimental study on laser cleaning of surface pollutants of cultural relics in Yungang Grottoes. Safety and Environmental Engineering, 2015, 22(2): 32–38. (in Chinese)
36. Rivas T, Pozo S, Fiorucci MP, et al. Nd: YVO4 laser removal of graffiti from granite. Influence of paint and rock properties on cleaning efficacy. Applied Surface Science, 2012, 263(15): 563–572.
37. Chen KX, Feng GY, Deng GL, et al. Mechanism of laser paint removal based on emission spectrum and component analysis. Spectroscopy and Spectral Analysis, 2016, 36(9): 2956–2959. (in Chinese)
38. Li X, Zhang Q, Zhou X, et al. The influence of nanosecond laser pulse energy density for paint removal. Optik, 2018, 156 : 841–846.
39. Antony K, Arivazhagan N, Senthilkumaran K. Numerical and experimental investigations on laser melting of stainless steel 316 L metal powders . Manuf. Process, 2014, 16 (3): 345–355.
40. Nepomnyashchii VV, Mosina TV, Radchenko AK, et al. Theory, manufacturing technology, and properties of powders and fibers-oxidation resistance of electrolytic ferromagnetic powders with an organic composite coating. Powder Metallurgy & Metal Ceramics, 2007, 46(7-8): 313–316.
41. Zhang H, Liu WW, Dong YY, et al. Mechanism and experimental study of paint removal with low frequency YAG pulsed laser. Laser and Optoelectronics Progress, 2013, 50(12): 114–120. (in Chinese)
42. Zhang ZY, Wang YB, Liang H, et al. Removal of paint with low thermal conductivity by high repetition frequency pulsed laser. Chinese Journal of Lasers, 2019, 46(1): 148–156. (in Chinese)
43. Yang JL, Zhou JZ, Sun Q, et al. Optimization of laser paint removal process parameters based on response surface analysis. Advances in Laser and Optoelectronics, 2019, 56(23): 183–190. (in Chinese)
44. Zhu YR, Zhu M, Shi Y, et al. Effect of laser defocus on removal mechanism of metal paint. Electric Welding Machine, 2020, 50 (1):29–33. (in Chinese)

45. Gaspar P, Rocha M, Kearns A, et al. A study of the effect of the wavelength in the Q-switched Nd:YAG laser cleaning of gilded wood. Journal of Cultural Heritage, 2000, 1(2): 133–144.
46. Shi SD, Li W, Du P, et al. Removing paint from a metal substrate using a flattened top laser. Chinese Physics B, 2012, 21(10): 295–301.
47. Zhao HC, Qiao YL, Du X, et al. Laser cleaning performance and mechanism in stripping of polyacrylate resin paint. Applied Physics A, 2020, 126(5): 360.
48. Lu Y, Yang L, Wang Y, et al. Paint removal on the 5A06 aluminum alloy using a continuous wave fiber laser. The Coatings, 2019, 9(8): 488.
49. Jin S, Wang JX, Yuan XD, et al. Laser paint removal technology for metal skin and composite surface of aircraft. Aeronautical Manufacturing Technology, 2018, 61(17): 63–70. (in Chinese)
50. Jiang YL, Ye YY, Zhou GR, et al. Research on laser paint removal technology of aircraft skin. Infrared and Laser Engineering, 2018, 47(12): 21–27. (in Chinese)
51. Zheng G, Tan RQ. Experimental study of TEA CO_2: laser paint removal. Laser Journal, 2005, 26(5): 82–84. (in Chinese)
52. Song GF, Li LC, Xia FJ, et al. Application exploration and experimental research of laser cleaning technology in ammunition repair. Laser & Infrared, 2017, 47(1): 29–31. (in Chinese)
53. Kono M, Baldwin KGH, Wain A, et al. Treating the untreatable in art and heritage materials: ultrafast laser cleaning of "cloth-of-gold". Langmuir, 2015, 31 (4):1596–1604.
54. Wang HJ, Liu WW, Yu Y, et al. Research status and prospect of laser cleaning of metal surface contaminants. Internal Combustion Engine & Parts, 2016, 8: 75–78. (in Chinese)

Chapter 7
Application of Laser in Rust Removal

7.1 Significance of Laser Rust Removal

In our daily lives and industrial production, we utilize numerous metal components and equipment. However, the corrosion of these metals not only poses inconveniences but also poses significant threats to life and production. Equipment that has been corroded often needs to be replaced regularly, leading to direct financial losses. While direct losses can be easily quantified, indirect losses are more challenging to calculate and often exceed direct losses by a wide margin, including waste of raw materials, impact on product quality, and environmental pollution. In the factory, changes in machine performance caused by corrosion will lead to a significant increase in subsequent production costs, and the corrosion of some parts may also cause fire and other safety accidents. Nowadays, reinforced concrete structures are widely used in buildings and bridges. The concrete is wrapped around the steel, if the steel corrodes, it will cause the concrete outside to peel off, resulting in serious structural problems and inability to bear weight. This is also one of the common failure modes in reinforced concrete buildings. Corrosion poses a significant threat to engineering equipment, critical structures, and infrastructure, potentially leading to catastrophic accidents. Especially in buildings closely related to life and factories related to petroleum and natural gas, corrosion is more likely to cause huge economic losses and even casualties. In 1983, the Mianus River Bridge in Connecticut, USA collapsed due to the rusting and breaking of its bearings, causing changes in the forces on the bridge deck and ultimately resulting in the deaths of three drivers on the bridge at the time. Similarly, the 2013 Qingdao pipeline explosion caused numerous casualties and extensive economic losses. The underlying reason of this disaster was corrosion-induced leaks that ultimately led to the explosion. The losses caused by corrosion are indeed greater than the sum of all natural disaster losses. According to a study conducted by the National Association of Corrosion Engineers (NACE),

F. Song and X. Lin, *Laser Cleaning*, https://doi.org/10.1007/978-981-97-6631-4_7

the global annual loss due to metal corrosion amounts to a whopping 2.5 trillion US dollars, which is equivalent to 3.4% of the global GDP. In fact, the economic losses caused by corrosion surpass 3% of the annual GDP in various countries. And the situation is even more severe in China, where the losses reach about 5%.

Therefore, protecting equipment and components from rusting is a crucial task in both production and daily life. To prevent equipment from corroding, it is essential to apply a layer of paint during the manufacturing process to isolate the metal from the atmosphere. However, in industrial settings, due to division of labor and process issues, metal materials may not immediately proceed to the painting stage after production. Instead, they may be stored in warehouses or even stacked outdoors, where they are exposed to the environment and are prone to rusting. We know that metal materials, particularly carbon steel materials, can rapidly (range from mere hours to several days) form corrosion products, known as floating rust, in a humid environment. This initial layer of rust appears at the onset of corrosion and is visible in Fig. 7.1. When paint is applied to a rusty surface, it often peels off along with the rust oxidation skin. Therefore, it is essential to ensure that the metal surface is clean and rust-free prior to painting. Any existing rust on the metal material must be removed. Even after painting, metal equipment may still develop rust due to wear, paint aging, and environmental factors such as oxygen and water vapor diffusion into the substrate, as demonstrated in Fig. 7.2. Generally, paint layer damage and rust appear initially in small patches. Over time, these areas of rust can become more severe, leading to widespread paint peeling. To prevent the spread of rust and ensure its containment, it is essential to conduct repairs promptly. During the repair process, it is necessary to remove both the paint and any generated rust. Therefore, rust removal is a crucial and essential step in both the pre-painting stage for component equipment and post-use maintenance of instruments.

I. **Physical Rust Removal Method**

In daily life or for minor rust removal tasks, mechanical friction is often employed. However, the shot peening method is commonly used for larger areas or more severe corrosion. This method is described in Chap. 1. During the shot peening process, abrasives (such as iron shot or iron sand) are sprayed onto the corroded metal surface at high speed. This effectively breaks down the weaker corrosion into smaller fragments, which then detach easily from the steel base surface.

Fig. 7.1 Rust on the surface of steel pipes stacked in a warehouse

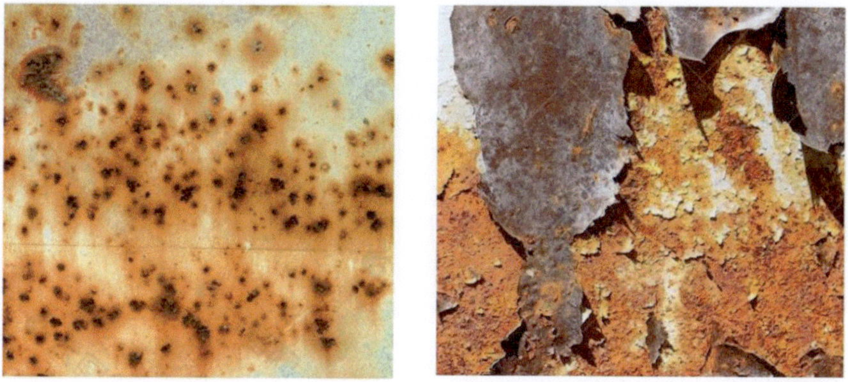

Fig. 7.2 Corrosion of substrate with painting

II. **Chemical Rust Removal Method**

Acid is often used to dissolve rust, such as iron hydroxide in acid solution, as it is a relatively low-cost and fast option. However, acid solutions can often corrode the metal substrate, so caution must be exercised. Alternatively, alkali can effectively remove rust without affecting the steel substrate, making it suitable for deep rust removal. Electrolytic rust removal is another technique that generally does not cause secondary corrosion of the metal surface and can even re-passivate rusted steel. However, it requires a specially designed electrolytic tank and is not suitable for on-site application.

When it comes to rust removal, shot peening and chemical cleaning are two widely and maturely used methods. However, these methods have significant drawbacks. Shot peening is noisy and dusty, creating environmental and health concerns. On the other hand, chemical cleaning often emits a pungent odor from acids and alkalis, which can be harmful to operators' health and has negative environmental impacts. In comparison, laser rust removal offers a more environmentally friendly alternative. It is a promising green technology that does not produce harmful by-products. After laser treatment, the surface undergoes passivation, which helps protect the substrate and slows down the rate of rusting.

7.2 Rust Formation and Basic Components

Rust, also known as corrosion, typically refers to the oxides that form on the surface of metals (including alloys), such as gold, silver, copper, iron, aluminum, etc. Some common rusts include iron rust, copper rust, aluminum rust, and silver rust. It's important to note that there can be multiple types of oxide for the same metal; for instance, iron rust can be Fe_2O_3, $Fe(OH)_3$, or a combination of both. Additionally, the chlorides and sulfides of these metals can also be classified as rust. If "rust" is used to describe the oxides of metals, then "corrosion" is used to describe various chemicals on the metal surface. Rust or corrosion is a chemical reaction between the metal and other components in the environment that can cause erosion of the metal substrate. Oxide rust is the most common type.

7.2.1 Corrosion Formation

According to the reaction mechanism, metal corrosion can be categorized into various types, including chemical corrosion, electrochemical corrosion, and physical corrosion.

I. **Chemical Corrosion**

The corrosion that occurs through a pure chemical reaction between a metal and its surrounding medium is referred to as chemical corrosion. This type of corrosion only occurs when the metal contacts with the medium directly, such as the meeting of steel and gases like O_2, H_2S, SO_2, Cl_2, etc. Over time, corresponding compounds can be formed on the metal surface, such as oxides, sulfides, chlorides.

II. **Electrochemical Corrosion**

Corrosion formed by the electrochemical reaction between a metal and an electrolyte is called electrochemical corrosion. In this process, the corrosion anode undergoes an oxidation reaction and releases electrons, while the corrosion cathode undergoes a reduction reaction and gains electrons. Taking iron products as an example, the standard electrode potential of $Fe \rightarrow Fe^{2+}$ is − 0.44 V, which is lower than the standard electrode potential of carbon (+ 0.3 V). When the two contact each other, corrosion will occur because of the potential difference. The reaction of the corrosion process is as follows [1, 2]:

Anode process:

$$Fe \rightarrow Fe^{2+} + 2e \tag{7.2.1}$$

Cathode process:

$$2H^+ + 2e \rightarrow H_2 \uparrow \tag{7.2.2}$$

(Hydrogen evolution corrosion reaction in acidic solution)

$$4H^+ + O_2 + 4e \rightarrow 2H_2O \tag{7.2.3}$$

(Oxygen reduction reaction in acidic solution)

$$O_2 + 2H_2O + 4e \rightarrow 4OH^- \quad (7.2.4)$$

(Oxygen reduction reaction in neutral or alkaline solution)

The electrochemical corrosion is a major reason for the corrosion of metals. In the life and industrial processes, iron is one of the most widely used metals, but also the most susceptible to corrosion. Taking cast iron as an example, after completing the basic cathode and anode reactions, an insoluble product known as rust can be generated through a series of secondary processes. These reactions mainly involve the following processes:

$$Fe_2 + 2OH^- \rightarrow Fe(OH)_2 \quad (7.2.5)$$

(Secondary products are generated when the products of the cathode and anode migrate and meet each other)

$$4Fe(OH)_2 + 2H_2O + O_2 \rightarrow 4Fe(OH)_3 \quad (7.2.6)$$

(Ferrous hydroxide is oxidized to iron hydroxide in the presence of sufficient oxygen and water)

$$6Fe(OH)_2 + O_2 \rightarrow 2Fe_3O_4 + 3H_2O \quad (7.2.7)$$

(Direct oxidation of ferrous hydroxide to ferric tetroxide when oxygen is in short supply)

$$4Fe(OH)_3 \rightarrow \gamma\text{-FeO O H} + H_2O \quad (7.2.8)$$

(Active ferruginic acid is formed after dehydration and decomposition of iron hydroxide)

$$\gamma\text{-FeO OH} \rightarrow \alpha\text{-FeO OH} + Fe_3O_4 + Fe_2O_3 \quad (7.2.9)$$

(Active ferruginic acid further decomposes to form other components of rust)

III. **Physical Corrosion**

The process of metal corrosion caused by physical dissolution is referred as physical corrosion. At elevated temperatures, numerous metals dissolve or crack in molten salts, alkalis, and molten liquid metals, which is metal corrosion caused by physical dissolution.

7.2.2 *Rust*

I. Main Components of Rust

In many components and devices, carbon steel is widely used, which contains iron as its primary metal element. As a result, rust is the most common and pervasive form of corrosion. The composition and formation of rust are greatly influenced by the surrounding environment. Environmental ions can significantly impact the composition of rust, and water can also adsorb on the oxidation products of iron. Left untreated, the lack of a dense structure in rust allows water and oxygen to continuously penetrate the surface rust layer, seeping into the material and causing the internal iron to continue to rust. This can lead to severe rusting, ultimately corroding the entire material, as illustrated in Fig. 7.3.

Iron oxide is generally referred to as rust, which is usually red and is produced by the oxidation–reduction reaction of iron and oxygen. Different types of rust can be produced under different circumstances. Rust is mainly composed of ferric oxide hydrate ($Fe_2O_3 \cdot nH_2O$) and iron hydroxide ($FeO(OH)$, $Fe(OH)_3$). Given sufficient oxygen, water, and time, iron will fully oxidize into rust. The combination of iron and other substances is sometimes referred to as rust, such as FeS_2, $FeSO_4 \cdot 7H_2O$, $FeCO_3$, $Fe_3(PO_4)_2 \cdot 8H_2O$, $NaFe_3(OH)_6(SO_4)_2$, $(NH_4)_2Fe(SO_3)_2 \cdot H_2O$, and so on. However, the most common and significant form of rust is iron oxide.

Fig. 7.3 Severely rusted iron

When the oxygen concentration increases or iron oxide is exposed to an oxygen environment for an extended period, the degree of iron oxidation and the chemical valence of iron can be increased. This transition occurs in the following order: Fe^{2+} (brown-green) → $Fe^{2+/3+}$ (The intermediate blending state has both Fe^{2+} and Fe^{3+}) → Fe^{3+} (red). Due to the presence of water vapor, rust formation in the atmosphere will result in the incorporation of water molecules or OH^- ions. The main components of rust include FeO(OH), $Fe_2O_3 \cdot nH_2O$, FeO, and Fe_3O_4. Typically, the side in contact with the metal is composed of FeO, while the side exposed to air is primarily FeO(OH) or Fe_2O_3, and the middle section consists mainly of Fe_3O_4.

II. **Thermodynamic Data of Rust Components**

According to "the Handbook of Practical Inorganic Thermodynamic Data" [1], Table 7.1 presents the thermodynamic data of certain rust components.

7.2.3 Copper Rust

Copper has been utilized in numerous applications because of its low electrical resistance and exceptional thermal conductivity. However, its surface can be corroded after a period of using. In ancient China, numerous bronze artifacts were produced. Over the course of thousands of years, the patina that accumulated on their surfaces has transformed their appearance, as can be observed in Fig. 7.4.

The layer of oxidation on the surface of copper, commonly referred to as patina, is a product of the reactions between copper and the oxygen, carbon dioxide, and water present in the atmosphere. The chemical process that leads to the formation of patina on copper is as follows:

$$2Cu + H_2O + O_2 + CO_2 = Cu_2(OH)_2CO_3$$

The main component of copper rust is basic copper carbonate, which has a chemical formula of $Cu_2(OH)_2CO_3$. It has a molecular weight of 221.12, a relative density of 3.8525, and refractive indices of 1.655, 1.875, and 1.909. This compound is insoluble in water and ethanol, but soluble in ammonia, acids, cyanides, and aqueous solutions of ammonium salts and alkali metal carbonates. It can be decomposed at 220 °C, and the chemical reaction formula for this process can be expressed as [3]:

$$Cu_2(OH)_2CO_3 = 2CuO + CO_2 \uparrow + H_2O$$

In atmospheric conditions, copper primarily corrodes uniformly, exhibiting a change in surface color due to corrosion products. This color change reflects the varying components of the corrosion products that form under different environmental conditions. Initially, copper oxidizes in the atmosphere, producing reddish-brown CuO and Cu_2O. In environments polluted with sulfur oxides, further oxidation

Table 7.1 Thermodynamic Data of Certain Rust Components

Chemical formula	T_{tr1}/K	T_{tr2}/K	T_{sb}/K	T_M/K	T_B/K	T_{DP}/K	ΔH_{tr1} (J mol^{-1})	ΔH_{tr2} (J mol^{-1})	ΔH_{sb} (J mol^{-1})	ΔH_M (J mol^{-1})	ΔH_B (J mol^{-1})
Fe	1184	1665		1809	3135		900	837		13,807	349,573
FeO				1650		3687				24,058	
Fe_3O_4	866			1870						138,072	
Fe_2O_3	953	1053				1735	669	0			

Note

1. T_{tr} represents the phase transition temperature of substance, T_{sb} is the sublimation temperature, T_M is the melting temperature, T_B is the boiling temperature, and T_{DP} is the thermal decomposition temperature. All temperatures are given in Kelvin (K). ΔH_{tr} represents the heat of phase transition, ΔH_{sb} is the heat of sublimation, ΔH_M is the heat of fusion, ΔH_B is the heat of vaporization. All heats are given in J/mol
2. The table lists two phase transition temperatures for the given substance: T_{tr1} for $\alpha \rightarrow \beta$ transition and T_{tr2} for $\beta \rightarrow \gamma$ transition
3. The temperature range for the existence of $Fe_2O_3 \cdot H_2O$ is 298–400 K
4. The decomposition temperature range of $FeCO_3$ is 298–800 K. FeOOH and α-FeOOH dehydrates to form α-Fe_2O_3 at a temperature of about 300 °C or 573 K. γ-FeOOH dehydrates and transforms into a metastable state of γ-Fe_2O_3 at approximately 300 °C or 573 K. As the temperature continues to rise, it transforms into α-Fe_2O_3, a stable phase, at about 450 °C or 723 K. The color of rust is as shown in Table 7.2

Table 7.2 Color of Certain Components of Rust

Chemical formula	Color
Fe	Sliver
FeO	Black
Fe_2O_3	Black
γ-Fe_2O_3	Dark brown
α-Fe_2O_3	Bright red
α-FeOOH	Red/brown/yellow
β-FeOOH	Orange
γ-FeOOH	Orange red

Fig. 7.4 Copper rust

occurs, generating Cu_2S. Cu_2S is a transient intermediate that is rapidly oxidized to form black CuS or blue-green $CuSO_4 \cdot 3Cu(OH)_2$. In marine atmospheres, chloride ions react with copper ions to produce blue-green $CuCl_2 \cdot 3Cu(OH)_2$. Under marine conditions, the primary corrosion products of brass include $CuCl_2$, Cu_2O, $ZnCl_2$, and $CuSO_4 \cdot 3Cu(OH)_2 \cdot 2H_2O$; the corrosion products of pure copper involve $CuCl_2$, Cu_2O, $Cu_2SO_4 \cdot H_2O$, $CuCl_2 \cdot 3Cu(OH)_2$, and $CuSO_4 \cdot 3Cu(OH)_2 \cdot 2H_2O$; the corrosion products of bronze are $CuCl_2$, Cu_2O, $CuCl_2 \cdot 3Cu(OH)_2$, and $CuSO_4 \cdot 3Cu(OH)_2 \cdot 2H_2O$; and those for beryllium copper also include $CuCl_2$, Cu_2O, $CuCl_2 \cdot 3Cu(OH)_2$, $CuSO_4 \cdot 3Cu(OH)_2 \cdot 2H_2O$, and Al_2O_3.

The corrosion products on the surface of bronze can be roughly divided into harmless rust and harmful rust. Harmless rust mainly refers to the ancient spots and crusts on the surface of bronze, which is characterized by a hard and dense rust layer. In contrast, harmful rust is characterized by a loose structure, which is similar to powder and commonly called as "powdery rust" [4]. The corrosion layer of bronze without "powdery rust" generally has a simple layered structure, and the outer layer is a salt compound of divalent copper. XRD analysis results show that the rust products are mainly malachite, azurite, hematite, quartz, and cassiterite, etc., without basic copper chloride. On the other hand, the corrosion layer of bronze with "powdery rust" is generally divided into three to four layers. The outer green layer is mostly a basic compound of divalent copper (including atacamite). The second layer is a reddish-brown layer of cuprous oxide. The third light-colored layer is mainly a layer of cuprous chloride. The fourth layer is the alloy substrate layer. Methods for removing copper rust include mechanical friction (manual friction, shot blasting, etc.), chemical reagent method, and thermal cleaning method. The chemical reagents used in cleaning mainly include acids and carbonate solutions.

7.2.4 *Oxide Skin on Aluminum Plate*

Objects made of aluminum or aluminum alloys may rust at a slower rate than iron. Due to their low density and lightweight nature, aluminum and its alloys have become increasingly prevalent in daily life. Nonetheless, long-term utilization can lead to the development of aluminum rust.

The chemical process that triggers the formation of aluminum oxide is as follows:

$$4Al + 3O_2 = 2Al_2O_3$$

The generated aluminum rust possesses a robust structure, effectively impeding the ingress of oxygen and water vapor into the substrate, thereby offering protection. This phenomenon is known as passivation. Consequently, the corrosion of aluminum is less severe compared to copper and iron. However, due to the frequent presence of mud and oil stains during the rusting process, the material not only appears unsightly but may also compromise its usability. As Fig. 7.5 illustrates, car tire aluminum alloy wheel hubs and industrial castings emerge noticeably clean and presentable after a cleaning.

7.2.5 *Silver Rust*

Silver is an essential ornamental metal, which has widespread applications in industry and research. Characterized by its inertness and exceptional corrosion resistance, silver remains unaffected in most environments. However, prolonged exposure to air

Fig. 7.5 Aluminum alloy wheel hub and industrial castings of a car tire

can initiate chemical reactions that lead to the formation of oxides on its surface, causing a loss of luster and a noticeable dulling of its appearance, as illustrated in Fig. 7.6. The primary component of silver rust is silver sulfide, which is generated by the interaction between silver and sulfides in the air. Furthermore, it can also react with acidic gases and nitrogen monoxide. To effectively remove silver rust, heating reduction reactions or soaking in ammonia water can be employed.

Fig. 7.6 Silver sulfide of silver rust

7.3 Laser Rust Removal Mechanism

For removing rust from the surface of metal substrates, laser cleaning offers a promising and effective alternative to traditional mechanical and chemical cleaning methods. Metal substrates suitable for laser cleaning mainly include carbon steel, austenitic stainless steel, martensitic stainless steel, titanium alloy, and aluminum alloy. The contaminants that can be removed through laser cleaning include rust layers, sulfide layers, oxide layers, paint layers, Al-Si coatings, and corrosion-resistant coatings. The mechanism of laser rust removal is similar to that of laser paint removal. In both cases, the contaminant layer and the substrate absorb the laser energy. This absorption causes a rise in temperature, which can reach the melting or vaporization point. Alternatively, it can generate thermal stress, plasma, or a shock wave. Ultimately, these effects lead to the separation of the contaminant layer from the substrate.

7.3.1 Absorption of Laser by Metals and Rust Layers

In Chap. 3, we mentioned that when a laser beam hits the surface of a material during propagation, the energy transfer of the laser beam satisfies the following relationship:

$$E_{reflected} + E_{absorped} + E_{transmitted} = E_0 \tag{7.3.1}$$

where E_0 represents the incident energy, $E_{\text{reflected}}$, E_{absorbed} and $E_{\text{transmitted}}$ represent the reflected, absorbed, and transmitted laser energy by the material, respectively. During rust removal, the substrate and corrosion are both opaque solids, and the laser transmittance is close to 0, which means $E_{\text{transmitted}} = 0$, substituting into Eq. (7.3.1), we can get

$$1 = \frac{E_{reflected}}{E_0} + \frac{E_{absorped}}{E_0} = R + A \tag{7.3.2}$$

where R is the reflectivity of the material, A is the absorption rate of the material.

Due to their unique electronic band structure, metals are opaque to visible light. Within this structure, there are numerous unoccupied energy levels situated above the Fermi level. When exposed to light, electrons in metals can easily absorb the energy of incident photons and become excited to these unoccupied energy levels. The presence of numerous such levels means that metals can absorb visible light of various frequencies, i.e., photons with different energies (ΔE). In reality, metals block all low-frequency electromagnetic waves (ranging from radio waves to ultraviolet light) and are only transparent to high-frequency electromagnetic radiation such as X-rays (0.1–10 nm) and γ-rays (1 pm–0.1 nm) [5].

During laser propagation, the intensity decays exponentially, and the incident laser intensity I at a distance x from the surface can be described as follows:

$$\mathrm{I} = I_0 e^{-\alpha x} \tag{7.3.3}$$

where I_0 represents the intensity of laser incident on the surface; α represents the absorption coefficient of the material to the incident laser.

The values of R, A, and α can be calculated based on the complex refractive index N' of the substance,

$$\mathrm{N'} = \mathrm{N} - \mathrm{iK} \tag{7.3.4}$$

where N is the refractive index of the material, which is determined by the speed of laser in the absorptive medium, and K is determined by the attenuation (absorption of laser energy) when laser propagates in the absorptive medium.

The absorption coefficient α can be expressed as:

$$\alpha = \frac{4\pi K}{\lambda} \tag{7.3.5}$$

where λ represents the wavelength of the incident laser.

7.3.2 Temperature Distribution During Laser Rust Removal

When the laser beam strikes the surface of the cleaning object, the material's surface absorbs a portion of the laser's energy. This absorption results in an increase of electron energy, leading to a transition from a lower energy state to a higher one. The laser energy is then converted into heat energy through intricate interactions involving phonons and photons. Subsequently, this heat energy is transferred to the surrounding medium via thermal diffusion, elevating the material's temperature. We take laser dry rust removal as an example to solve the temperature distribution by heat conduction equation.

For dry laser rust removal, a Cartesian coordinate system is established and the thickness of the sample is L. When the laser incidents on the rusted surface of the sample, the heat conduction equation [6] can be written in a steady state:

$$\alpha \frac{\partial T(x, y, z, t)}{\partial t} = \left(\frac{\partial^2 T}{\partial^2 x} + \frac{\partial^2 T}{\partial^2 y} + \frac{\partial^2 T}{\partial^2 z} \right) \tag{7.3.6}$$

where ρ, c, k, α are respectively the density, specific heat, thermal conductivity, and thermal diffusivity of rust, $\alpha_t = \rho c/k$. Since the lasers used are mostly Gaussian beams and the spot area is very small, we approximate it as uniformly heated, so it can be simplified by the Fourier heat conduction model [7]:

$$\begin{cases} \alpha \frac{\delta T}{\delta t} = \frac{\partial^2 T}{\partial z^2} & 0 \leq z \leq L, t \geq 0 \\ \alpha \frac{\delta T}{\delta z}\Big|_{z=0} = q \\ T(z,t)|_{t=0} = T_0 \; 0 \leq z \leq L \\ \frac{\delta T}{\delta z}\Big|_{z=L} = 0 & t \geq 0 \end{cases} \tag{7.3.7}$$

where q is the volumetric energy density of the laser. The above equation can be solved by the Gauss–Seidel iteration method. α_t can be measured by experiment. By solving the above equation, the temperature distribution can be obtained. Generally speaking, only its numerical solution can be obtained.

The numerical simulation of the temperature field distribution on the metal substrate surface during laser rust removal was conducted by Ren [8] using the Gaussian surface heat source and finite element grid partitioning strategy. This simulation was then compared with the experimental results. When the average power of the irradiated laser was set to 8 W, the laser energy density was relatively low, with the highest temperature on the metal substrate surface reaching approximately 700 °C. No molten pool was observed on the cross-section of the metal substrate. As the average power of the laser increased to 12 W, the temperature on the metal substrate surface exceeded 1200 °C, rapidly rising above the melting point of low carbon steel. This resulted in the formation of a molten pool morphology. When the average power of the pulsed laser was further increased to 16 W, the highest temperature on the surface of the metal substrate reached approximately 1700 °C, exceeding the melting point. At this power level, a molten pool with a concave center of the light spot and small molten beads distributed at the edge of the light spot formed on the surface of the metal substrate.

7.3.3 *Main Mechanism of Laser Rust Removal*

Due to the different laser energy absorbed by the corrosion layer and the substrate, the thermal expansion coefficient is different, resulting in different temperatures at different locations. The temperature gradient will produce thermal stress, which makes the rust layer vibrate. At high temperatures, chemical reactions are triggered. When the temperature reaches the melting and vaporization points, ablation effects occur. Additionally, the laser may interact with air in the affected area, generating a plasma shock wave. For wet laser cleaning, there can also be a phase explosion effect due to liquid evaporation, but this method is less commonly used in laser rust removal. These are the primary mechanisms involved in laser rust removal, and we will explore them in further detail.

I. **Thermal Effect and Thermal Stress**

When the material rapidly absorbs laser energy, the different thermal expansion rates of the substrate and the rust layer result in a rapid pressure build-up at the junction.

This pressure exerts force on the adjacent material, generating a corresponding reaction force. This leads to the formation of a two-way pressure wave between the rust layer and the substrate, known as thermal stress. Once the thermal stress reaches a critical level, it exceeds the bonding force between the rust layer and the substrate, causing the rust layer to detach from the substrate. The primary force acting between the rust layer and the substrate is viscous force, the magnitude of which is [9]:

$$\mathrm{F(z, t)} = \left(\mathrm{B} + \frac{4}{3}\mathrm{G}\right)\frac{\partial \mathrm{u(z, t)}}{\partial \mathrm{z}} - B\gamma T(z, t)\mathrm{c} \tag{7.3.8}$$

The bulk modulus and shear modulus of the basement are denoted as B and G, respectively. The expansion coefficient of the basement is given by γ. The intermolecular gap is represented by z, while t denotes the action time. The displacement function and temperature function of the basement are given by $u(z, t)$ and $T(z, t)$, respectively. When the thermal stress exceeds the viscous force, the rust layer detaches from the basemen.

II. **Chemical Reaction**

During laser rust removal, the main targets are metals and metal oxides, both of which are prone to undergo various chemical reactions at high temperatures. The heat absorption and release of these reactions have a significant impact on the temperature and properties of each component involved. As such, it is essential to consider their possible reactions during the rust removal process. The possible reactions of the main corrosion products of iron during the rust removal process are shown below [10, 11]:

(1) FeO:

$$\mathrm{2Fe + O_2 \xrightarrow{1873-1923\ K} 2FeO} \tag{7.3.9}$$

$$\mathrm{FeO + C \xrightarrow{Above\ 1223\ K} Fe + CO \uparrow - Q} \tag{7.3.10}$$

$$\mathrm{FeO + nCO \xrightarrow{Above\ 1223\ K} Fe + CO_2 + (n-1)CO + Q} \tag{7.3.11}$$

(2) Fe_3O_4:

$$\mathrm{Fe_3O_4 + 4C \xrightarrow{1123-1173\ K} 3Fe + 4CO \uparrow} \tag{7.3.12}$$

$$\mathrm{Fe_3O_4 + CO \xrightarrow{when\ it's\ higher\ than\ 853\ K} 3FeO + 4CO_2} \tag{7.3.13}$$

$$\mathrm{Fe_3O_4 + 4CO \xrightarrow{when\ it's\ lower\ than\ 853\ K} 3Fe + 4CO_2} \tag{7.3.14}$$

$$\mathrm{2Fe_3O_4 \xrightarrow{2573\ K} 6FeO + O_2 \uparrow} \tag{7.3.15}$$

(3) Fe_2O_3:

$$6Fe_2O_3 \xrightarrow{1735\ K} 4Fe_3O_4 + O_2 \quad (7.3.16)$$

$$Fe_2O_3 + CO \xrightarrow{\Delta} 4FeO + CO_2 \quad (7.3.17)$$

$$FeCO_3 \xrightarrow{1735\ K} FeO + CO_2 \uparrow \quad (7.3.18)$$

III. **Plasma Shock Waves**

When light is irradiated on the metal surface, plasma is excited, and the wavelength of this plasma is much smaller than the original wavelength. When laser cleaning metal materials, the laser passes through the contaminants to reach the surface of the metal substrate. A plasma is formed between the substrate and the contaminants. Once the laser energy reaches a certain value, it triggers the expansion of the plasma, causing an explosion and generating plasma shock waves. These shock waves and explosions occur between the surface of the substrate and the layer of contaminants, effectively removing the contaminants. Han et al. [12] studied the process of thin film damage by lasers, analyzed various factors that contribute to thin film damage, and believed that the laser plasma effect played a crucial role in the laser cleaning process. Additionally, if the temperature can reach the melting point of rust, melting will occur. For metal rust, this temperature is relatively high.

Many researchers have studied the mechanism of laser rust removal. Zhou [13] conducted research on the mechanism of laser rust removal using a pulsed fiber laser. This laser had a maximum power output of 100 W, a pulse width of 100 ns, a wavelength of 1064.5 nm, and a repetition frequency of 100 kHz. He used laser energy densities of 30.6 J/cm^2 and 10.2 J/cm^2 to perform step-by-step laser cleaning of the rust layer on the surface of AH32 marine steel. The original rust layer was found to be loose and porous with various blocky oxides present. After the first cleaning with a 60 W laser, the thick rust layer was removed, and the rusted surface had many cracks and pits; after the second cleaning with a 20 W pulsed laser, the remaining rust on the sample surface was removed, revealing uniformly arranged laser action traces. This study suggests that the mechanisms for step-by-step laser rust removal mainly include hole blasting mechanism and ablation evaporation mechanism. During the first cleaning, the loose porous rust structure absorbed a large amount of incident laser energy. The air and moisture enriched in this structure rapidly expanded under the thermal effect of the pulsed laser, generating a large blasting force that caused the rust layer to crack and peel off. During the second cleaning, due to increased surface temperature after the first laser irradiation, the base rust's absorption rate of the laser increased. The rust removal mechanism was primarily ablation evaporation [14].

Pan [15] studied thicker rust layers on steel substrates. They divided the rust layer into three distinct layers, as illustrated in Fig. 7.7, and attributed the cleaning of the first and second layers mainly to the phase explosion mechanism.

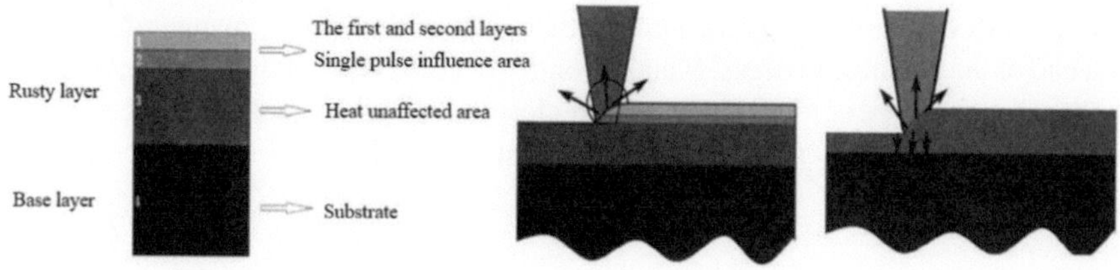

Fig. 7.7 Schematic diagram of layered model and cleaning mechanism: **a** layered model, **b** explosion mechanism of the first and second layers, **c** expansion and ablation mechanism of the third layer

The first, second and third layers are made of rust, while the fourth layer is the surface of the steel substrate. When the laser beam is vertically applied from the first layer, the first and second layers become single-pulse thermal influence zones, while the third layer (closer to the basement) forms a thermal end influence zone. The first and second layers are divided by the isotherm of the rust decomposition temperature, and the third layer experiences thermal elastic expansion and ablation effects. Furthermore, because the rust layer has a loose, porous structure, when the laser beam passes through it, refraction occurs, producing a focusing effect similar to a lens and enhancing the near field.

7.4 Laser Rust Removal Equipment and Characterization Methods

7.4.1 Laser Rust Removal Equipment

The complete laser rust removal system typically comprises a laser, a mobile transmission device, a beam adjustment device, a control device, a monitoring device, auxiliary devices such as a dust removal device and so on. Pulsed CO_2 lasers, Nd:YAG lasers, or fiber lasers are commonly utilized in these systems. The CO_2 laser and Nd:YAG laser are commonly utilized in the mechanical field, whereas excimer lasers are often employed for ultra-precision cleaning. The beam adjustment device, consisting primarily of optical lenses and other components, enables adjustment of the spot shape and size of the laser output. This adjustment in turn controls its energy density of radiation, enabling it to meet different rust removal thresholds under varying circumstances and ultimately enhancing cleaning efficiency [16–18]. The mobile transmission device, propelled by a stepper motor, drives the cleaning sample or laser output end. It must consider the high flexibility required for actual rust removal while ensuring that the spot covers the sample position. Fiber optic transmission can be employed. The monitoring device, equipped with detection instruments, monitors sound, image, spectrum, and other signals generated during the cleaning process. These signals are promptly fed back to the control device. The control device coordinates and controls other parts of the device, managing the start and end of the cleaning process. Manual control or computer control can be employed for rust removal operations. For laser wet cleaning, an additional device is required to deposit a liquid film on the substrate surface.

7.4.2 Characterization Methods of Laser Rust Removal Effects

During the laser rust removal process, a plasma is generated, and its use for elemental analysis is a crucial method for characterizing laser rust removal. Yao et al. [19] employed a Nd:YAG pulsed laser to clean rusted iron blocks and subsequently conducted plasma elemental spectroscopy analysis. They used laser output energy ranging from 0 to 10 J, a wavelength of 1.06 μm, a pulse width of 20 ns, a spot diameter of 4 mm, and a frequency of 1 pulse/1.5 min. The initial iron mass fraction in the iron block was 82.32%. After cleaning, the plasma spectrum generated by the pulsed laser cleaning of the iron block sample was collected using a step grating spectrometer, and the iron atomic spectrum was marked. It was observed that the relative intensity of the plasma spectrum after cleaning had significantly increased. Subsequently, the elemental mass fraction was measured using chemical methods for verification. It was discovered that when the characteristic spectral line intensity at 563.40 nm increased by 443.11%, the cleaning rate reached 99.84%, indicating that the iron block had been thoroughly cleaned. Additionally, the fluctuation of the plasma spectral line also significantly decreased. Since the proportions of nitrogen and oxygen in the air are fixed, their impact on the spectral line was not significant. Therefore, plasma spectroscopy analysis can provide an important basis for determining whether laser rust removal is complete.

In the study of elemental content analysis after laser rust removal, it is not only possible to analyze the content of substrate elements, but also to simultaneously analyze the content of impurity elements. In addition to using a grating spectrometer for plasma spectroscopy analysis, various instruments can be employed for elemental analysis, including scanning electron microscopes. Qiao and their colleagues [20] utilized a high repetition rate, high-energy laser with a wavelength of 1064 nm to examine laser rust removal on carbon steel surfaces. This laser had a power ranging from 200 to 500 W, a pulse width of 80 ns, a pulse repetition rate between 5 and 15 kHz, a scanning width of 5 cm^2/s, and an overlap rate of 5%. The researchers used the Nova NanoSEM50 scanning electron microscope to analyze the relative content of surface elements after laser rust removal.

It was observed that the Fe element content of the original rust layer was low, while the O element content was high. When the scanning speed was set at 6.0 mm/s, the relative content of Fe element was at its highest, and the relative content of O element was at its lowest, indicating that this was the optimal condition for rust removal. Qiao et al. [20] also reported that the carbon element content was 6.23 wt%. Compared to oxygen, this indicated that there was adsorbed carbon on the surface, cleaning the rust layer and it had not been oxidized. The study also examined the relative content and elemental composition of micro-areas within the laser spot trace, non-smooth micro-areas and micro-areas of the overlapped areas. The results showed that the iron content in both micro-areas within the laser spot trace and non-smooth micro-areas was consistent, indicating successful cleaning. However, the oxygen content within the non-smooth micro-area was higher, possibly indicating that some rust dust had not been fully removed. The relative carbon content at the overlap was low due to a lower degree of inert reaction in areas with fewer number of scanning. Overall, this study demonstrates that rust on carbon steel surfaces can effectively be removed by laser treatment through comprehensive elemental analysis.

7.5 The Impact of Different Laser Parameters on Laser Rust Removal Effect

The impact of different laser parameters on the effectiveness of laser rust removal varies significantly. Previous studies have shown that laser power, repetition frequency, number of scanning, scanning speed, and scanning line spacing have great effects on laser rust removal. Of course, the severity of corrosion on the cleaning object is the primary determining factor for selecting laser cleaning parameters.

I. Laser Power and Energy

When investigating the effects of different laser parameters on laser rust removal, it is crucial to first consider laser power density, which is the laser power per unit area. Here, laser power includes both average power and peak power (single-pulse energy). Power density depends on laser power, spot size, repetition frequency, and other parameters. Let's first discuss the impact of average laser power. Chen et al. [10] utilized a pulsed laser with a wavelength of 1064 nm to explore the effect of laser power on the rust removal effect of stainless steel surfaces. The laser parameters were as follows: single pulse energy of 50 mJ, single pulse duration of 100 ns, repetition frequency of 10 kHz, spot diameter of 3 mm, number of scanning of 5. Three different average output powers were selected, specifically 300 W, 400 W, and 500 W. The surface morphology and composition of the stainless steel before and after cleaning were analyzed using SEM (Scanning Electron Microscopy) and EDS (Energy Dispersive Spectroscopy) imaging techniques. At 300 W power, large pieces of oxide on the surface decomposed, forming spherical oxide fragments. Elemental analysis revealed an increase in the atomic percentage of Cr and Ni elements, while Fe and O elements still dominated. The cleaning effect was general. At 400 W power, the bulk oxide on the stainless steel surface decreased, and the surface changed from rough to smooth. Elemental analysis revealed a shift from predominantly Fe and O elements to predominantly Fe, O, and Cr elements, indicating that the surface oxide layer has been almost completely removed, the sub-surface layer is exposed. The surface is smoother and the laser cleaning effect is better than before. At 500 W power, blocky oxides increase, and the surface becomes rough. Elemental analysis indicates a further increase in the proportion of Cr elements. The oxides on the stainless steel surface have been effectively cleaned, but there is some damage to the substrate. From this analysis, it can be concluded that as the laser power increases, the thickness of laser cleaning increases, resulting in better cleaning effectiveness. The cleaning threshold of laser rust removal on stainless steel surface is approximately 3.96×10^3 W/cm^2, while the substrate damage threshold is about 5.52×10^3 W/cm^2.

Qiao et al. [20] used the elemental analysis method to explore the impact of laser average power on cleaning effect. A high repetition rate, high-energy laser with a wavelength of 1064 nm was used for laser rust removal on the surface of carbon steel. The pulse width was 80 ns, scanning speed was 5 cm^2/s, and overlap rate was 5%. Power settings were 200 W, 300 W, and 500 W, with pulse repetition frequencies of 5 kHz, 10 kHz, and 15 kHz. The experimental results indicate that higher power correspond to higher peak power densities of the laser, resulting in increased Fe element content after substrate cleaning. Zhang et al. [21] compared the cleaning effects and mechanisms of the oxide film on the surface of aluminum alloy under different laser energy densities. The results showed that laser cleaning could completely remove the oxide layer. The pulsed fiber laser operated at a wavelength of 1064 nm with a pulse width of 100 ns. It has a maximum repetition frequency of 500 kHz and a maximum average power of 100 W, emitting laser in Gaussian mode. The initial cleaning threshold (i.e., when cleaning began to take effect) was 12.7 J/cm^2, and the complete cleaning threshold was 25.5 J/cm^2. At low energy densities, laser ablation causing phase explosion is the primary cleaning mechanism, where substrate evaporation squeezes out the molten oxide layer, forming pulse craters. At high energy densities, besides phase explosion, collision effects also occur, leading to splashing and removal of the oxide layer. Transient energy absorption induces thermal stress coupling effects, resulting in the separation of the substrate and oxide layer. Impact effects from evaporation lead to the removal of the oxide layer. Aluminum vapor absorbs laser energy, forming high-energy density plasma, thereby enhancing the coupling between the laser and plasma.

Chen et al. [22] studied the effect of laser cleaning on the surface morphology of A7N01 aluminum alloy by varying the laser energy density. Using a fiber laser with a wavelength of 1064 nm, a frequency of 600 kHz, and a pulse width of 150 ps to clean the oxide film on the surface of the aluminum alloy. The study found that when the energy density is exceeded 1.53 J/cm^2, plasma gas was generated, which increases the contact area and time between the oxygen in the air and the high-temperature aluminum alloy. This resulted in more intense oxidation phenomena, leading to an increase in both laser cleaning marks and secondary oxidation phenomena on the surface of the aluminum alloy with the increasing energy density.

II. **Repetition Frequency**

Cao et al. [23] utilized an RFL-P50Q Q-switched fiber laser to clean the oxide layer on the copper cathode of lithium-ion batteries. The laser parameters were as follows:

wavelength of 1064 nm, power of 50 W, and spot diameter of 54 μm. The experiment investigated the effect of different pulse frequencies on the cleaning effectiveness. The scan line spacing was 0.06 mm, the scanning speed was 5000 mm/s, and the scanning was performed twice, with repetition frequencies of 40 kHz, 50 kHz, and 60 kHz. The results showed that at a frequency of 40 kHz, the original oxide layer was removed, but a new oxide layer formed afterward. At a frequency of 50 kHz, the oxide layer was essentially removed without regeneration, while at a frequency of 60 kHz, some parts of the oxide layer could not be completely removed. A higher repetition frequency can increase the cleaning speed, but it needs to be combined with the overlap between scan lines and number of scanning to effectively clean the rust on the metal surface.

III. The Number of Scanning

The number of scanning will affect the rust removal effect. Sun et al. [24] conducted research on laser removal of surface oxide films from TiNi alloy using the MoDel RQM-0100 Nd:YAG pulsed laser cleaning machine. The laser parameters included a repetition frequency of 20 Hz, a power percentage of 80%, a scanning speed of 200 mm/s, and a scanning interval of 0.03 mm. The TiNi alloy samples consisted of 44.34% titanium and 55.66% nickel. Sample 1 and 2 (Sample 1 to 4 were identical) were scanned 6 times and 10 times, respectively. It was found that the surface of sample 2, scanned 10 times, was brighter than that of sample 1 scanned 6 times. Additionally, the cross-section of sample 2 was more uniform than that of sample 1. However, the oxide films on both samples were not completely removed. After reaching a certain number of scanning, the changes in laser cleaning effectiveness became less noticeable with the increase in the number of scanning. Photographs of the surfaces before and after cleaning for both types of samples are shown in Figs. 7.8 and 7.9.

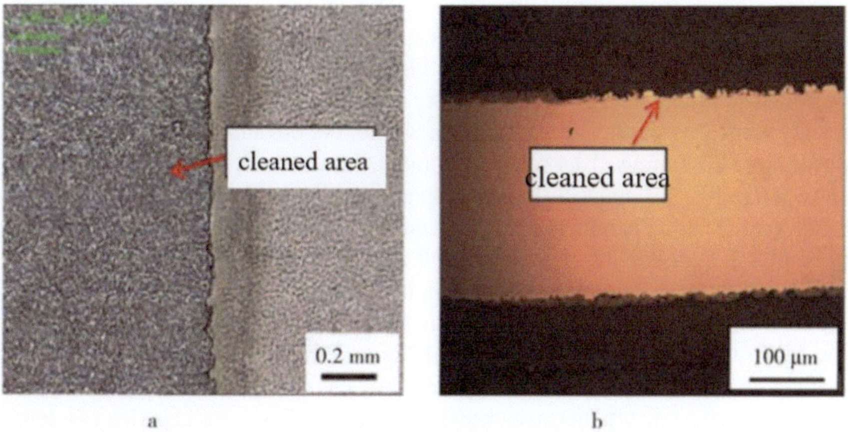

Fig. 7.8 Comparison of surface morphology of sample 1 before and after laser cleaning. **a** Surface topography; **b** Cross section

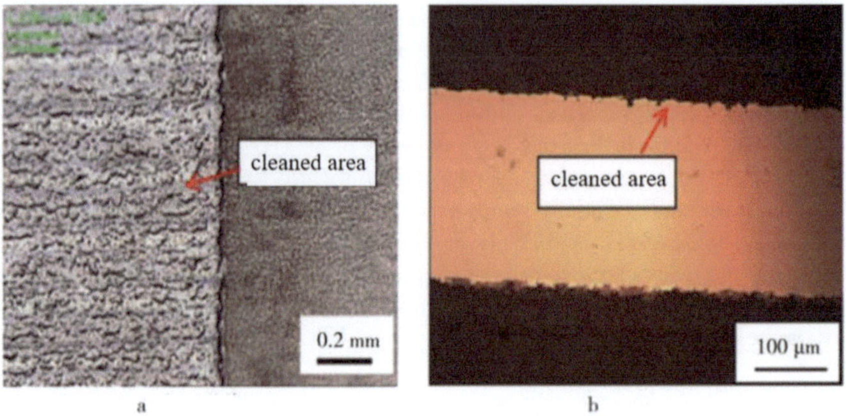

Fig. 7.9 Comparison of surface morphology of sample 2 before and after laser cleaning. **a** Surface topography; **b** Cross section

Changing the laser parameters to a repetition frequency of 100 Hz, power percentage of 100%, scanning speed of 500 mm/s, scanning spacing of 0.09 mm, laser rust removal was performed on TiNi alloy sample 3 and 4, with the number of scanning of 3 and 5 respectively. It was observed that the surface of sample 3 was brighter than that of samples 1 and 2. The unscanned area have reduced, and from the cross-section, it can be seen that sample 3 has less residual oxidation film; while sample 4 exhibited a more uniform surface after laser cleaning, and the oxide film was more thoroughly removed, as shown in Figs. 7.10 and 7.11. Additionally, it was observed during the experiment that removing the oxidation film from TiNi alloy was more challenging compared to rust due to its denser and more robust nature. Moreover, since there was no protective gas in the laser cleaning experiment, and TiNi alloy has a strong affinity for oxygen, secondary oxidation was easy to occur.

Zhu et al. [25] used acousto-optical Q-switched diode-pumped Nd:YAG laser to clean the rust on the surface of 5A12 aluminum alloy. The cleaning speed was controlled by changing the number of laser scans at the same position. The number of laser scans was set to 5, 10, 15, 20, 25, 30, 40, 60, 80, and 100, and the cleaning speeds were 20.7 mm/s, 10.4 mm/s, 6.9 mm/s, 5.2 mm/s, 4.1 mm/s, 3.5 mm/s, 3.0 mm/s, 2.6 mm/s, 1.7 mm/s, 1.0 mm/s respectively. By testing the cleaned samples, the optimal cleaning parameters were found to be a power of 98 W and a scanning speed of 4.1 mm/s. If the scanning speed is too fast, the cleaning effect is poor; if the speed is too slow, the cleaning efficiency is low. According to the research results, it can be found that if the power is below the cleaning threshold, no matter how many times it is scanned, there will be no effect. If the power reaches the cleaning threshold but is still relatively low, then increasing scan times can improve the cleaning effect due to thermal accumulation effect. However, cleaning efficiency

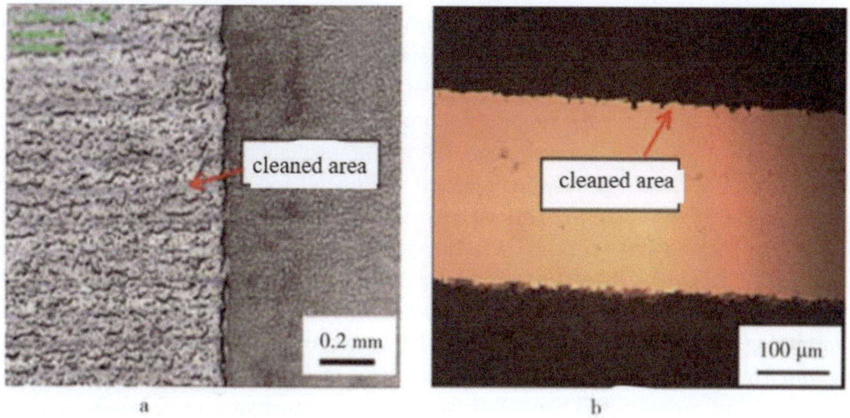

Fig. 7.10 Comparison of surface morphology of sample 3 before and after laser cleaning. **a** Surface topography; **b** Cross section

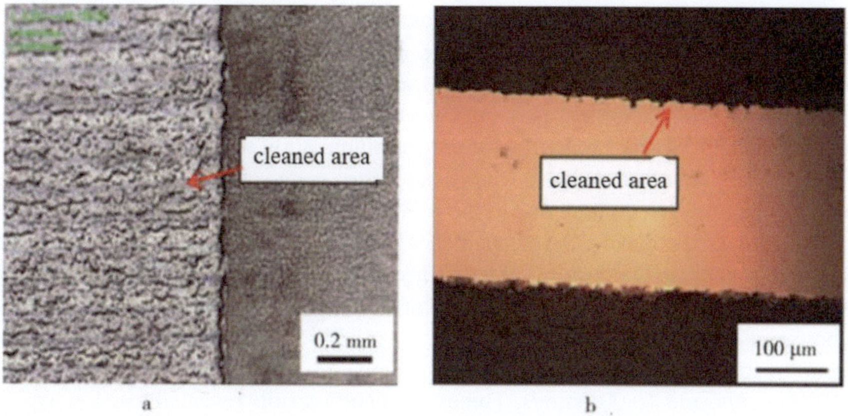

Fig. 7.11 Comparison of surface morphology of sample 4 before and after laser cleaning. **a** Surface topography; **b** Cross section

is low when using multiple scans. It is only used for stubborn rust that is strongly bonded to the substrate, or for thick rust, or when high-power lasers are not available, or when careful cleaning is needed to prevent substrate damage.

IV. **Scanning Speed**

The faster the scanning speed, the higher the cleaning efficiency, and the smaller the vice versa. Some metals, such as aluminum alloys, have a lower melting point relative to iron. Thus, with a slower scanning speed and a longer dwell time of the laser in the same area, it may cause melting phenomena on the substrate. Luo et al. [11] employed the POWER-LASE Rigeli400 nanosecond pulse solid-state laser for cleaning the oxide layer on the surface of aluminum alloy anodes. The laser parameters included a wavelength of 1064 nm, peak power of 400 W, minimum pulse width of 15 ns, maximum pulse frequency of 10 kHz, and a spot size of 0.8 mm. They conducted experiments on 5 identical samples with scanning speeds of 0.5 m/min, 1.5 m/min, 2.5 m/min, 3.5 m/min, and 4.5 m/min, respectively. After laser cleaning, the interface morphology between cleaned and uncleaned areas were characterized by an optical microscope, the surface morphology post-cleaning was observed by a scanning electron microscope (SEM). Energy spectrum analysis (EDS) was used for phase identification. The results showed that the surface of sample 1 was a molten morphology; the surface of sample 2 was mainly molten morphology, with tearing morphology appearing; sample 3 exhibited both molten and tearing morphology; sample 4 was mainly tearing morphology; while the tearing strength decreased on the surface of sample 5. These observations were attributed to excessive energy accumulation and subsequent melting of the aluminum substrate when the scanning speed was too low (0.5 m/min). Conversely, when the scanning speed was 3.5 m/min, the laser mainly peeled off the oxide layer, and further increasing the speed led to a decrease in cleaning efficiency. Hence, it was evident that scanning speed also affected the cleaning threshold and damage threshold. Moreover, they further analyzed the influence of scanning speed on the distribution of surface elements of the substrate. Taking samples 1 and 4 as examples, they found an increase in oxygen content for sample 1 and a decrease in oxygen content for sample 4. This was attributed to surface oxidation of the aluminum alloy when the scanning speed exceeded the damage threshold, whereas an appropriate scanning speed resulted in a decrease in oxygen content due to the peeling off of the oxide layer.

Cao et al. [23] utilized a Q-switched fiber laser to investigate the effect of different scanning speeds on the removal of oxide layers from lithium-ion battery electrodes via laser cleaning. The laser parameters were as follows: a wavelength of 1064 nm, a power of 50 W, a spot diameter of 54 μm, a repetition frequency of 50 kHz, and a scanning line spacing of 0.06 mm. The scanning speeds were adjusted to 2500 mm/s, 5000 mm/s, and 10,000 mm/s. The experimental results revealed that low-speed scanning could remove the oxide layer but might result in secondary oxidation. Moderate-speed scanning did not induce secondary oxidation, but a single scan was insufficient to completely remove the oxide layer, necessitating a second scan. High-speed scanning left a significant portion of the oxide layer uncleared. There exists an optimal value for scanning speed. Additionally, the experiments found that the influence of scanning speed on cleaning efficacy was generally higher than that of repetition frequency.

Qiao et al. [20] conducted an experiment of laser rust removal on carbon steel surfaces using a high-repetition-rate, high-energy laser with a wavelength of 1064 nm and set the laser power to 500 W, repetition frequency to 10 kHz, pulse width to 80 ns, overlap rate to 5%, the scanning speed from 3.0 to 9.0 cm^2/s, and an interval of 1.0 cm^2/s. The experimental results revealed that at a scanning speed of 6 cm^2/s, the iron content was highest and the best cleaning effect. It is evident that the scanning speed generally plays a more important role on cleaning efficacy than the repetition frequency. The cleaning effect will decrease if the scanning speed is too low or too high.

V. **Scanning Line Spacing or Overlap Amount**

If the scanning line spacing is too small or the overlap is large, the scanning path will be too long, resulting in extended cleaning time and reduced efficiency. Conversely, if the scanning line spacing is too large or the overlap is too small, the power density in some areas will be too small, leading to a decrease in cleaning effectiveness. Cao et al. [23] used a Q-switched pulsed fiber laser to explore the effect of scanning spacing on the effect of laser cleaning the oxide layer of lithium-ion battery electrodes. The laser parameters were as follows: a wavelength of 1064 nm, power of 50 W, spot diameter of 54 μm, repetition frequency of 50 kHz. The experiment was conducted at a scanning speed of 5000 mm/s, two scans and the scanning line spacing of 0.05 mm, 0.06 mm, and 0.07 mm. The experiments found that with a scanning line spacing of 0.05 mm, the original oxide layer was removed, but secondary oxidation occurred. At a scanning line spacing of 0.06 mm, the oxide layer was essentially removed without any signs of secondary oxidation. However, with a frequency of 60 kHz, some portions of the oxide layer remained.

Zhou [13] used a pulsed fiber laser with a maximum average power of 100 W, a pulse width of 100 ns, a wavelength of 1064 nm, and a pulse repetition frequency of 100 Hz as the light source to perform step-by-step laser cleaning of the rust layer on the surface of AH32 marine steel. The scanning speed and overlap amount were controlled by the scanning galvanometers, a fixed platform, and a control system. The focused laser spot diameter was 50 μm, and the maximum scanning speed of the high-speed galvanometers was adjustable up to 8000 mm/s. At a scanning speed of 3000 mm/s, laser energy densities of 30.6 J/cm^2 and 10.2 J/cm^2 were used respectively. The scanning path is shown in Fig. 7.12. The overlap amount was inversely proportional to the scanning speed. According to the experimental results, when the overlap amounts were set to 50%, 40%, and 30%, the scanning speeds were 2500 mm/s, 3000 mm/s, and 3500 mm/s respectively. After laser cleaning, the surface of the substrate exhibited a partially molten state, with a smooth and uniform interior within the spot, while dendritic protrusions were distributed at the edges.

Deschênes [26] compared the cleaning conditions at different overlap amounts. The schematic diagram of the laser spot overlap used is shown in Fig. 7.13. It was found that too much overlap would cause damage to the substrate, while too little overlap might result in incomplete cleaning.

VI. **Laser Mode**

Deschênes [26] also compared the cleaning effects of single-mode and multi-mode lasers. It was found that for removing oxide layers and rust, the multi-mode laser was almost twice as effective as the single-mode laser. This is due to the more uniform light spot of the multi-mode laser, while the single-mode laser is Gaussian distributed.

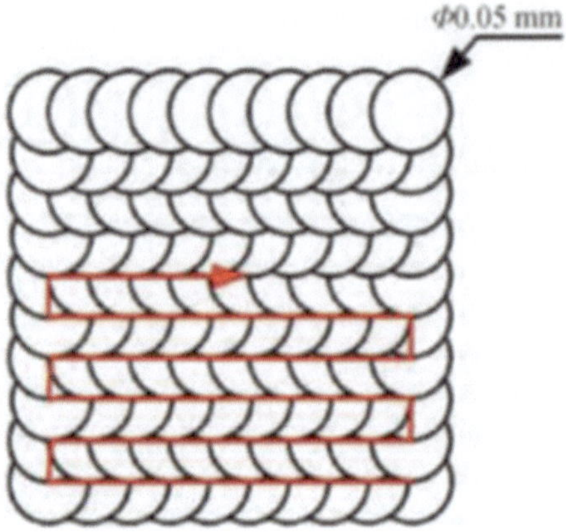

Fig. 7.12 Spot scanning route and lap amount

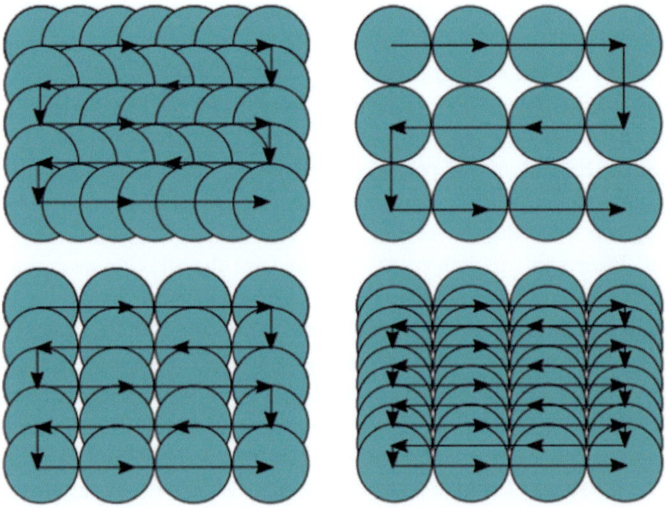

Fig. 7.13 Several different degrees of overlap ratio used in the experiment

VII. **Degree of Corrosion of the Sample**

In addition to the parameters of the laser itself, the degree of corrosion of the sample has a decisive effect on the cleaning effectiveness. Tian et al. [16] used a Q-switched pulsed Nd:YAG laser to perform laser rust removal experiments on iron samples. The laser parameters included a wavelength of 1064 nm, a pulse width of 10 ns, a repetition frequency above 20 Hz, and a maximum single pulse energy of 220 mJ. In the experiment, different levels of corrosion were classified for various samples, ranging from mild to severe, denoted as grades A, B, and C respectively. It was found that for grade A samples, the laser cleaning threshold was very low, For grade

B samples, the threshold was higher compared to grade A samples. For grade C samples, due to the thick rust layer, a single cleaning session was insufficient to effectively remove the rust. And simply increasing the energy could not improve the cleaning effect well, it was necessary to further increase the repetition frequency.

Yang et al. [27] conducted a study on laser rust removal of deeply corroded steel plates. They utilized a pulsed fiber laser with a maximum average power of 20 W to clean severely corroded Q235 steel substrates. They found that employing a slow/ fast multiple-scanning method could achieve better laser rust removal effects. When the repetition frequency was too high, the peak power did not reach the cleaning threshold, and the rust could not be removed; conversely, a repetition frequency that was too low could lead to secondary oxidation on the sample surface due to the high peak power; An appropriate scanning spacing can improve the quality of rust removal.

Deschênes et al. [26] conducted cleaning experiments on rust of different degrees on iron substrates. Photographs of the rust are shown in Fig. 7.14. The rust in Figure (a) is the mildest, predominantly surface rust, designated as Grade A. The rust in Figure (b) is of moderate severity, designated as Grade B. The rust in Figure (c) is the most severe, designated as Grade C. Cleaning was performed using a Q-switched fiber laser (1064 nm) with different modes and average powers, as shown in Table 7.3. It can be seen that regardless of the laser parameters used, the cleaning speed was fastest for Grade A rust (surface rust) and slowest for Grade C rust..

VIII. **Comparison of Dry and Wet Laser Cleaning**

For metal rust removal, dry cleaning is generally adopted. The reason is that in wet cleaning, typically using water, liquid molecules can penetrate into the substrate and

Fig. 7.14 Corrosion of iron surfaces to varying degrees

Table 7.3 Laser cleaning speed under different corrosion grades (unit: cm^2/s)

Corrosion grades	Cleaning speed			
	100 W single mode	200 W single mode	500 W single mode	2 kW single mode
A	12–21	25–42	50–75	324
B	4–12	10–25	15–50	204
C	1–4	2–10	5–15	102

cause new corrosion. Therefore, there is relatively less research on wet laser cleaning of rust. Lei Zhenglong [28] compared dry and liquid film-assisted wet laser cleaning methods for removing rust layers from high-strength steel surfaces. They conducted a comparative analysis of the rust removal effects of both methods using a nanosecond fiber pulsed laser with a wavelength of 1064 nm, a pulse width of 30 ns, and an adjustable frequency ranging from 1 to 10 kHz, with a maximum average power of 400 W. The results indicated that both methods could effectively remove rust layers from the sample surfaces, and under low power conditions, the liquid film-assisted laser cleaning demonstrated better results compared to dry laser cleaning.

The relative effect of laser cleaning can be demonstrated when used in conjunction with other cleaning methods. Li [29] mentioned a method of laser-assisted chemical cleaning. This method combines laser-induced oscillation waves with commonly used chemical corrosion techniques for rust removal. Specifically, the method involves focusing Q-switched laser pulses into an HCl aqueous solution (hydrochloric acid) to generate plasma and thereby produce oscillations. In the experiment, a Q-switched laser with a wavelength of 1064 nm, a spot radius of about 10 μm, and pulse energies ranging from 300 to 500 mJ were used.

The laser-induced plasma is difficult to completely remove the oxide films under aerobic conditions, but it can effectively remove other contaminant particles, especially carbon-containing pollutants. Using laser-assisted hydrochloric acid can effectively remove the oxide layer. The combination of hydrochloric acid and laser cleaning compared to hydrochloric acid alone shows varying effects on cleaning depending on the position of the laser spot. Overall, laser assistance enhances the cleaning effect. The improvement is not significant when the pulse energy is less than 300 mJ, but it becomes significant when the pulse energy exceeds 300 mJ, leading to a noticeable improvement in the efficiency of oxide film removal. Moreover, the laser-assisted cleaning does not damage the base.

7.6 Research and Practical Applications of Laser Rust Removal

I. Laser Removal of Oxide Film on Lithium-Ion Battery Poles

Currently, the electrode poles of lithium-ion batteries are mostly made of copper. Due to environmental factors, oxidation films can form on these poles after long-term use. Cao [23] used an RFL-P50Q type Q-switched pulsed fiber laser to study the laser cleaning of the oxide layer on the lithium-ion battery electrode poles. After laser cleaning, the connectivity and corrosion resistance of the electrode poles were tested. The results showed that laser cleaning can remove the oxides from the electrode poles, reduce the contact resistance between the electrode pole and the conductive strip, improve the contact performance, and the more severe the oxidation, the greater the reduction in resistance after laser cleaning. The oxidation layer between the batteries can cause voltage between the pole and the conductive strip. The worse the battery consistency, the greater the dynamic voltage fluctuation. Laser cleaning can improve the consistency of the battery pole connections. Observations of electrode poles placed in a constant temperature and humidity environment showed that samples cleaned with lasers were more corrosion-resistant than those polished with sandpaper, as the remelting after laser cleaning can form a corrosion-resistant structure. It is evident that laser cleaning of lithium-ion batteries has a good cleaning effect.

Ni [30] utilized a pulsed laser with an average power of 100 W and a scanning speed of 10 mm/s to clean the 5 μm thick anodic oxide film on the edge of the aluminum alloy welds. This process completely removed the anodic oxide film layer, and the welds were uniformly formed with a metallic luster on the surface. No clustered pores, impurities, or other internal defects were observed. The tensile strength of the laser-cleaned aluminum alloy joint was 298–303 MPa, the tensile fracture elongation at break was 6.2–6.5%. The performance range of the laser-cleaned welds was consistent with that of the mechanically scraped welds.

With the widespread use of electric vehicles, a large number of waste lithium-ion batteries not only seriously pollute the environment but also cause significant energy and resource wastage. Liu [31] used Nd:YAG solid-state pulsed laser to clean the solid electrolyte interphase films accumulated on the surface of the lithium iron phosphate battery cathodes causing battery capacity degradation. By scanning electron microscopy and Fourier infrared spectroscopy analysis, the pulse laser energy density for the best cleaning conditions was determined to be 0.142 J/mm^2. Through laser cleaning, the remanufacturing of waste battery electrodes was achieved.

II. **Application of Laser Rust Removal in EMU Maintenance**

In the maintenance of EMUs, severe corrosion of aluminum alloy axle boxes is a common issue. Laser cleaning has proven effective in removing rust from aluminum alloy axle boxes, while also leading to improvements in their mechanical properties. In a study conducted by Niu et al. [32], a 100 W laser cleaning machine was used to investigate the laser cleaning of rusted parts of aluminum alloy axle boxes. The laser parameters were as follows: power of 100 W, repetition frequency of 70 kHz, scanning width of 20 mm, and scanning frequency of 260 Hz. The results, as shown in Fig. 7.15, showed that laser cleaning could reduce the oxygen content to 1/3 of its original level and effectively remove the rust layer from the surface of the aluminum alloy axle box. The damage degree to the substrate from single cleaning was measured, and the loss thickness of substrate was approximately 17.5 μm, which was negligible. Moreover, as cleaning times increased, the substrate loss decreased. Surface roughness measurements showed almost no change.

Examination of the microstructure of the aluminum alloy surface after laser cleaning by a metallographic microscope revealed the presence of a hardened layer that slightly increased surface hardness. Measurements using a Vickers microhardness tester showed a slight increase in hardness on the surface of the axle box body components after laser cleaning. Additionally, this hardened layer could enhance the oxidation resistance of the substrate. Mechanical properties such as tensile strength and bending performance of the aluminum alloy after laser cleaning were also measured, showing slight improvements in both tensile strength and bending strength

Fig. 7.15 Laser cleaning effect of component mounting surface

of the aluminum alloy substrate. The experiments demonstrated that laser cleaning could effectively remove corrosion from aluminum alloy components of high-speed train axle boxes and improve material properties.

III. **Application of Laser Rust Removal in Shipbuilding and Maintenance**

China, as a major shipbuilding nation, requires extensive rust removal treatment for large quantities of steel plates during ship construction. Additionally, annual ship maintenance involves paint and rust removal operations. Xie et al. [33] cleaned the rust on the surface of the ship plate with a fiber laser and proposed a method to determine the parameters of overlapping scanning rust removal process based on the geometric characteristics of single-line scanned groove profiles. Experimental results revealed the relationship between the geometric characteristics of groove profiles and laser energy density: within the range of 0.5–5 J/mm^2 energy density, the depth, width, and cross-sectional area of the single-line scanned groove profile exhibited an approximate linear relationship with the energy density. The research indicated that laser rust removal can meet the cleanliness standards for shipboard coatings and the requirements of satisfy surface roughness. Liu et al. [34] employed a 1070 nm laser with a peak power of 1550 W, a minimum pulse width of 0.2 ms, and a maximum repetition frequency of 2 kHz to clean the rust of AH32 ship-grade B steel plates. The thickness of the rust laye is 20 μm. Figure 7.16 illustrates the cleaning process.

As a comparison, a shot blasting rust removal experiment was also conducted, and the effects of the two cleaning methods were compared in terms of adhesion and anti-corrosion through coating film adhesion test (GB/T5210.9-2006) and artificial atmosphere corrosion test, namely salt spray test (GB/T10125-1997). The surface after laser cleaning met the national standards.

IV. **Application of Laser Rust Removal in Ammunition Repair**

Chapter 6 discusses the application of laser paint stripping in ammunition repair. In fact, the pollutants to be cleaned not only include paint but also rust. Song et al. [35] conducted experiments on the cleaning of copper-made ammunition casings using a 20 W pulsed fiber laser. Laser parameters were as follows: wavelength of 1064 nm, average power of 20 W, pulse energy of 1 mJ, pulse width of 80 ns, and spot diameter of 0.03 mm.

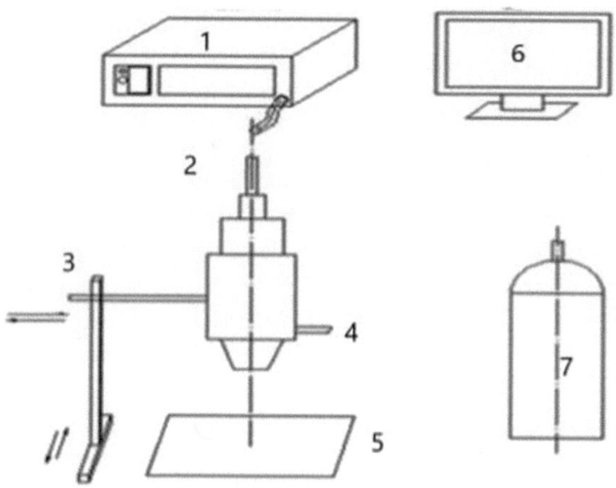

1--pulsed laser, 2--fiber, 3--two-dimensional regulation, 4--process head, 5-- steel plate, 6--industrial computer, 7-- gas protection system.

Fig. 7.16 Schematic diagram of experimental device for laser rust removal of Marine steel plate. 1—pulsed laser, 2—fiber, 3—two-dimensional regulation, 4—process head, 5—steel plate, 6—industrial computer, 7—gas protection system

The experimental results found that the rust layer of the shell peeled off smoothly, revealing a metallic luster and a relatively uniform texture. Additionally, the surface temperature of the material changed minimally, thus mitigating the risk of ammunition ignition due to the thermal effect. However, the drawback was the relatively low operation efficiency, at 0.67 s/mm^2. When a copper ammunition casing has both paint and rust, a stronger laser can be used for initial cleaning, followed by a weaker laser to further clean the remaining rust and paint layers, achieving the desired cleaning effect. The efficiency of combined rust and paint removal is slightly higher than that of simple rust removal, about 0.083 s/mm^2, but still relatively low. It suggests that laser rust removal holds promise for ammunition repair, but further research is needed to improve efficiency.

V. **Application of Laser Rust Removal in the Protection of Metal Cultural Relic Preservation**

Laser cleaning can be utilized to remove rust layers from the surface of metal artifacts, offering a safer alternative to traditional methods. Zhang et al. [36] used the THUN-DER type IV Nd:YAG laser from Quanta System Company to conduct laser rust removal experiments on gilded bronze artifacts. Laser parameter included a wavelength of 1064 nm, pulse energy of 1 J, pulse width of 5–8 ns, energy density of 0.30–0.50 J/cm^2, maximum repetition rate of 40 Hz, spot diameter of 10 mm, and divergence of 0.5 mrad. Experimental results demonstrated significant corrosion cleaning effects, with artifacts restoring their gilded metallic luster. Three-dimensional microscopy observations revealed no residues, no traces, or new scratches on the artifact surfaces. It is evident that laser cleaning yields excellent results while ensuring safety and reliability.

VI. **Application of Laser Cleaning in Mechanical Industry**

Over prolonged usage, the surfaces of mechanical product components tend to accumulate layers of paint, rust, oil and water stains, carbon deposits, scale, and other contaminants. Chapter 6 elucidates how laser cleaning can be employed for both paint stripping and rust removal in mechanical products. Apart from facilitating the removal of surface corrosion, laser cleaning also enables the formation of a corrosion-resistant protective layer on metal components, thereby enhancing the durability of mechanical products [18].

7.7 The Impact of Laser Rust Removal on Material Properties

Many studies have indicated that laser rust removal not only effectively removes corrosion like other methods but also enhances the properties of materials to some extent.

I. **Improvement of Material Corrosion Resistance**

Wang et al. [37] investigated the effects of laser energy density, pulse repetition frequency, and beam scanning speed on the removal of corrosion from the surface of A3 steel. The corrosion resistance of the A3 steel surface was found to be improved after laser cleaning. At the same time of laser cleaning, laser treatment of metal surfaces often change the chemical composition or physical properties of the metal, achieving the purpose of modification. For steel products, corrosion resistance is the most important and crucial characteristic that needs improvement. Kwok et al. [38] employed kilowatt-level continuous Nd:YAG lasers to irradiate the surface of stainless steel materials in an argon environment. Laser irradiation induced surface melting, while simultaneously improving cavitation erosion resistance and pitting corrosion resistance. It is particularly worth noting that a uniform, very thin, cellular dendritic structure was formed on the surface of the steel. This surface material, separated from large precipitates during the laser surface melting process, plays a crucial role in improving the corrosion resistance of steel.

For thick rust layers, after laser treatment, regardless of the wavelength and pulse width of the chosen laser, there will be a significant darkening reaction on the substrate surface. This darkening reaction is related to the heat-induced dehydration reaction of rust:

$$2\alpha\text{-FeOOH [or } \gamma\text{-FeOOH]} + h\nu \rightarrow \gamma\text{-Fe}_2\text{O}_3 + \text{H}_2\text{O},$$
$$6\gamma\text{-Fe}_2\text{O}_3 + h\nu \rightarrow 4\text{Fe}_3\text{O}_4 + \text{O}_2$$

During the process of laser treatment of metal products, physical and phase change processes will occur under certain conditions:

$$Fe_3O_4 \xrightarrow{200\ ^\circ C} \gamma\text{-}Fe_2O_3 \xrightarrow{400\ ^\circ C} \alpha\text{-}Fe_2O_3$$

These physical and phase changes are all related to the process of pulsed laser heating, and the formed darkened surface has passivity, which can stabilize the surface of the metal products and prevent further chemical reactions from occurring. After a long period of time, the darkened surface does not undergo further discoloration reactions, meaning it does not transition into a corroded dark gray layer, indicating the absence of new rust formation.

Zhang [39] employed a diode-pumped nanosecond pulsed fiber laser to clean hot-rolled AA5083-O aluminum alloy. The laser had a wavelength of 1064 nm, pulse width of 10 ns, and repetition frequency of 10 kHz. The corrosion behavior of the aluminum alloy after laser cleaning was studied using electrochemical impedance spectroscopy. The results indicated that the surface exhibited higher corrosion resistance after laser cleaning in a 3.5 wt% NaCl solution, with significantly increased impedance and reduced capacitance. Changes in the state of surface oxides were determined through glow discharge optical emission spectroscopy, X-ray photoelectron spectroscopy, and high-resolution transmission electron microscopy, indicating improvements in microstructure.

II. **Impact on Welding Performance**

In recent years, there have been many articles on research regarding laser cleaning pretreatment for metal welding, and laser rust removal has also been found to effectively improve the welding performance of metals. Luo et al. [40] used a 1064 nm nanosecond pulsed solid-state laser to remove the oxide layer from TC11 titanium alloy. The laser parameters were as follows: power of 150 W, maximum frequency of 10 kHz, minimum pulse width of 15 ns, maximum pulse energy of 40 mJ, spot diameter of 0.8 mm, zero defocus, and cleaning speed of 10 mm/s. The sample thickness was 6 mm. After cleaning, the interface was observed using an upright reflection microscope, the height differences were measured using a step gauge, the morphology was analyzed with SEM, and the energy spectrum was analyzed with EDS. The results showed that the oxide layer was completely removed after cleaning. No significant changes were observed in the microstructure under a metallographic microscope. SEM analysis showed that a large number of pollutant particles were removed. The cleaned sample was welded with a vacuum electron beam, as shown in Fig. 7.17, revealing Grade I weld seams without impurities, pores, or other defects, demonstrating that laser cleaning can effectively improve welding quality.

Luo et al. [40] used the POWER—LASE Rigel i400 nanosecond pulsed solid-state laser to clean anodized aluminum alloy. The laser parameters were as follows: output power of 400 W, spot movement speed of 15 mm/s, and spot length of 25 mm. The experiment revealed that the weld seam surface without laser cleaning was black and had multiple impurity defects; while the weld seam surface after laser cleaning

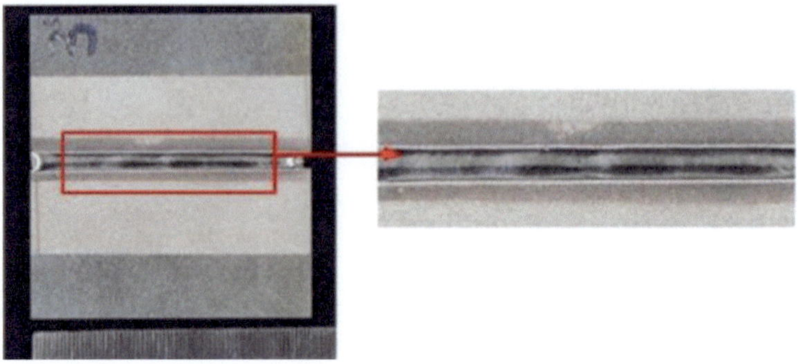

Fig. 7.17 Microstructure of titanium alloy before and after cleaning

exhibited a silver-white without obvious impurity defects. X-ray inspection of the weld seam defects revealed only two small pores, which did not exceed the tolerance limit. Thus the laser-treated weld seam was classified as grade I.

Song et al. [41] utilized a Nd:YAG frequency-doubled laser to conduct experiments on aluminum alloy (5A06) sheets. The laser parameters were: wavelength of 532 nm, power density of 5.31×10^8 W/cm^2, pulse width 10 ns, repetition rate of 10 kHz, spot diameter of 30 μm, and scanning speed of 15 mm^2/s. The effect of laser cleaning on the quality of weld seams formed by the TIG welding method was investigated. If the oxide layer of the base material is not completely removed, it will absorb moisture during welding, resulting in forming of pores in the weld seam and making it prone to cracking. After laser cleaning, not only were the contaminant particles reduced, but also the surface oxygen distribution significantly decreased. And there was a noticeable reduction in the number of pores, as shown in Fig. 7.18. Consequently, the quality of the weld seam was improved.

Chen [42] employd a laser with wavelength of 1064 nm, pulse width of 150 ps, repetition rate of 600 kHz, and power of 15 W (the cleaning equipment and schematic of the spot scanning are illustrated in Fig. 7.19). The oxide layer on the surface of the aluminum alloy was cleaned, the laser spot diameter was about 70 μm, with scanning spacing of 0.03 mm and scanning speed v ranging from 50 to 4000 mm/s.

Observing the shape after laser cleaning through SEM, it was found that micro-nanostructures generated, which had certain effects on the wettability, roughness, and optical absorption rate of the material surface. Compared to the original material surface, different laser cleaning parameters resulted in different hydrophilicity and hydrophobicity on the material surface. (1) When the scanning speed of the laser spot was less than 100 mm/s, the roughness in the x and y directions increased compared to the original surface and mechanically polished surfaces. (2) As the scanning speed gradually increases, the roughness decreased and stabilized. (3) Compared to the original surface and the surface treated by mechanical polishing, the weld penetration and weld quality of the material surface obtained after laser cleaning were both significantly improved.

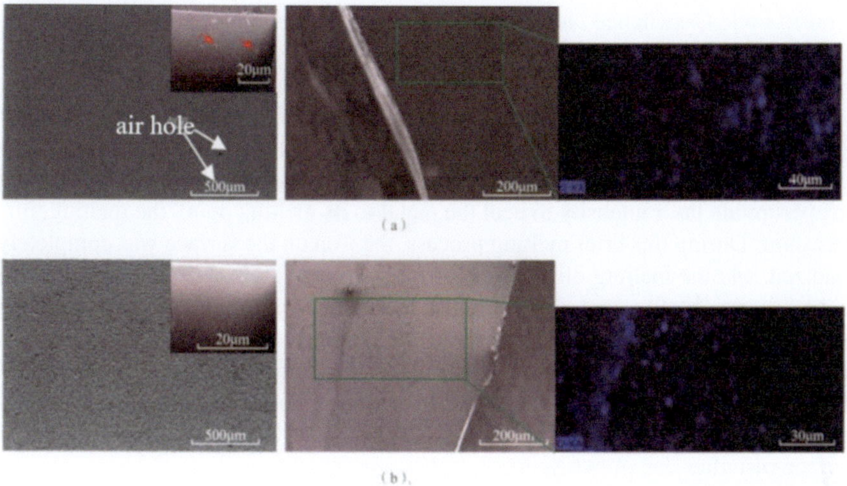

Fig. 7.18 Microstructure of weld obtained by TIG welding after traditional grinding and laser cleaning: **a** traditional grinding; **b** laser cleaning 气孔: air hole

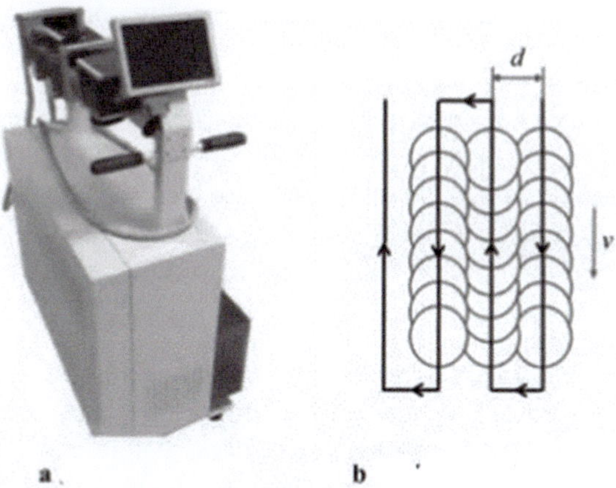

Fig. 7.19 Laser cleaning equipment and cleaning spot diagram. **a** Laser cleaning equipment; **b** cleaning spot

III. **Passivation**

The oxide layer on the surface of the aluminum alloy is relatively dense, which can prevent oxygen and moisture in the air from entering the substrate, thus acting as a passivation role. As for rust, it is generally a porous structure, and moisture and oxygen will continuously infiltrate the substrate, causing increasingly severe corrosion.

Research by Li et al. [43] indicates that laser cleaning can form a dense oxide layer on the surface, thereby passivating it. They utilized a self-built quasi-continuous wave acousto-optic Q-switched Nd:YAG laser, with pulse width ranging from microseconds to over a hundred nanoseconds. At a frequency of 3 kHz, the maximum average output power was 30 W with a pulse width of 150 ns. The thickness of the rust layer on the steel plate was a few micrometers. The experiment found that while the laser was removing rust, the metal substrate underwent oxidation. When the laser intensity was low, the metal surface was not heated to the melting point of iron. By selecting an appropriate laser intensity to heat the metal to its melting point, the melting time was short. During this brief melting process, the iron on the surface was completely oxidized, and the melting of the iron surface helped homogenize the concentration and temperature of oxygen in the molten iron. Under this condition, a passivating oxide layer was formed. When the laser intensity was large enough, the diffusion time of oxygen in the melt pool was longer. During laser rust removal, a thick, layered macroscopic structure oxide layer was formed on the metal matrix. Samples cleaned with this method, when exposed to the air environment, slowly developed a layer of surface rust after one or two years.

Zhang et al. [44] used a diode-pumped pulsed fiber laser with a wavelength of 1064 nm, a pulse width of 10 ns, and a repetition rate of 10 kHz. The laser intensity could be set to 3.5 J/cm^2, 5.7 J/cm^2, 7.1 J/cm^2, 8.5 J/cm^2, and 11.3 J/cm^2. The corrosion behavior of laser-cleaned AA7024-T4 aluminum alloy was studied using electrochemical impedance spectroscopy (EIS) and scanning vibrating electrode technology (SVET). The results indicated that the corrosion resistance of the surface after laser cleaning in a 3.5 wt.% NaCl solution was higher than that of the original hot-rolled alloy. Active anodic spots appeared on the surface, and a dense protective oxide layer containing Al_2O_3 and MgO was formed, which contributed significantly to the enhancement of corrosion resistance.

IV. **Other**

Laser rust removal also affects other properties of the material, including surface roughness, microstructure, hardness, and other mechanical properties. Ren et al. [45] used a Yb-dopped pulsed fiber laser with a wavelength of 1064 nm to study the mechanical properties of the surface of low carbon steel after laser rust removal. Measurements of surface roughness before and after cleaning revealed that the surface roughness of the cleaned samples increased compared to the original samples. During the cleaning process, some pulses may interact with the substrate through the rust layer, potentially affecting the microstructure of the substrate, such as partial melting or phase changes due to heat conduction. Observations using an INSPECT F50 field-emission scanning electron microscope showed that there was no re-melting of the substrate surface after laser claeaning. Subsequently, the microstructure of the original and cleaned samples was observed using a Zeiss Observer. Z1m metallographic microscope, as depicted in Fig. 7.20, indicating no significant changes in the microstructure of the substrate.

Ren et al. [45] also characterized the hardness of the sample. They used a Rockwell hardness tester with a C-type indenter and found no significant changes in hardness before and after laser rust removal. Then they measured the surface hardness of the samples using an HR15T Rockwell surface hardness tester and found a slight

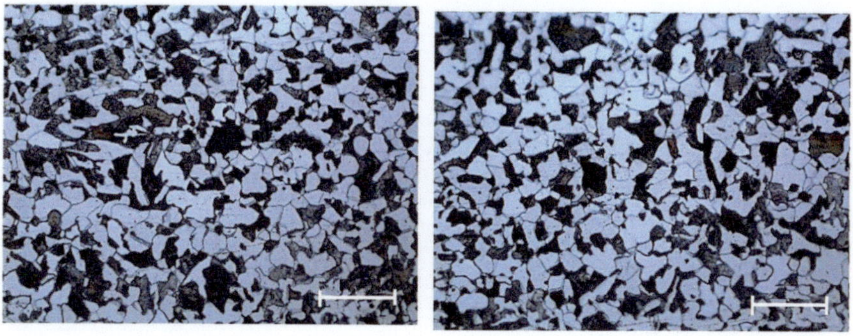

Fig. 7.20 Microscope image of sample surface before and after laser cleaning. **a** Surface microstructure of the original sample; **b** Surface microstructure of the sample after laser rust removal

increase in cleaned surface hardness a, but the change was not significant. They also characterized the mechanical properties of the sample, and obtained various mechanical property parameters such as elongation, strain hardening index, tensile strength, yield strength, and elastic modulus of the tested material through their self-developed IBIS-2 type ball pressure method mechanical property rapid detection device. It was found that there was no significant change in various mechanical property parameters before and after rust removal. Shi [46] used a 120 W MOPA pulsed fiber laser to clean aluminum alloy, characterized its surface morphology with a scanning electron microscope (SEM), measured its microhardness with a Vickers hardness tester, and characterized its tensile and bending properties with a BT1-FR020-A50/Zwick universal testing machine. The results showed that laser cleaning can effectively remove the oxide layer from the surface of the aluminum alloy and improve its tensile and bending properties.

From the perspective of rust removal effectiveness, laser rust removal demonstrates excellent cleaning results. rust removal effect. As analyzed above, it can enhance weld quality, improve material corrosion resistance, and offer controllable strength. In comparison, shot blasting and high-pressure water jetting methods provide relatively good cleaning results, but controlling the intensity is challenging. Manual/mechanical friction and chemical agent methods yield general results. Regarding rust removal efficiency, the current speed of laser rust removal can reach more than 1 m^2/h, which is not yet high enough. Additionally, apart from the initial investment in equipment, the cost of a single laser cleaning operation is comparable to or even lower than that of other methods. Moreover, laser rust removal has certain advantages over most other methods as it can collect peeled metal rust easily and environmentally.

References

1. Ye DL. Handbook of Practical Inorganic Thermodynamics Data. Beijing: Metallurgical Industry, 1981. (in Chinese)
2. Jiang L. Engineering Chemistry. 2nd edition. Beijing: Chemical Industry, 2006. (in Chinese)
3. Ren PY, Li RX, Wu X. Corrosion behavior of copper materials in marine atmosphere under the action of magnetic field. Corrosion & Protection, 2019, 40(6): 419–423. (in Chinese)
4. Zhou H, Zhu HF, Cai LK. Morphology and composition of the rust on bronze artifacts. Sciences of Conservation and Archaeology, 2005, 17(3): 22–27. (in Chinese)
5. Zheng J, Liang H, Ma WB, et al. Physical properties of materials. Tianjin: Tianjin University, 2008. (in Chinese)
6. Cui XM, Liu DY, Huai XL. Temperature fields on materials surface under high-energy pulse laser irradiation. Journal of University of Science and Technology Beijing, 2000, 22(5): 470–474. (in Chinese)
7. Yang B, Gu JH, Gui RS. Study on model about the material heated by pulsed laser. Journal of Soochow University, 2008, 24(4): 41–44. (in Chinese)
8. Ren ZG, Chen J, Chen HN. Surface temperature field on metal substrate during laser derusting based on finite element simulation. Surface Technology, 2018, 47(12): 333–339. (in Chinese)
9. Cui JJ, Guo Q, Wang CY, et al. Experimental research on the dry laser cleaning process. Journal of Zhongyuan University of Technology, 2018, (4): 29–32. (in Chinese)
10. Chen GX, Lu HF, Zhao Y, et al. Effect of power on laser cleaning result of stainless steel surface. Opto-electronic Engineering, 2017, 44(12): 1217–1224. (in Chinese)
11. Luo Y, Wang X, Zhao HF, et al. Effect of laser cleaning on the surface morphology and welding performance of 2219 aluminum alloy. Applied Laser, 2017, 37(4): 544–549. (in Chinese)
12. HAN JH, DUAN T, GAO XH, et al. The influence of laser plasma effects on the characteristics of thin film damage. Spectroscopy and Spectral Analysis, 2012, 32(5): 1162–1165. (in Chinese)
13. Zhou JZ, Li HT, Sun Q, et al. Laser derusting mechanism of AH32 steel based on cleaned surface topography. Optics and Precision Engineering, 2019, 27(8): 1754–1764. (in Chinese)
14. Wei CH, Wang LJ, Liu WP, et al. Thermal response of 45# steel coupling with multi-layer oxide film evolution by laser irradiation. Optics and Precision Engineering, 2014, 22(8): 2061–2066. (in Chinese)
15. Pan Y, Wang MD, Liu JC, et al. Process and mechanism of HT200 laser de-rusting for 100 W fiber laser. Applied Laser, 2019, 39(2): 269–274. (in Chinese)
16. Tian B, Zou WF, Liu S J, et al. Introduction of rust removed by dry laser cleaning. Cleaning World, 2006, 22(8): 33–38. (in Chinese)
17. Song F, Liu SJ, Zou WF. Typical application of laser cleaning——removal of rust and paint. Cleaning World, 2005, 21(11): 38–41. (in Chinese)
18. Wang HJ, Liu WW, Yu Y, et al. Research status and prospect of laser cleaning of metal surface contaminants. Internal Combustion Engine & Parts, 2016, 8: 75–78. (in Chinese)
19. Yao HB, Yu WL, Li YR, et al. Analysis of laser cleaning of iron in air based on characteristics of plasma spectra. Acta Photonica Sinica, 2013, 42(11): 1295–1299. (in Chinese)
20. Qiao YL, Zhao JX, Wang SJ, et al. Laser cleaning and elemental composition analysis of rusty surface. Laser & Infrared, 2018, 48(3): 299–304. (in Chinese)
21. Zhang GX, Hua XM, Huang Y, et al. Investigation on mechanism of oxide removal and plasma behavior during laser cleaning on aluminum alloy. Applied Surface Science, 2020, 506: 144666.
22. Chen HG, Chen JW, Chen H. Effect of energy density on laser-cleaned surface morphology of A7N01 aluminum alloy. Electric welding machine, 2020, 50(3): 97–101, 125. (in Chinese)
23. Cao NF, Feng WF, Wen CG. Method for the removal of terminal post oxides on power Li-ion battery. Battery Bimonthly, 2017, 47 (6): 343–346. (in Chinese)
24. Sun SW, Chen YH, Chen W, et al. Laser cleaning process of TiNi alloysurface rolling oxide film. Journal of Netshape Forming Engineering, 2018, 10(5): 132–136. (in Chinese)
25. Zhu G, Wang S, Cheng W, et al. Investigation on the surface properties of 5A12 aluminum alloy after Nd:YAG laser cleaning. Coatings, 2019, 9: 578.

26. Deschênesj M, Fraser A. Empirical study of laser cleaning of rust, paint, and mill scale from steel surfac. Materials Processing Fundamentals. New York: Springer, 2020: 189–201.
27. Yang MK, Zhou FR, Ma Y, et al. Laser rust removing process of deep rusting steel plates. Applied Laser, 2018, 38(6): 85–90. (in Chinese)
28. Lei ZL, Sun HR, Chen YB, et al. Elimination of rusting layer from high-strength steel surface using different laser cleaning methods. Chinese Journal of Lasers, 2019, 46(7): 78–83. (in Chinese)
29. Li SG. Laser-assisted chemical cleaning of thin oxide films on carbon steel surfaces. Cleaning World, 2005, 21(5): 37–41. (in Chinese)
30. Ni JM, Guan YJ, Hu MH, et al. Influence analysis of laser cleaning anodic oxide coatings on aluminum alloy joint property. Electric Welding Machine, 2017, 47(7): 102–105. (in Chinese)
31. Liu WW, Liu LH, Zhang H, et al. Theoretical and experimental study on laser cleaning of cathode from lithium-ion batteries. Cleaning World, 2016, 32(6): 17–23. (in Chinese)
32. Niu FJ, Qi XS. Research on the application of laser cleaning technology in EMU maintenance. China High and Nes Technology, 2017, 1(11): 75–77. (in Chinese)
33. Xie YF, Liu HW, Hu YX. Determining process parameters for laser derusting of ship steel plates. Chinese Journal of Lasers, 043(4): 103–110. (in Chinese)
34. Liu HW, Zhou YM. Technology and application of optical fiber laser derusting of ship steel plates. Marine Technology, 2016, (6): 87–93. (in Chinese)
35. Song GF, Li LC, Xia FJ, et al. Application study on laser washing technology in ammunition repair. Laser & Infrared, 2017, 47(1): 29–31. (in Chinese)
36. Zhang XT, Zhang PY, Yang C, et al. Laser cleaning technology in the conservation of gilt bronze. Sciences of Conservation and Archaeology, 2013, 25(3): 98–103. (in Chinese)
37. Wang Z, Zeng X, Huang W. Parameters and surface performance of laser removal of rust layer on A3 steel. Surface&Coatings Technology, 2003, 166(1): 10–16.
38. Kwok CT, Man HC, Cheng FT. Cavitation erosion and pitting corrosion of laser surface melted stainless steels. Surface&Coatings Technology, 1998, 99(3): 295–304.
39. Zhang SL, Suebka C, Liu H, et al. Mechanisms of laser cleaning induced oxidation and corrosion property changes in AA5083 aluminum alloy. Journal of Laser Applications, 2019, 31(1): 012001.
40. Luo Y, Wang X, Zhao HF, et al. Effect of laser cleaning pretreatment before welding on the welding performance of TC11 titanium alloy. Surface Engineering & Remanufacturing, 2017, 17(2): 26–28. (in Chinese)
41. Song Y, Wang HP, Wang Q, et al. Research and application of laser fine surface processing technology: cleaning and polishing. Aeronautical Manufacturing Technology, 2018, 61(20): 78–86. (in Chinese)
42. Chen JW, Peng ZZ, Wang FS, et al. Influences of laser cleaning on surface morphology and laser-MIG hybrid welding characteristics of aluminum alloy. Electric welding machine, 2019, 49(8): 66–71. (in Chinese)
43. Li W, Du P, Zhang J, et al. Passivation process in quasi-continuous laser derusting with intermediate pulse width and line-scanning method. Applied Optics, 2014, 53(6): 1103.
44. Zhang FD, Liu H, Suebka C, et al. Corrosion behavior of laser-cleaned AA7024 aluminum alloy. Applied Surface Science, 2018, 435: 452–461.
45. Ren ZG, Wu CZ, Chen HN, et al. Mechanism of laser derusting and surface properties of low carbon steel. Opto-electronic Engineering, 2017, 44(12): 1210–1216. (in Chinese)
46. Shi TY, Wang CM, Mi GY, et al. A study of microstructure and mechanical properties of aluminum alloy using laser cleaning. Journal of Manufacturing Processes, 2019, 42:60–66.

Chapter 8
Application of Laser in Mold Cleaning

8.1 Significance of Mold Cleaning

In the process of tire production, molds are essential mechanical equipment. Tire molds can be divided into segmented molds and two-piece molds. The former consists of the tread ring, mold sleeve, and upper and lower side plates, while the latter consists of the upper mold and lower mold. Segmented molds can further be categorized as conical surface guided segmented molds and inclined plane guided segmented molds.

The quality of molds directly impacts the quality of tires. Tire molds include patterns and branding designs, as shown in Fig. 8.1, which require intricate carving techniques. Therefore, after a certain period of use, molds must be cleaned to ensure the cleanliness of their surfaces. This is essential to guarantee the quality of tires as well as the lifespan of the molds [1].

Other industrial productions, such as the toy and footwear industries, also rely on the use of molds, as shown in Fig. 8.2. Cleaning the molds is also necessary in these industries to ensure product quality and extend the lifespan of the molds [2].

When pollutants accumulate to a certain extent, they first affect the surface shape of products such as tires, toys, and shoes, resulting in them becoming defective or even scrap. Secondly, pollutants can corrode molds, thereby impacting their lifespan. The residues in molds differ across different industries. For example, tire molds are repeatedly used under high pressure and high temperature conditions, accompanying the vulcanization process of rubber. As a result, they inevitably accumulate a combination of rubber, additives, and mold release agents, leaving behind pollutants such as sulfides, inorganic oxides, silicone oil, and carbon black on the mold surface and grooves. In the case of production materials containing resins like polyvinyl chloride, gases can be produced that corrode various types of mold steel. Residues separated from flame retardants and antioxidants can also corrode steel. Coloring agents can cause rust on steel, and rust stains are difficult to remove. Moisture and oxygen in the air can also damage molds [3–5].

F. Song and X. Lin, *Laser Cleaning*, https://doi.org/10.1007/978-981-97-6631-4_8

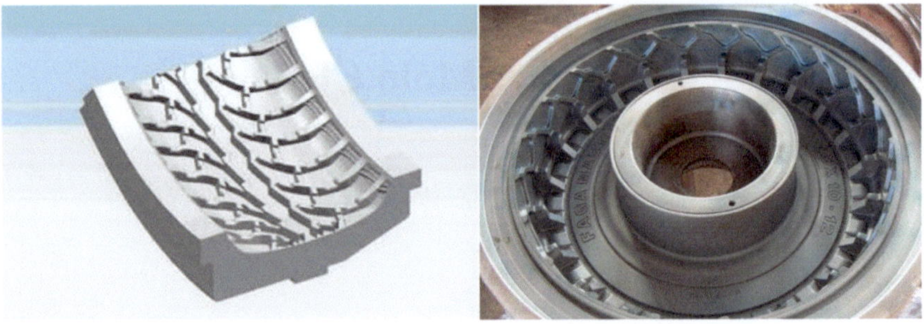

Fig. 8.1 Tire mold

Mold cleaning is a crucial process in maximizing the lifespan of molds, improving their reliability, and ensuring product quality. As a key step in mold maintenance, it is essential to perform cleaning quickly and promptly in industrial production. In many factories, molds awaiting cleaning occupy almost all the limited operating space in corridors and workshops [6]. Therefore, mold cleaning has always been a matter of concern, and the development of efficient, cost-effective, and non-polluting mold cleaning techniques is of great importance.

8.2 Commonly Used Mold Cleaning Methods

Similar to paint stripping and rust removal, traditional cleaning methods for molds include mechanical cleaning, chemical cleaning, and dry ice cleaning.

I. **Mechanical Cleaning**

Mechanical cleaning is a mature technology widely used in the tire industry and other industries for mold cleaning. It is a simple and practical method that does not require high equipment requirements. Depending on the specific needs, manual friction using sandpaper or wire brushes can be employed, as well as machine shot blasting. The former has lower efficiency but is suitable for mild contamination, small molds, or cleaning scenarios with fine requirements. The latter has higher efficiency but may cause mechanical damage to the mold, shorten its lifespan, and easily clog the mold's vent holes. Clearing the vent holes can be a labor-intensive and time-consuming process, often requiring the mold to be removed.

Fig. 8.2 Other types of molds

II. **Chemical Cleaning**

Chemical cleaning methods primarily involve the use of chemical agents. Initially, alkaline solutions were commonly used, along with high temperatures, by soaking the mold in hot alkaline water. However, this method has low efficiency and is only suitable for cleaning small molds. It is ineffective for large molds. Acid cleaning, on the other hand, is convenient, cost-effective, and efficient. However, it can easily damage the mold, especially when acid molecules enter the interior, causing corrosion and reducing the mold's lifespan. Additionally, these chemical agents can pollute the environment and pose health risks to workers.

III. **Dry Ice Cleaning**

Dry ice cleaning utilizes a dry ice blasting machine to convert liquid CO_2 into dry ice pellets of specific sizes (ranging from 2 to 14 mm) using a dry ice pelletizer, as shown in Fig. 8.3. These dry ice pellets have a temperature of – 78 °C and are propelled onto the mold surface at high speeds using the dry ice blasting machine, as illustrated in Fig. 8.4. The impact force instantly causes the dry ice pellets to vaporize, absorbing a significant amount of heat and triggering intense heat exchange on the mold surface. This rapid temperature change causes contaminants to cool, shrink, and become brittle, leading to their detachment. Additionally, the dirt on the mold surface and the mold itself have different thermal expansion coefficients. The resulting thermal stress, caused by the temperature difference between the surface and the interior, can overcome the adhesive forces between the pollutants and the mold, thereby breaking the bond between the two materials. Moreover, within a fraction of a second, dry ice expands in volume by approximately 800 times at the point of impact, creating a "mini-explosion" that effectively dislodges dirt particles [7]. Dry ice cleaning is a typical thermodynamic process, and since the dirt layer on the inner surface of tire molds is mainly composed of rubber, which exhibits significant cold brittleness, dry ice cleaning is more effective when the mold is in a high-temperature state [8, 9]. Dry ice cleaning offers several advantages, such as no wastewater generation, minimal abrasion and corrosion, and the ability to clean molds while they are in operation.

III. **Laser Cleaning Method**

Laser cleaning technology is a rapidly developing cleaning technique in recent years. It involves using high-energy lasers to irradiate the surface of the mold, causing the

Fig. 8.3 3 mm dry ice pellets

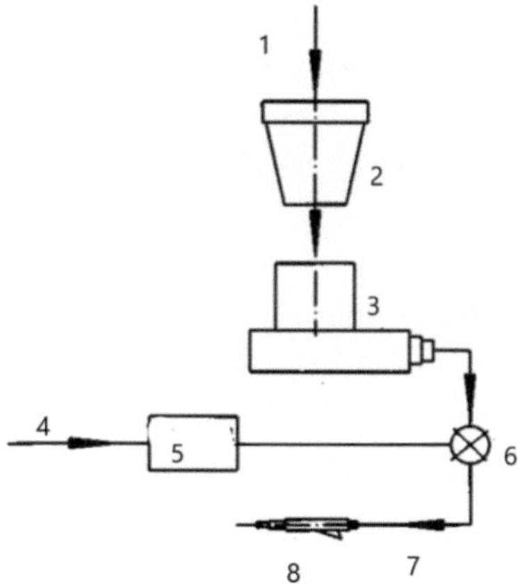

Fig. 8.4 Schematic diagram of dry ice cleaning equipment. 1—dry ice pellets, 2—storage tank, 3—pellets transport, 4—compressed air, 5—dryer, 6—mixer, 7—insulation line, 8—spray gun

pollutants adhering to the mold to be removed through the effects of ablation, vibration, and plasma shockwaves. Laser cleaning is characterized by being pollution-free, low-noise, and safe for cleaning purposes.

Laser cleaning and dry ice cleaning are both innovative methods of mold cleaning. In comparison, laser cleaning offers the following advantages [10, 11]:

(1) Dry ice cleaning generates significant noise, with many cleaning systems exceeding 100 dB, which does not comply with government environmental regulations. As a result, operators are required to wear heavy protective headgear. Additionally, according to noise level regulations, simultaneous cleaning with dry ice is challenging to achieve during regular production. In contrast, laser cleaning produces minimal noise, posing no harm to human health.
(2) Laser cleaning has low labor intensity. The dry ice cleaning process requires manual control, while laser cleaning operates automatically. Dry ice mold cleaning can be a labor-intensive task that takes a long time. In contrast, laser cleaning technology is advanced and highly automated. It requires minimal labor and time, as it only requires the installation and activation of the laser.
(3) Laser cleaning offers a relatively high safety factor. Manual dry ice cleaning requires operators to climb onto heated sulfuration machines and work between the open halves of the mold. There is a certain level of risk involved when the energy is released and the sulfuration machine closes. Additionally, placing the dry ice nozzle onto the hot mold cavity can potentially lead to accidents such as hose bursts. In contrast, laser cleaning can be controlled remotely, making it a safer option.
(4) The dry ice cleaning process requires the temperature of the hot mold surface to undergo a sudden change when impacted by ice particles. Therefore, this method is most effective for cleaning hot molds. On the other hand, laser cleaning can be used for both hot and cold equipment without any change in performance. As a result, laser cleaning does not require the molds to be heated specifically for the cleaning process [8].

(5) The operational costs of dry ice cleaning (dry ice powder and electricity consumed) are generally higher compared to laser cleaning.
(6) Dry ice cleaning can potentially cause damage to complex ventilation systems.
(7) Laser cleaning can be done online, with low cleaning costs and minimal mold wear.

Laser cleaning has certain limitations, such as high initial equipment investment and a long investment recovery period. Currently, the majority of tire factories in China do not have laser cleaning equipment. Only a few suppliers offer high-power cleaning machines (200 W and above) for the tire industry, with initial equipment investments typically exceeding millions of RMB. However, with advancements in domestic technology and gradual implementation in production practices, laser cleaning shows greater advantages and promising prospects. Jetter [9] provides an example: In 2001, a machine producing 20,000 tires per day with eight vulcanizers required daily cleaning of the tire molds. If three vulcanizers were used for cleaning per shift or nine vulcanizers were used daily (some factories clean twice), it would take approximately 15 h of operation and 10 h of downtime to remove the molds from the vulcanizers for offline cleaning. By using lasers to clean both halves of the molds, only 0.3 h of operation time and 3 h of downtime would be needed. Thus, cleaning one vulcanizer would save 14 h of operation time and 7 h of downtime. Assuming only 10 cleanings (5 vulcanizers) were done, with 5 using offline cleaning in the mold workshop, a total of 70 h of operation time and 35 h of downtime could be saved each day. Over 320 working days, this would increase the total operating time by 22,400 h and the machine time by 11,200 h. It is evident that using laser cleaning would yield significant returns.

The market for laser cleaning of molds, particularly in the field of tire molds, has gained attention. Equipment, technology, and research for laser cleaning of tire molds have emerged. It is foreseeable that laser cleaning will become the mainstream method for mold cleaning in the future and will play a vital role in the mold industry.

8.3 Laser Cleaning Device for Molds

The laser cleaning machine used for mold cleaning consists of components such as a laser, control system, cooling system, vacuum cleaner, and inspection system [12], as shown in Fig. 8.5. The control system serves as the central control unit for the entire laser cleaning machine, coordinating, and managing the operation of other subsystems to accomplish the cleaning tasks. Many molds also have tiny exhaust holes, and the laser system can clean exhaust holes with a diameter of 1.5 mm. These exhaust holes are distributed in the tread area of the mold, and laser cleaning can effectively clean them to a depth of a few millimeters [9].

The operator can manually control the laser cleaning machine by simply setting a certain distance between the laser cleaning system's head and the mold to start the cleaning process. The operator often needs to set instrument parameters at the control

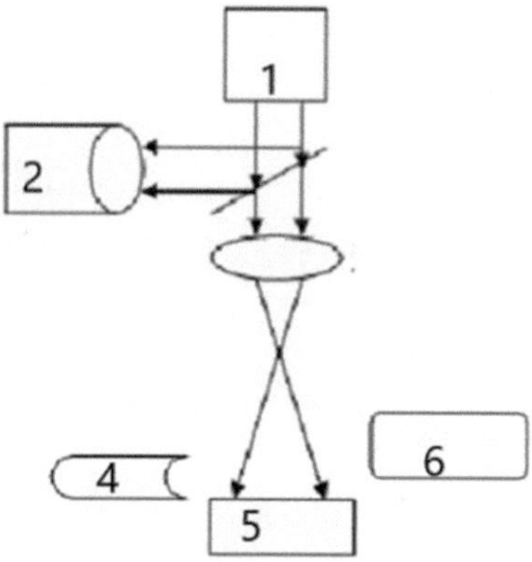

Fig. 8.5 Schematic diagram of laser cleaning system. 1—laser system, 2—energy meter, 3—lens, 4—adsorption device, 5—sample, 6—detection device

panel in front of the equipment. The time required for cleaning the mold depends on factors such as the frequency of cleaning, rubber compound formulation, and the condition of the mold surface. The total time for cleaning one mold, including setup time and cleaning time, typically ranges from about 30–50 min [3]. If the time for loading and unloading the mold and the components on both sides of the mold for cleaning are added, it can take an additional 15–20 min.

The cleaning machines used for laser cleaning of molds and paint stripping and rust removal are similar. Here, we will only focus on the selection of lasers, optical systems, and beam delivery systems.

8.3.1 Selection of Laser System

The process of laser removing rubber adhered to the surface of molds is essentially an interaction between the laser and the material. It largely depends on the material's absorption coefficient and the pulse energy of the laser, and the absorption coefficient and penetration capability of the laser are closely related to its wavelength. Generally, the shorter the wavelength, the greater the energy of the photons ($h\nu$, where h is the Planck constant and ν is the frequency), and the stronger the effect of atomic excitation. On the other hand, if the wavelength is longer, the energy of the photons gradually decreases, but the thermal effect increases, and the laser's radiation penetration capability also increases. When selecting a laser, it is important to consider the wavelength and choose a laser with a wavelength near the material's absorption peak as the irradiation light source, so that the material has a strong absorption capacity for the laser.

In practical applications, it is necessary to choose a laser with high single-pulse energy and affordable price. CO_2 lasers, Nd:YAG lasers, and fiber lasers are the most commonly used laser sources for laser cleaning. Table 8.1 shows the laser parameters used for a certain CO_2 laser tire mold cleaning machine. Once the laser type, power,

Table 8.1 Main technical parameters of laser

Wavelength	10.6 μm
Maximum power	2000 W
Power instability	$\leq \pm 3\%$
Seal operating life	$\geq$ 8 h
Working air pressure	70–90 kPa [13]
Operating noise	$\leq$ 70 dB
Laser size (length × width × height)	1890 mm × 920 mm × 1710 mm
Laser weight	2000 kg
Electronic console weight	20 kg

and wavelength are determined, the selection of pulse count, pulse width, pulse shape, and laser repetition frequency becomes crucial factors in the process.

8.3.2 *The Structure of the Light Guide System and the Design of the Optical Path System*

If a CO_2 laser is chosen, a beam delivery system using a beam delivery arm is required as there is no suitable fiber optic for CO_2 lasers. The optical path of the beam delivery system is shown in Fig. 8.6. The laser emitted by the laser source is directed towards the output mirror, enters the beam delivery arm, and then reflects off the mirrors on the arm to reach the focusing lens at the top of the laser head. After focusing, the laser beam reaches the surface of the mold to complete the cleaning process.

In the laser cleaning process, ensuring a good cleaning quality requires careful control of the beam diameter. Changes in the beam diameter can lead to variations in the focal point. Therefore, to achieve optimal cleaning results and high efficiency, it is important to minimize the changes in the beam diameter. Due to the natural divergence of the laser beam during propagation, the divergence angle is typically around 2 mrad, which means that for every 1 m of propagation, the beam diameter

Fig. 8.6 Schematic diagram of optical path transmission of laser cleaning mold device. 1—laser, 2—reflector, 3—focusing lens

increases by approximately 2 mm. In situations where the laser has to travel over long distances, the beam diameter can change significantly. Therefore, in such cases, it is necessary to compensate for the length of the optical path to maintain a consistent beam diameter.

8.4 Mechanism of Laser Cleaning Rubber Mold

Taking the cleaning of tire molds as an example, rubber is the main component of tires, and during the tire production process, rubber particles easily adhere to the surface of the molds, affecting the quality of the products. The adhesion between the rubber particles and the mold surface is mainly due to van der Waals forces, electrostatic forces, and capillary forces. For micron-sized particles, van der Waals forces play a major role in the adhesion. There are three main methods for removing adhered particles: detachment, rolling, and sliding. For laser dry cleaning, the primary method of removing adhered particles is detachment. For more detailed analysis on the force acting on the contaminants and the dynamics of contaminant removal, please refer to Chap. 3.

I. Calculation of Adhesion Force F_{ad}

According to the analysis in Chap. 3, assuming the mold is made of 45 steel and has rubber particles adhered to its surface, the van der Waals force (F_{ad}) between the rubber particles and the 45 steel in air can be calculated based on the Hamaker constants of rubber particles and 45 steel with air [13]. This force represents the adhesion between the rubber particles and the 45 steel.

$$F_a = \frac{AR}{12h^2} \tag{8.4.1}$$

where h represents the distance between the particles, R is the particle radius, and A is the Hamaker constant. The parameters used for the calculation can be found in Table 8.2.

Based on the calculations, the adhesion forces between a 5 μm rubber particle and 45 steel are estimated to be approximately 2.49×10^{-8} N [14, 15]. If the rubber particle sizes are 0.2 μm and 1 μm, the adhesion forces between them and 45 steel are approximately 9.97×10^{-10} N and 4.98×10^{-9} N, respectively [14, 15].

Table 8.2 Physical parameters of particles and matrix

Parameters	Hamaker constant of rubber particles A_{11} (J)	Hamaker constant of 45 steel A_{33} (J)	Hamaker constant of air A_{22} (J)	A_{123} (J)	Density of natural rubber ρ (kg m^{-3})
Value	16.20×10^{-20}	15.38×10^{-20}	1.94×10^{-20}	5.98×10^{-20}	1150

II. Calculation of Cleaning Force F_{clean}

Due to the temperature dependence of the main thermophysical properties parameters of 45 steel, such as specific heat capacity (c_p), thermal conductivity (λ), and thermal diffusivity (α), it is necessary to determine the corresponding values of c_p, λ, and α at different temperatures for accurate calculations. Table 8.3 provides the calculated values of c_p, λ, and α_t at different surface temperatures of 45 steel under various laser energy densities, using the interpolation method [16].

The laser energy is absorbed by the surface of the mold substrate, causing a rapid increase in temperature on the surface. At the same time, heat is transferred into the interior of the substrate. Due to the transient thermal effect of the laser, the surface of the substrate experiences an instantaneous thermal expansion acceleration perpendicular to the surface. This acceleration acts on the particles adhered to the surface, resulting in a cleaning force (F_{clean}) in that direction. When the cleaning force is greater than the adhesion force between the particles and the substrate ($F_{clean} > F_{ad}$), the particles separate from the substrate. Taking the adhesion of rubber particles to the mold surface as an example, a one-dimensional transient heat conduction equation is established for the laser-heated solid surface. The finite element method is then used to solve the temperature field on the mold surface and inside, enabling the calculation of the cleaning force [13, 17–19].

Assuming that the mold substrate absorbs laser energy and the surface temperature increases by ΔT, the thermal expansion displacement generated on the surface of the mold is given by $H = \beta\delta\Delta T$, where β is the thermal expansion coefficient of the mold, which is 1.2×10^{-5}/°C for 45 steel. Δ represents the thermal penetration depth of the mold within time t, given by $\delta = 4\alpha_t t$. Assuming that the particles experience the same instantaneous thermal expansion acceleration as the substrate, the cleaning force acting on the particles at the end of a laser pulse can be calculated as F_{clean}.

$$F_{clean} = \frac{m\text{H}}{\tau^2} \tag{8.4.2}$$

Table 8.3 Thermophysical parameters of No. 45 steel at various temperatures

Power density ($\times 10^2$ W/cm^2)	1.5	3	4.5	6	7.5	9	10.5
Temperature (°C)	153.44	289.59	422.45	558.58	697.18	836.67	978.51
Specific heat capacity c_p (J/(kg °C))	489.62	521.30	572.35	664.79	849.66	744.54	610.57
Thermal conductivity λ (W/(m °C))	45.46	38.37	37.71	32.88	28.75	26.28	24.49
Thermal diffusivity α_t (10^{-6} m^2/s)	12.30	9.40	8.41	6.31	4.32	4.51	5.12

Table 8.4 Calculation results of mold surface cleaning force under different laser power densities

Power density ($\times 10^2$ W/cm^2)	0	1.5	3	4.5	6	7.5	9	10.5
Cleaning force ($\times 10^{-8}$ N)	–	0.95	1.67	2.36	2.73	2.84	3.50	4.36

where m represents the mass of the particles, and τ represents the pulse width of the laser. When the pulse period is 20 μs, the calculation results of the cleaning force for 5 μm rubber particles at different power densities are shown in Table 8.4.

Using the same method, it is possible to determine the cleaning force of 0.2 and 1 μm rubber particles at different power densities.

III. **Dry Laser Cleaning of Particles**

According to the different thermal expansion of the materials, laser dry cleaning can be divided into three modes: substrate thermal expansion to remove particles, particle self-expansion to detach from the substrate, and joint absorption of laser energy by the substrate and particles to remove particles. Existing theoretical and experimental studies have shown that the cleaning effect of particle self-expansion detachment from the substrate is much lower than that of substrate absorption, and the size of the removed particles is generally above the micrometer level. The third mode combines the first two modes. When the size of the particles is much larger than the wavelength of the incident light, irregular shape, surface unevenness, and uneven distribution of the particles cause the incident light to be reflected from various directions. Generally, the diameter of the particles to be removed is larger than the wavelength of the laser used, so the focus should be on strong laser absorption by the mold substrate.

8.5 Impact of Laser Cleaning Parameters on Molds Cleaning Effect

8.5.1 Laser Source

The commonly used laser types for mold cleaning are TEA CO_2 laser and Nd:YAG laser. Generally, pulsed Nd:YAG lasers have lower average power and smaller pulse energy compared to CO_2 lasers. However, CO_2 lasers are difficult to transmit using flexible optical fibers, making practical operation inconvenient. This is especially true for mold substrates with curved surfaces, such as tire molds. Precise focusing control is required during cleaning. Cleaning using beam delivery systems or direct irradiation is not as convenient as cleaning through fiber-optic beam delivery. Moreover, using fiber-optic beam delivery allows for easy cleaning of small areas within the mold [9].

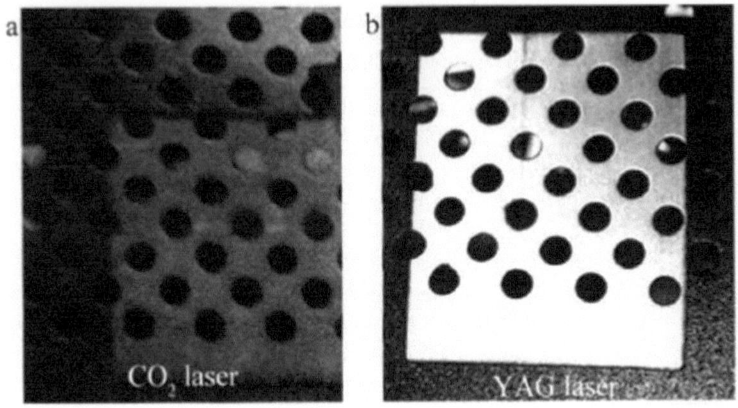

Fig. 8.7 Comparison of the absorptivity of rubber to different lasers

Zhang et al. [20] compared the absorption rates of rubber on steel plate surfaces for CO_2 laser and Nd:YAG laser. The results showed that the absorption rate of rubber for Nd:YAG pulsed laser was significantly higher than that for CO_2 laser, as shown in Fig. 8.7.

8.5.2 Laser Energy Density

Zhang et al. [20] conducted laser cleaning experiments on tire molds using a YAG pulsed laser generated by a solid-state semiconductor-pumped Q-switched laser. The laser parameters were as follows: pulse width of 200 ns, single pulse energy of 20 mJ, peak power of 100 kW, actual spot diameter of about 0.4 mm, and scanning speed of 2000 mm/s. The experiments showed that increasing the scanning speed, reducing the average power, or using a lens with a longer focal length would result in a decrease in the energy density of the laser cleaning. As a result, the cleaning effect deteriorated significantly, and the rubber on the mold surface was difficult to remove completely. It can be concluded that the cleaning effect of tire molds is related to the laser energy density, and the threshold for complete cleaning is 250 mJ/mm^2.

Wang et al. [21] conducted experiments on the removal of rubber from hard aluminum surfaces using an HGM-50 Q-switched pulsed laser. The laser parameters were as follows: wavelength of 1.06 μm, pulse width of 200 ns, repetition frequency of 3.7 kHz, spot diameter of 0.09 mm, and scanning speed of 300 mm/s. The cleaning efficiency was analyzed using metallographic microscopy and a CCD camera. The cleaning effects at different laser energy densities are shown in Fig. 8.8. The experiment found that there was almost no cleaning effect when the energy density was less than 4.42 J/cm^2. As the energy density increased, cleaning occurred on the substrate surface, and the cleaning efficiency continued to increase until the energy density reached 25.1 J/cm^2, where the cleaning efficiency reached 100%. Therefore, 25.1 J/

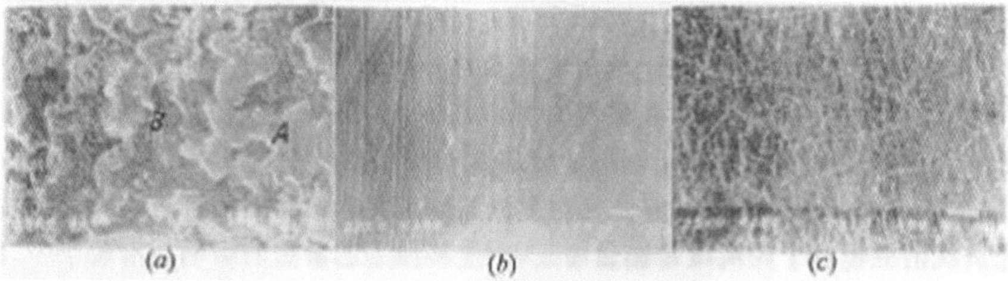

Fig. 8.8 Cleaning effect of sample surface under different laser energy densities

cm^2 was identified as the threshold for complete cleaning. When the energy density increased to 29.7 J/cm^2, surface ablation occurred on the substrate, indicating that the threshold for substrate damage was 29.7 J/cm^2.

Ye et al. [22] conducted a study on the cleaning of a No. 45 cast steel sample mold using a pulsed laser with a wavelength of 1064 nm. They investigated the effects of different laser pulse energies on the substrate and the cleaning efficiency, as shown in Table 8.5. With increasing pulse energy, no clear pattern was observed in the changes on the substrate surface. The initial cleaning threshold for the tire mold was determined to be 50.9 mJ, while the threshold for complete cleaning in single-pulse mode was found to be 180.9 mJ.

After laser irradiation, the surface roughness of the sample was observed to be 0.80 μm, while the control group, which underwent shot blasting cleaning, had a surface roughness of 1.43 μm. This indicates that the impact of laser on the surface of cast steel is smaller than that of shot blasting cleaning. According to industrial standards, laser irradiation will not affect the use of cast steel.

8.5.3 Laser Repetition Frequency

During the experiment of laser cleaning on tire molds using YAG pulsed laser, it was observed that when the laser power was kept constant and the laser repetition frequency was increased, the peak power of the laser pulses decreased. As a result, the cleaning effect deteriorated, and it became difficult to completely clean the mold. Additionally, the cleaning speed decreased. Conversely, when the repetition frequency was decreased, the cleaning effect improved, and the cleaning speed increased.

8.5.4 Beam Quality

For laser systems that require fiber transmission, the beam quality of the laser may deteriorate after exiting the fiber. The use of beam shaping and focusing lenses can

Table 8.5 Effect of laser power on cleaning effect

No.	Power (W)	Pulse frequency (Hz)	Scanning speed (mm/s)	Defocusing distance (mm)	Sample condition	Experimental phenomena	Experiment analysis
1	30	10	5	– 3	Flat surface with a rubber layer of about 0.1 mm	No obvious change in the rubber part and no change in the mold	The power is too low to remove rubber
2	60	10	5	– 3	Flat surface with a rubber layer of about 0.1 mm	The rubber particles begin to partially vaporize with pungent smell	The power is too low to remove rubber
3	80	10	5	– 3	Flat surface with rubber layer removed	The rubber vaporization is obvious, the rubber begins to burn, and the cleaning effect is 5904 obvious. T1205 here are burn marks on the surface, but there is no damage after wiping	The power is appropriate, and the mold has no trace of damage
4	90	10	5	– 3	Flat surface with rubber layer removed	The rubber vaporization is obvious, the rubber begins to burn, and the cleaning effect is obvious. There are burn marks on the surface, but there is no damage after wiping	The power is appropriate, and the mold has no trace of damage
5	120	10	5	– 3	Flat surface with rubber layer removed	The vaporized carbon falls off from the surface of the mold, causing slight damage to the surface of the mold and turning it black	The power is slightly higher, which has caused damage to the mold surface

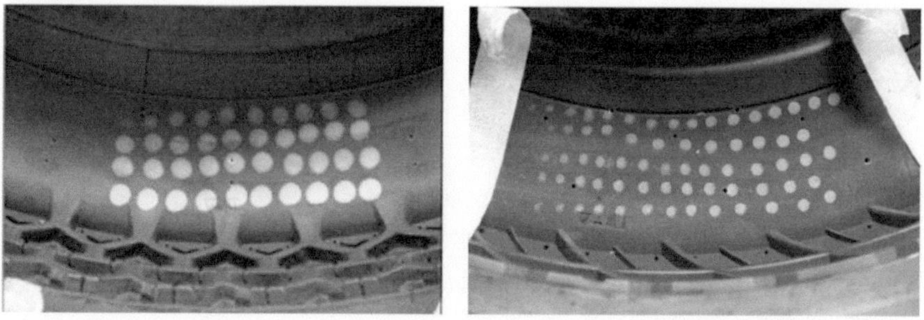

(a) (b)

Fig. 8.9 Photos of tire molds after **a** single laser pulse experiment and **b** multiple pulse experiment

affect the divergence angle and spot radius. When using a long focal length lens, the divergence angle is smaller, and the focal depth is longer, which is beneficial for cleaning operations. However, the spot size is larger, resulting in lower energy density. On the other hand, when using a short focal length lens, the divergence angle is larger, and the focal depth is shorter, which is not conducive to cleaning operations. However, the spot size can be made smaller, resulting in higher energy density. Therefore, a balance needs to be achieved between improving beam quality and increasing power density.

8.5.5 *Pulse Number*

In the experiment conducted by Ye et al. [22] on the cleaning of mold samples made of 45# cast steel using a 1064 nm pulsed laser, they also investigated the influence of different pulse numbers on the experiment. The experimental results are shown in Fig. 8.9. The results indicate that when the pulse numbers are set at 20, 40, 60, 80, and 100, the initial cleaning threshold and complete cleaning threshold remain the same, which are 48 mJ and 118.7 mJ, respectively. It can be observed that the complete cleaning threshold for multiple pulses is lower compared to a single pulse, while the initial cleaning threshold is slightly lower, but not significantly. For laser cleaning with the same number of pulses but different pulse numbers, the initial cleaning threshold and complete cleaning threshold are the same.

References

1. Kochan A. In-press mould cleaning in the tyre industry. Industrial Robot, 2001, 28(2): 12–113.
2. Jia XS, Zhang YD, Chen YQ, et al. Laser cleaning of slots of chrome-plated die. Optics and Laser Technology, 2019, 119:105659.

3. Xie L. Rubber mould cleaning technology and its developing trend. Rubber and plastic technology and equipment, 2001,27:18-22. (in Chinese)
4. Xie L. Laser cleaning mold technology. China Rubber, 1999, 7:22–23. (in Chinese)
5. Zhou GL, Zhao HX. New technology of laser cleaning tyre mould. Special Purpose Rubber Products, 2003, 5:39–41. (in Chinese)
6. Buchter E. Mould cleaning with laser radiation-mild, economical process for cleaning moulds. Kunststoffe-plast Europe, 1999, 89(11):72. (in Chinese)
7. Zuo H. Summarization of dry ice purge technology. Low Temperature and Specialty Gases, 2005, 23(2):12–14. (in Chinese)
8. Wang CQ. Cleaning technology for tire mold with laser and dry ice and its application. Tire Industry, 2018, 38(10):579–582. (in Chinese)
9. Jetter J, Hu P, Ma MM. Mold laser cleaning technology. China Rubber/Plastics Technology and Equipment, 2001, 27(11):30–32. (in Chinese)
10. Gong NM. New cleaning technology of the tire pattern. Cleaning World, 2004, 20(4):29–31. (in Chinese)
11. Lin Q, Shi MQ, Zhang M, et al. Laser cleaning and its applications. Guangzhou Chemical Industry, 2010, (6):23–25. (in Chinese)
12. Zhou GL, Sun HY, Wang CS. Research on laser cleaning process to rubber mold. Special Purpose Rubber Products, 2008, 29(6):34–36, 44. (in Chinese)
13. Sun HY. Research on the process of laser cleaning rubber product molds. Qingdao: Qingdao University of Science and Technology, 2009. (in Chinese)
14. Wang ZY. High-power CO_2 laser beam transmission and focusing and its impact on processing quality. Beijing: Beijing University of Technology, 1997. (in Chinese)
15. Lv BD. Laser optics: beam characterization, propagation and transformation. Resonator Technology and Physics. Beijing: Higher Education Press, 2003. (in Chinese)
16. Sun LB. Research on the mechanism of laser cleaning of rubber product molds. Qingdao: Qingdao University of Science and Technology, 2008. (in Chinese)
17. Zhou GL, Kong LB, Sun HY. Finite element analysis of temperature field based on ANSYS in laser cleaning mold. Manufacturing Automation, 2008, 30(9):90–92. (in Chinese)
18. Wang MC. Finite element method. Beijing: Tsinghua University Press, 2003. (in Chinese)
19. Tan DH, Lu DS, Song WD, et al. A finite element analysis of the temperature distribution on a substrate surface during laser cleaning. Journal of Huazhong University of Science and Technology, 1999, 6:50–53. (in Chinese)
20. Zhang ZH, Yu XC, Wang Y, et al. Experimental study about cleaning of tire molds with pulse YAG laser. Laser Technology, 2018, 233(1):131–134. (in Chinese)
21. Wang ZM, Zeng XY, Huang WL. Parameters and mechanisms of laser cleaning rubber layer on a tyre mould. Chinese Journal of Lasers, 2000, 27(11):1050–1053. (in Chinese)
22. Ye YY, Jia BS, Chen J, et al. Laser cleaning of the contaminations on the surface of tire mould. International Journal of Modern Physics B, 2017, 31(16–19):9–13.

Chapter 9
Laser Cleaning of Historical Buildings and Artworks

9.1 Significance of Cultural Relics Cleaning and Restoration

Cultural relics are the remnants and traces of human activities throughout the course of societal development, representing invaluable treasures that are non-renewable in human society. Various types of cultural relics reflect human social activities, social relationships, social ecology, and ideologies during different historical periods from different perspectives. They display humanity's capacity to utilize and transform nature. The protection, management, and scientific research of cultural relics are significant. They contribute to people's understanding of their own history and creativity, unveil the objective laws of human social development, and promote the progress of contemporary and future societies.

China, with a history of five thousand years of civilization, boasts a vast quantity of cultural relics. However, due to environmental pollution and inadequate preservation, many cultural relics and artworks are gradually suffering from corrosion and contamination. Factors such as climate change, sunlight radiation, air pollution, insect infestation, and mold proliferation contribute to the weathering and detachment of stone carvings, corrosion of bronze artifacts, crumbling and efflorescence of bricks and tiles, fading and blistering of murals, cracking and decay of wooden materials, silk and fabric deterioration, paper damage from insects and mold, cracking and warping of bones and tortoiseshell, and fur becoming brittle and shedding. These threats are constantly diminishing the lifespan of cultural relics. Taking the Capital Museum as an example, the total number of unearthed cultural relics in its collection has exceeded 200,000. However, due to unfavorable exhibition conditions, most of these artifacts are stored in warehouses and have never been displayed to the public. Most of these cultural relics need restoration [1].

The restoration of cultural relics involves a significant amount of work focused on cleaning various types of surface contaminants (such as oil, rust, dust, decayed plants, etc.) to restore them to their original appearance as much as possible. For

F. Song and X. Lin, *Laser Cleaning*, https://doi.org/10.1007/978-981-97-6631-4_9

instance, the ancient buildings within the Summer Palace feature a considerable number of stone components, some adorned with intricate patterns. However, due to acid rain, pollution, and the touch of visitors, certain architectural elements like railings, steps, as well as the surfaces of stone steles and carvings accumulate a layer of dirt, hastening the natural weathering of the stone. Removing these adhered pollutants from the surfaces not only restores the color of the stone but also slows down the pace of weathering.

Regarding the cleaning methods for cultural relics, there are different viewpoints. Taking the example of the stone cleaning at the Summer Palace mentioned earlier, on October 24, 2004, more than 10 Chinese and foreign stone conservation experts, including Lu Shoulin, Chairman of the Chinese Society for Cultural Relics Conservation Technology, gathered to provide advice on cleaning the building materials—bricks and stones—of structures like the Paiyun Hall and the Foxiang Pavilion in the Summer Palace. Chinese experts, represented by Mr. Lu Shoulin, suggested using physical methods such as scraping, wiping, and brushing for cleaning, while American experts advocated for chemical methods.

However, whether it is physical cleaning methods or chemical cleaning methods, both can cause certain damage to cultural relics. These traditional cleaning methods have their drawbacks in terms of cultural relic protection. Cultural relics are non-renewable, so the requirements for cleaning them are more stringent.

In the field of cultural relic cleaning, some commonly used traditional cleaning methods include mechanical cleaning, chemical cleaning, and ultrasonic cleaning. Mechanical cleaning involves using mechanical means such as scraping, wiping, brushing, or sandblasting to remove surface contaminants. Chemical cleaning utilizes organic cleaning agents and employs methods like spraying, dripping, soaking, or high-frequency vibration to remove surface deposits. Ultrasonic cleaning involves placing the object to be cleaned into a cleaning agent and using the vibration effects generated by ultrasonic waves to remove dirt.

However, these cleaning methods have their respective limitations when applied to cultural relic cleaning. On one hand, each cleaning method has its drawbacks. For instance, mechanical cleaning cannot meet the requirements of delicate cleaning and may easily damage the surface of relics. Some fragile artifacts such as silk, paper, murals, and fabrics cannot be cleaned using mechanical methods. Chemical cleaning methods can lead to environmental pollution, provide limited cleanliness, and may result in unforeseen damage to the relics themselves. Some relics, having endured for thousands of years in specific environments, have complex surface contaminants that are not easily removed by single cleaning agents. Ultrasonic cleaning is ineffective against sub-micron-level contaminants, and the size limitations of cleaning tanks restrict the size of the artifacts that can be cleaned. Moreover, a crucial point is that many artifacts cannot undergo wet cleaning, rendering chemical and ultrasonic methods ineffective.

With the research and application of laser cleaning, people have proposed using laser cleaning in the restoration and preservation of cultural relics. Laser cleaning technology offers advantages such as selectivity, non-contact operation, real-time controllability, environmental friendliness, and reliability.

In 2003, Teule et al. [2] developed a controllable laser cleaning machine specifically designed for artworks. This machine utilized a KrF excimer laser and four right-angle prisms arranged to form an optical arm. By using motors to control the horizontal and vertical directions, it achieved two-dimensional controllability, resulting in successful cleaning of artworks. Kane et al. [3] from Macquarie University in Australia utilized a high repetition frequency frequency-doubled copper vapor laser (UV-CVL) and a medium-power excimer laser to clean nanoparticles from the surfaces of optical components such as glass and silicon substrates. They achieved complete cleaning of particles larger than 0.3 μm on small areas.

The National Museum and Painting Collection Center in Merseyside, UK, is one of the world-class cultural relic collection sites. The research group led by Martin [4] has done fruitful work in the restoration and cleaning of cultural relics. They have published a monograph on laser cleaning of cultural relics—"Application of Laser Cleaning in Cultural Relic Protection", which delved into theoretical and experimental research on laser cleaning technology, particularly in cultural relic preservation. They also developed laser cleaning machines and conducted extensive research and practical applications on many British artifacts. The results proved that laser cleaning technology is an excellent cleaning method that surpasses other cleaning methods.

In Europe, countries such as the United Kingdom, Germany, Italy, France, and Greece have successfully applied laser cleaning technology in cultural relic preservation and achieved remarkable results.

Excimer lasers and Nd:YAG lasers are commonly used laser sources in cultural relic cleaning. To minimize the potential damage caused to artifacts by thermal effects during laser cleaning, picosecond and femtosecond lasers have been introduced into the field of cultural relic cleaning.

The powerful ultrafast femtosecond pulses possess a unique capability for material ablation with minimal accompanying damage, which can eliminate the surface contamination layers while preserving artworks and heritage items. The primary advantages of using short-pulse, high repetition rate lasers for laser cleaning include high-precision surface treatment and minimal invasiveness, achieving submicron precision when removing pollutants. Moreover, high repetition rate pulses can increase scanning speed and avoid thermal accumulation. With the rapid development of femtosecond laser technology, ultrafast laser cleaning has the potential to become a key technology for treating artworks and heritage items that were previously difficult to restore [5].

9.2 Laser Cleaning of Stone Artifacts

Stone sculptures or ancient buildings, among other stone artifacts, are often exposed to the natural environment for extended periods. Over time, the surface of these artifacts undergoes changes, resulting in the formation of a black, hard shell. This shell not only alters the original appearance of the artifacts but can also pose a threat to the artifacts themselves or interfere with their maintenance. Therefore, cleaning becomes very necessary.

9.2.1 Laser Cleaning Applied to Buildings

In Europe, laser cleaning projects have been used for the cleaning of historical buildings including: the Amiens Cathedral in France, St. Stephen's Cathedral in Austria, the Tomb of the Unknown Soldier in Poland, Rotterdam City Hall in the Netherlands, St. Frederik's Cathedral in Denmark, the Cathedral of Santa Maria Assunta in Italy, and the famous Westminster Palace in London, etc. [6–8].

Taking the example of the Amiens Cathedral in France, in September 1992, to commemorate the twentieth anniversary of the establishment of the World Heritage Protection Organization under UNESCO, a restoration project was carried out for the renowned Amiens Cathedral. The exquisite marble carvings on the west facade's Portal of the Virgin Mary were crucial to the project. During the one-year restoration project for the Portal of the Virgin Mary, restoration personnel used laser beams to remove a few millimeters thick black dirt covering the marble carvings. This revealed the original color of the marble surface, restoring the beauty of the intricate carvings.

9.2.2 Stone Artifacts and Cultural Relics

I. **Research on Laser Cleaning of Stone Cultural Relics**

For artworks such as many stone sculptures made of limestone, When SO_2 and NO produced by fuel combustion in the atmosphere drift onto these stone sculptures, they react chemically to form a black shell that deposits on the surface. Gypsum black shells are one of the most severe forms of degradation affecting architectural cultural heritage worldwide. They not only affect the aesthetic appearance but also accelerate the weathering of the stone sculptures, thereby shortening their lifespan. Therefore, removing them from the granite substrate is necessary.

In the past, traditional methods such as high-pressure water jet washing or mechanical friction were primarily used for cleaning, but these methods often caused damage to the samples, and some finely carved places are not easy to clean. However, using laser cleaning for contaminated samples can avoid these issues and achieve excellent cleaning results. Generally, the cleaning speed for pollutants on brick and stone buildings is about 0.6 m^2/h, while the cleaning speed for stains on limestone sculptures is about 0.2 m^2/h. Increasing the energy density of the laser can increase the cleaning efficiency, but it is crucial to ensure that the building or artwork is not damaged.

Different wavelengths of lasers have varying effects on cleaning stone artifacts. Taking the example of the commonly used white marble substrate found in European architectural monuments, a study conducted by Ortiz et al. [9] used two types of Q-switched Nd:YAG pulsed laser systems. One had a wavelength of 1064 nm, pulse width of 8 ns, and repetition rate of 20 Hz, while the other was the fourth harmonic with a wavelength of 266 nm, pulse width of 10 ns, and repetition rate of 10 Hz. The base material of the cleaning object is white marble, with dimensions of 10 cm × 5 cm × 2 cm. The pollutants are ash and iron oxide deposits, as well as coatings sprayed with spray paint and markers. The primary component of the graffiti ink was acrylic polymer. High-pressure water jet cleaning was used for comparison. The cleanliness was quantified using CIE color parameters. The experiments revealed that using UV 266 nm laser pulses for cleaning effectively removed weathered surfaces and performed well in cleaning green and red graffiti and oxide residues. High-pressure water jet cleaning was also effective in removing surface deposits. When using 1064 nm laser cleaning, regardless of dry or wet cleaning methods, the substrate turned yellow when cleaning deposits and black spray stains. However, when the laser wavelength was 266 nm, no yellowing was observed on the marble surface. Similar phenomena have been reported in other studies [10, 11]. This yellowing effect induced by laser cleaning may be due to the presence of trace amounts of iron oxide in the stone or residues. Ortiz et al. [9] used surface X-ray micro fluorescence mapping to determine the presence of specific elements. They found that before cleaning, only Si signals (components of black spray) were detected. After cleaning, strong signals of Ca and Mg (main components of dolomite white marble) and weak Si signals were observed, as shown in Fig. 9.1.

The experiments showed that there was no observed structural damage after cleaning with a 1064 nm laser. However, cleaning with a 266 nm laser or high-pressure water jet resulted in surface damage to the stone. Additionally, cleaning with a 266 nm laser caused surface erosion exceeding 100 μm in size.

In the study conducted by Ortiz et al. [9], laser cleaning of surface deposits, iron oxide stains, and various graffiti (black, red, green) on white marble was investigated. Both near-infrared (IR) and ultraviolet (UV) pulse lasers were utilized, and the cleaning effects were evaluated through colorimetric measurements, X-ray micro fluorescence, as well as optical and electron microscopy. The results revealed that the 266 nm UV pulsed laser ablation effectively removed all stains except for the tip ink, causing some surface damage at a microscopic scale. On the other hand, the

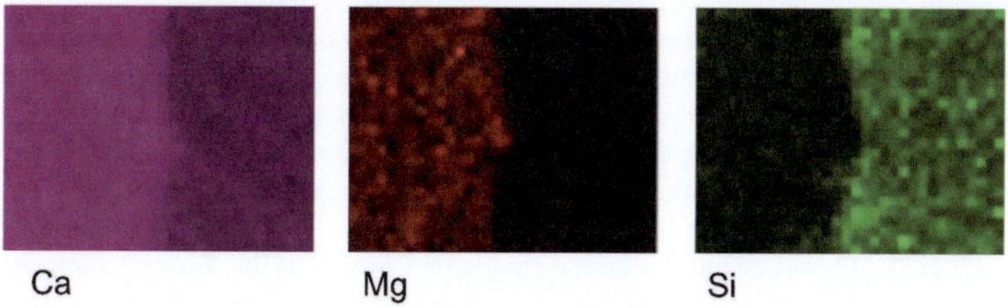

Fig. 9.1 The X-ray micro fluorescence of black spray on stone and the original interface

1064 nm IR pulsed laser could remove black deposits, but it led to yellowing of the stone surface afterward. Fluorescence analysis of X-ray samples indicated that the residual surface weathering agent was the main cause of yellowing. In summary, it can be concluded that pulse laser cleaning technology is beneficial for the rapid cleaning of small particles.

Pozo-Antonio [12] conducted a study on laser cleaning of gypsum black shells on granite in the northwest Iberian Peninsula using both dry and wet methods. These black shells form a thick solid covering on the surface of granite, primarily composed of needle-like, flake-like, and desert-like gypsum, with the main minerals being quartz (29%), potassium feldspar (25%), sodic plagioclase (24%), muscovite (13%), and biotite (4%). The laser was a combination of two wavelengths (1064 and 355 nm) from an Nd:YAG Q-switched laser system to overcome the discoloration issue when cleaning black shells with a single wavelength laser. The pulse width of the laser was 6.5 ns, with a repetition rate of 13 Hz. Dry and wet laser cleaning methods with different laser parameters were studied. It was found that under wet methods, the cleaning effect was better.

When both wavelengths have an energy density of 0.3 J/cm^2, under pre-wetted surface conditions, the removal rate of carbonaceous contamination particles and gypsum is high, with no damage to the granite minerals. The mechanism of laser cleaning involves the melting, spalling of biotite and potassium feldspar, and cleavage fracture of quartz.

China has also conducted research on the laser cleaning of stone artifacts. Qi [13], Ye [14], used a pulsed laser with a wavelength of 1064 nm, a pulse width of 10 ns, and a spot area of 7 mm^2 to conduct laser cleaning experiments on the surfaces of rock carvings at the Yungang Grottoes, targeting various organic pollutants. The repetition frequency was set at 1 Hz. Laser cleaning for artifact preservation first requires ensuring that the artifacts are not damaged. Therefore, the experiment began by determining the laser's damage threshold, with the upper limit of laser energy used in the cleaning experiments set at 80% of the damage threshold for the sandstone substrate. Samples of sandstone from the Yungang Grottoes area were used, with contaminants including ink stains, lime, smoke residues, and colored paint. The experiment continuously adjusted the pulse energy and the number of pulses used to find the optimal parameters for cleaning different pollutants. A comparison between dry and wet laser cleaning was also conducted. The results showed that the greater the laser energy, the more significant the cleaning effect. For dry laser cleaning, with 5–6 pulses, ink can be removed when the energy reaches 60.8 mJ, smoke stains can be removed at 46.6 mJ, paint can be removed at 64.3 mJ, and lime can be removed at 192.3 mJ. For ink and lime, wet laser cleaning shows a more significant cleaning effect, with energy requirements of 45.9 mJ and 81.2 mJ respectively, with the same number of pulses. For smoke stains and paint, there is not much difference in the cleaning effect between dry and wet laser cleaning.

Qi et al. [13, 14] conducted laser cleaning on Cave 43 of the Yungang Grottoes to remove ink stains and smoke soot. The laser energy density used was approximately 0.66 J/cm^2, with a pulse repetition rate of 10 Hz. Dry cleaning revealed that the damage threshold of sandstone in the Yungang Grottoes was 73.5 mJ, while the

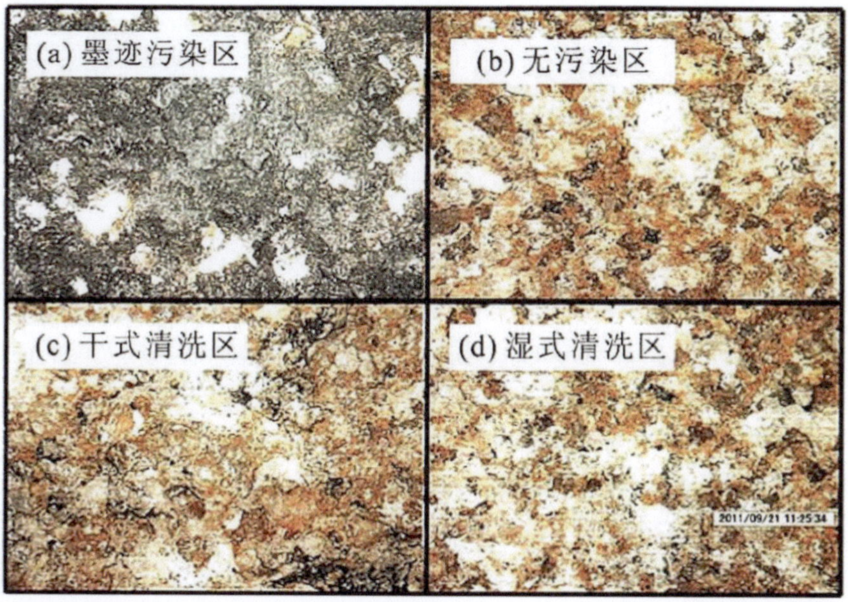

Fig. 9.2 Cleaning effect of ink stains on the rock surface of Cave 43 in the Yungang Grottoes: **a** area with ink stain contamination; **b** area without contamination; **c** area cleaned with dry method; **d** area cleaned with wet method

dry-cleaning thresholds for ink stains, smoke soot, paint, and lime contaminants on the sandstone surface were 21 mJ, 20.5 mJ, 32.5 mJ, and 49.5 mJ respectively. The thresholds for wet cleaning were 17.7 mJ, 20.5 mJ, 32.5 mJ, and 33.5 mJ, respectively. As shown in Fig. 9.2, dry laser cleaning effectively removed contaminants, but compared to the uncontaminated area, there were still small amounts of ink residue. Conversely, after cleaning with the wet cleaning method of pure water and ethanol, the removal of contaminants was excellent, with virtually no residue remaining.

Based on existing research, the mechanism of laser removal varies for different types of hard shells and even during different stages of the cleaning process. For instance, in the case of multi-layer coatings, the top black layer may be removed through thermal evaporation and laser-induced fragmentation, whereas the primary mechanism towards the end of the cleaning process may involve selective decomposition and evaporation. This selectivity is determined by the absorption characteristics of different materials to the laser, for example, black particles absorbing 3–4 times more of the 1064 nm wavelength laser compared to gypsum layers or the substrate of cultural artifacts. The main mechanisms in the nanosecond laser stripping of hard shells are as follows:

(1) Selective decomposition and evaporation: The surface-attached hard shell absorbs much more laser energy of a certain wavelength than the substrate. After the hard shell absorbs most of the laser energy, it expands due to heat and

separates from the substrate, or directly vaporizes and evaporates, thus achieving the purpose of cleaning.

(2) Splitting induced by shockwaves: Laser irradiation on the surface of the hard shell generates shockwaves. When the shockwave reaches the middle and lower layers of the hard shell, it returns partially and interferes with the incident sound wave, creating mechanical resonance that causes the dirt to break and fall off.

Laser cleaning of surface hard shells on sculptures and buildings is currently widely used, and the technology is relatively mature. Generally, the hard shell absorbs laser energy much more strongly than the underlying substrate, and its decomposition or damage threshold is usually much lower than that of the substrate. Therefore, by selecting an appropriate laser fluence value between the thresholds of the hard shell and the substrate, the coating can be removed without affecting the substrate.

II. **Practical Application of Laser Cleaning of Stone Cultural Relics**

Laser cleaning of stone cultural relics has been practically applied. To illustrate the successful application of laser cleaning in the preservation of sculptures and ancient architectural relics, the following are some examples of laser cleaning.

Figure 9.3 shows a comparison before and after laser cleaning of the Hermes statue. Figure 9.3a depicts the statue before cleaning, exhibiting significant contamination. Figure 9.3b shows the statue after cleaning, fully restored to its original appearance.

Figure 9.4 is a marble sculpture with a history of nearly 2000 years. The original appearance and carved details of the sculpture were obscured by a black hard shell on the surface. The situation after partial cleaning is shown in the figure, where the lower half remains uncleaned, while the upper half displays the appearance after Nd:YAG laser cleaning. Under the action of the laser, the black hard shell pollution layer is removed, revealing the original details of the substrate.

Fig. 9.3 Comparison before and after cleaning of the Hermes statue

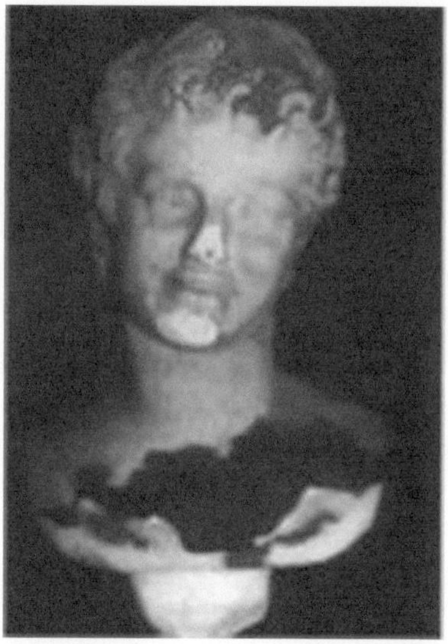

Fig. 9.4 Comparison before and after laser cleaning of the marble statue

Figure 9.5 illustrates a gypsum relief contaminated with soot, with the right half successfully cleaned. Figure 9.6 depicts a stone carving on the exterior wall of a church, where the right side has been cleaned while the left side is the original appearance of the pollution.

Laser-assisted cleaning of hard shells on buildings and sculptures has been applied in reality. In 1992, the World Heritage Protection Organization conducted repairs on the renowned Amiens Cathedral in France. During the one-year restoration of the Virgin Mary's Gate, maintenance personnel used laser beams to remove several millimeters thick black dirt layer covering the marble carved patterns. Similarly, in 2002, laser cleaning of the sculptures of saints on the Cologne Cathedral in Germany was also a successful example.

Fig. 9.5 Comparison of laser cleaning on relief (Part I)

Fig. 9.6 Comparison of laser cleaning on relief (Part II)

Due to the non-renewable nature of cultural relics, utmost caution must be exercised during the cleaning process. Over hundreds or thousands of years, external pollutants have formed uneven black shells on stone sculptures, with complex compositions. Generally, when the thickness and irregular shape of the shells and their composition are unclear, careful selection of laser fluence is essential to avoid damaging the cultural relics. It is advisable to start with a low laser fluence and opt for 2 or 3 cleaning sessions, rather than choosing a higher fluence that may exceed the damage threshold of the substrate. During the initial stages of cleaning, inconspicuous areas should be selected for testing. In some cases, it may be prudent to prepare mock samples and conduct preliminary experiments to accumulate data before proceeding with formal cleaning procedures.

In existing reports, lasers used to remove hard shells from the surfaces of sculptures or ancient buildings in cultural heritage conservation mostly employ Q-switched Nd:YAG lasers ($\lambda = 1064$ nm, pulse width around 10 ns). In some specialized applications, the third harmonic of the Q-switched Nd:YAG laser ($\lambda = 355$ nm) is utilized, which has a higher damage threshold to the substrate compared to the fundamental frequency. Additionally, certain types of hard shells with specific compositions can only be removed using the third harmonic. And in the cases of bio-hard-shells formed by fungi and other organisms, the efficiency of using the third harmonic laser for cleaning is significantly higher. Therefore, a combination of the fundamental frequency and the third harmonic of Q-switched Nd:YAG lasers can be employed in cleaning procedures.

Laser cleaning incurs higher costs. Ortiz et al. [9] believe that the primary factor affecting the cost is the operation time. Laser cleaning requires applying pulses to small areas sequentially, with each pulse covering an area of approximately 4 ~ 6 mm^2. It typically requires at least 1 ~ 2 min per square centimeter for cleaning. Overall estimation suggest that laser cleaning is nearly 20 times more expensive than traditional techniques. However, due to the irreplaceability of laser cleaning, especially in the cleaning of non-renewable cultural relics, it remains great potential.

9.3 Laser Cleaning of Metal (Bronze) Cultural Relics

Bronze cultural relics hold significant importance in ancient cultural relics and artworks. Traditional methods of mechanical and chemical cleaning pose risks of residue and surface abrasions. Zhang et al. [15] conducted laser cleaning experiments on bronze gilded cultural relics provided by the Hebei Cultural Relics Protection Center. The right side of the bronze gilded statue that needs to be cleaned is covered by green corrosion, while red corrosion covered most areas of the left side of the body, head, and face, resulting in blurred original contours, as shown in Fig. 9.7a. The laser cleaning parameters were as follows: a wavelength of 1064 nm, pulse energy of 1 J, pulse width of 5 ~ 8 ns, energy density of 0.30 ~ 0.50 J/cm^2, maximum repetition rate of 40 Hz, spot diameter of 10 mm, and divergence of 0.5 mrad. After laser cleaning, the thick layer of red corrosion tightly adhered to the gilded layer was removed, revealing the gilded layer without damage. The statue regained its original luster, and the facial and body contours became clearer, reflecting the original craftsmanship value. Examination under a three-dimensional video microscope post laser cleaning revealed preserved traces of the original craftsmanship with no residue or new abrasions. After verification by the Hebei Cultural Relics Protection Center, the cleaning met the predetermined requirements, demonstrating excellent and safe results of laser cleaning for this gilded bronze cultural relic. The experiment provided applicable laser cleaning parameters and methods for gilded bronze cultural relics, accumulating experience for laser cleaning of gilded bronze cultural relics.

Laser cleaning of bronze artifacts can sometimes cause discoloration due to high temperatures. One reason for this is the effect of laser radiation on surface contaminants (such as lead particles) and metal corrosion, which can lead to their ablation, melting, and redeposition, causing changes in light reflection [16–18]. Alternatively, discoloration may occur due to the laser-induced transformation of copper corrosion

Fig. 9.7 Comparison photos of laser cleaning on bronze gilded cultural relics. **a** Before laser cleaning; **b** after laser cleaning

products into cuprous oxide and cupric oxide. Shen et al. [19] proposed a method involving the application of agar gel on the surface of bronze artifacts. This method allows for the adjustment of laser energy while reducing mechanical stress and minimizing the redeposition of particulate contaminants. The experiment utilized an Nd:YAG pulsed laser cleaning machine with a wavelength of 1064 nm and two modes: LQS mode with a pulse width of 100 ns and pulse energies of 150, 300, and 450 mJ, and SFR mode with a pulse width of 30 ~ 110 μs and pulse energies of 100 ~ 2000 mJ. The spot diameter was fixed at 3.5 mm, and the repetition rate was set at 1 and 5 Hz, with 20 pulses per experiment. The agar powder was prepared into 2, 4, 6, 8, and 10 mm thick gel layers at the concentrations of 1, 2 and 3%. A comparative observation was conducted using video microscopy and scanning electron microscopy, and some samples underwent analysis using Raman spectroscopy and X-ray diffraction. The experimental results showed that laser energy was attenuated after passing through the gel layer, and energy dissipation increased with the concentration and thickness of the gel layer. Surface micro-melting under SFR mode radiation was observed under scanning electron microscopy, consistent with the findings of Froidevaux et al. [17]. Then the water–ethanol wetting method was used again, and it was found that there was still a cooling effect, which was more obvious under multiple pulse conditions. The surface coated with gel and water–ethanol will protrude spherical structures. After energy spectrum analysis, the composition of the spherical structure is no different from the substrate, as shown in Fig. 9.8.

This spherical morphology conforms to the previously mentioned micro-melting state, indicating that the gel method reduces the spherical structure and decreases laser energy. Comparative experiments were conducted between gel-laser cleaning, mechanical methods, and wet laser cleaning, revealing that the gel-laser combined method resulted in a more uniform post-cleaning state and a more natural visual effect, with higher cleaning efficiency compared to mechanical cleaning. Afterwards, the gel-laser cleaning was used to clean different cultural relics, which had a good effect, but there were differences in the thresholds of different cultural relics.

Some precious bronze artifacts are typically plated with a layer of gold. During long-term storage, the formation of numerous pores can lead to corrosion of the metal underneath the gold layer. As a result, corrosion products accumulate on the surface

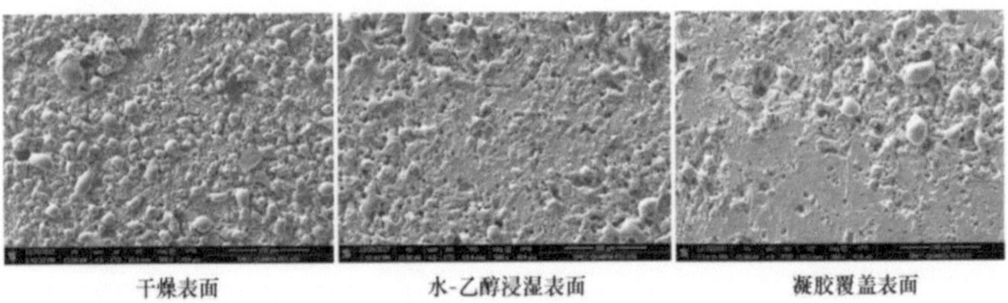

Fig. 9.8 Secondary electron images of sample surfaces after single-pulse irradiation in SFR mode. **a** Dry surface; **b** surface wetted with water–ethanol; **c** surface covered with gel

of the gold layer. Lee et al. [20] conducted an analysis of the surface of gold-plated bronze, investigating the efficacy of Nd:YAG laser irradiation in removing corrosion products before and after treatment, and compared to chemical cleaning. They utilized SEM-EDS, AFM, and XPS analysis to determine surface characteristics. The results indicated that chemical cleaning removed copper corrosion products through chemical reactions but did not uniformly eliminate them. In contrast, laser cleaning removed some corrosion layers, leaving behind a successfully treated surface of gold-plated bronze, as confirmed by various surface analyses, without any damage or residual contaminants.

9.4 Laser Cleaning of Ceramic Artifacts and Artworks

In the history of human civilization, the use of pottery is both ancient and widespread, making it an important part of artifacts and artworks. Ancient Chinese pottery, such as Han Dynasty terracotta warriors, exhibits high artistic, cultural, historical, and scientific value, particularly in areas such as lacquer work, sculpture, and painted decoration techniques. Zhang et al. [21] conducted a study on the laser cleaning of painted female figurines from the Han Dynasty using Nd:YAG infrared lasers. Figure 9.9 shows a photograph of the figurines. Before the laser cleaning experiment, the composition of surface contaminants on the pottery figurine was determined using a spectrometer. Qualitative analysis of XRF spectra indicated a significant peak for the Mn element in the black contaminant layer, suggesting the presence of Mn as the cause of its black coloration. Similarly, high Fe element peaks were observed in the spectra of the deep red mineral pigment layer, corresponding to the red coloration produced by Fe. Qualitative analysis of LIBS spectra revealed that the main elements detected in the black contaminant layer were Na, Mg, Al, Si, K, Ca, Mn, Fe, and Ba, while the deep red pigment layer mainly contained Na, Al, K, Ca, and Fe. The pink pigment layer was found to contain Na, Al, K, Ca, Hg, and S, while the white mineral layer contained Na, K, and Ca. It was determined that the black spots consisted of a chemically complex mixture. Through different pulse width and energy combinations, it is verified that the weakening effect of gel layers with different concentrations and thicknesses on the laser energy of different pulse and width energy combinations is similar in trend and amplitude, independent of the laser wavelength.

The pottery figurine needs to remove the surface's black contaminant layer, while the pigment layer acts as the substrate. The laser parameters used in the experiment were as follows: wavelength of 1064 nm, maximum output energy of 450 mJ, spot diameter ranging from 1.5 to 6 mm, and beam quality of 80–30 mm mrad. Two operational modes were employed: in LQS mode, the laser pulse width ranged from 100 to 200 ns; in SFR mode, the laser pulse width ranged from 30 to 110 μs. The experiment found that for the red mineral pigment layer, the cleaning effect was unsatisfactory when using the LQS mode. However, when choosing the SFR mode, the cleaning mechanism mainly involved thermal effects, and at higher energy

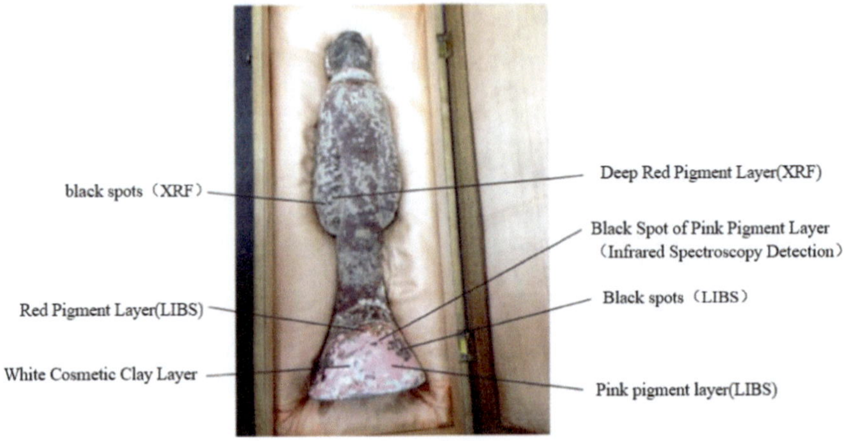

Fig. 9.9 Han Dynasty painted terracotta female figurine

levels, there was a risk of ablation. For the black spots on the white mineral pigment layer, the LQS mode was effective, resulting in reduced spot size, with no damage to the pigment at an energy level of 150 mJ. The cleaning mechanism primarily involved vibration effects, but pigment damage occurred when the energy reached 300 mJ. Regarding the black spots on the white mineral pigment layer, in the SFR mode, there was no significant cleaning effect at an energy level of 100 mJ, and although some effect was observed at 200 mJ, it was prone to ablation. It can be observed that the composition of the pollutants on the artifacts and the substrate pigment significantly influence the cleaning effect, determining the energy required for removal. The experimental results demonstrate the feasibility of laser cleaning for the black spots on Han Dynasty pottery figurines. By employing two different modes, within a certain range, surface contaminants on the figurines can be effectively removed. In the LQS mode, using optical parameters of an output energy of 150 mJ, a frequency of 3 Hz, and a spot diameter of 3–4 mm, black spots can be efficiently removed without damaging the pigment base.

Xu et al. [22] used laser cleaning to remove corrosion from the surface of ceramic products. The output power, repetition rate, and pulse duration (FWHM) of the fiber laser can be adjusted from 1 to 100 W, 100 to 400 kHz, and 58 to 240 ns, respectively, with the laser output being Gaussian distributed. The cleaning samples consist of two layers, including a ceramic substrate and a soil rust layer covering it. The experiment investigated the influence of repetition rate, pulse duration, laser power, and scanning speed on the cleaning effect. For high-repetition nanosecond pulse lasers, the cleaning effect is not significantly affected by the laser repetition rate and pulse duration (100–400 kHz and 58–240 ns). Laser power and scanning speed are the two main parameters affecting the cleaning effect, and reasonable laser power and scanning speed can achieve the desired cleaning effect. The results indicate that laser irradiation can directly remove corrosion from the surface of ceramic products, leaving the ceramic surface smooth after cleaning.

9.5 Laser Cleaning of Calligraphy and Painting Artifacts

Artifacts in the calligraphy and painting categories have unique substrates, typically composed of materials like paper, wood, or fabric, which are relatively fragile. Traditional methods such as mechanical or chemical cleaning may result in damage to the surface materials or harm the pigments, leading to irreparable damage to the artwork. Compared with these traditional methods, using laser cleaning protection technology can achieve good cleaning effects by controlling energy and action position.

The basic principle of laser cleaning for calligraphy and painting is generally understood as follows: The contaminants on the surface of the artwork absorb the energy from the laser beam, leading to a series of physical and chemical changes. These changes include thermal expansion-induced ablation and photodecomposition caused by the breaking of chemical bonds, ultimately resulting in the removal of contaminants from the painting surface. For the former, it requires the process to be quick, with minimal heat transfer to the substrate, to prevent damage. Conversely, for photodecomposition, the high energy of the ultraviolet laser's single photons is exploited. Since cultural relics and artworks are nonrenewable, the selection of laser parameters for cleaning, including wavelength, pulse width, and laser energy, must be carefully considered. The most used lasers for this purpose are KrF excimer lasers ($\lambda = 248$ nm) and third harmonic YAG lasers, known for their short wavelengths and high single-photon energy. In addition to wavelength, the most critical parameter is the laser energy per unit area, which can be determined by the efficiency of laser cleaning in experiments. Generally, the substrate of calligraphy and painting is quite sensitive to lasers. When using laser cleaning, allowing the laser to penetrate the substrate may affect the pigment layer, leading to long-term color degradation of the artwork and compromising its integrity. Therefore, thorough calculations and careful study of the penetration depth of light are necessary before cleaning. This involves considering at least two aspects. First, the threshold required for cleaning, which can be calculated based on the initial thickness of the contaminant layer, laser energy, and the thickness that can be removed per pulse. After calculation, a small inconspicuous part of the calligraphy or painting (such as a few square millimeters at the edge) can be selected for experimentation to study the removal rate and efficiency and determine the cleaning parameters. Second, the damage threshold, which is related to the thickness of the contaminant layer, laser energy, and substrate material. During cleaning, it is important to cautiously increase the laser energy to avoid damaging the cultural relics.

Paper cultural relics and artworks differ significantly from artifacts made of other materials. Paper is composed of plant fibers, making it soft, and the pollutants produced during long-term use and storage are quite different from the rust, paint, and pollutant particles on the surfaces of hard materials such as metal and stone. Long-term storage of paper may lead to mold growth, especially under favorable temperature and humidity conditions, where spores can develop into colonies. Enzymes produced by these molds can cause hydrolysis of the paper fibers, reducing the mechanical strength of the paper. Laser cleaning can be applied to remove mold effectively. The interaction between mold and paper differs from the typical van der Waals forces interaction between pollutants and substrates. It primarily involves chemical bonds between microbial molecules. Laser cleaning can achieve mold removal through photodecomposition, where high-energy photons break or weaken these bonds, leading to mold stripping Alternatively, it can work through thermal effects, rapidly increasing the temperature above the mold's melting or boiling point, causing instantaneous combustion and vaporization of the mold for cleaning purposes.

Zhao et al. [23] conducted a study on the cleaning of calligraphy and painting artworks, using a semiconductor-pumped third harmonic Nd:YAG laser for laser cleaning of mold-contaminated rice paper. The main parameters were as follows: the wavelength is 355 nm, the pulse width is (15 ± 3) ns, the maximum output average power is 10 W, the scanning speed is 1 mm/s, and the line spacing is 0.1 mm. Different experiments were conducted with varying repetition rates, power densities, and laser energy densities. It was found that at a repetition frequency of 20 Hz, an output power of 0.004 W, and a laser energy density of 0.85 J/mm^2, after one scanning, the surface only showed a slight color change; after two scans, most of the mold was removed, but some residue remained. At a repetition rate of 40 Hz, an output power of 0.019 W, and an energy density of 3.2 J/mm^2, after one scan, most of the mold was removed, but some residue remained; after two scans, almost all the mold was completely removed. However, when the laser pulse repetition rate was 50 Hz, the output power was 0.026 W, and the laser energy density was 4 J/mm^2, the high laser energy density caused significant damage to the rice paper substrate. Selecting appropriate parameters for laser cleaning can effectively remove mold from rice paper without damaging the substrate.

After most Western paintings (oil paintings) are completed, a layer of varnish with a thickness of 0.05–0.08 mm is applied to the painting. This varnish layer serves several purposes: it prevents direct contact of the pigments with the environment, protects the pigments from exposure to ultraviolet light, and enhances the glossiness of the painting. However, over time, the varnish layer undergoes oxidation, semi-oxidation, recombination of atomic groups, and polymerization under natural light exposure, resulting in the formation of a hard, blackened layer on the painting surface, known as aging. Additionally, some paintings, especially those exposed to natural environments for long periods (such as frescoes in churches), may accumulate dust, soot, and other pollutants on their surfaces. These accumulated substances and the aged varnish layer significantly reduce the glossiness of the painting, impairing its aesthetic appeal, and may potentially damage the artwork itself.

The ultraviolet laser, owing to its high single-photon energy, utilizes the photodecomposition effect to disrupt or weaken the interaction between contaminants and oil paintings, achieving the purpose of cleaning. This process is a "cold" process that produces almost no heat. Therefore, it does not cause significant impact on the substrate. Figure 9.10 illustrates a comparison of the oil painting of the Prophet Isaiah from Mount Athos, Greece, dating back to the seventeenth century, before and after laser cleaning and protection using an ultraviolet excimer laser [24].

Fig. 9.10 Comparison before and after laser cleaning of the portrait of the Prophet Isaiah

An eighteenth century Flemish egg tempera painting on a wooden panel, measuring 27 cm × 37 cm, was cleaned. Figure 9.11a shows the original appearance of the painting, with a small area at the lower left already laser-cleaned to obtain suitable laser cleaning parameters in a less visible edge area. Figure 9.11b displays the top part of the painting already cleaned with lasers. For comparison, a small area at the bottom of Fig. 9.11b (below the already cleaned area in Fig. 9.11a) was also chemically cleaned, achieving a removal effect similar to laser cleaning, but with contamination of the pigment layer by chemical solvents. Figure 9.11c contrasts the differences between the original surface (top) and the surface after laser cleaning (bottom). Figure 9.11d shows the result obtained by removing the remaining varnish with a solvent free of harmful substances after laser cleaning [25].

Figure 9.12 depicts a nineteenth century oil painting measuring 49 cm × 77 cm. An uneven layer of lime mortar is scattered across the painting, obscuring its original appearance (Fig. 9.12a). The varnish layer has also aged, being only a few micrometers thick. Such a thin aged varnish layer makes it a big problem for mechanical cleaning to control the precision, and chemical cleaning is also difficult to prevent the solvent from entering the pigment layer and causing irreparable damage. By employing KrF excimer laser assistance to remove the top layer of aged varnish, both lime particles and photo-induced fragments of the varnish can be cleared together. The removal depth of each pulse of the KrF laser can be as low as 0.1 μm, indicating no issues with precision. The final restoration result is depicted in Fig. 9.12b [25].

9.6 Partial Monitoring Methods for Laser Cleaning of Cultural Relics and Artworks

Laser cleaning for cultural relics and artworks places paramount importance on protecting the substrate compared to industrial products. It is crucial to rigorously control laser parameters during the cleaning of cultural relics and artworks to prevent

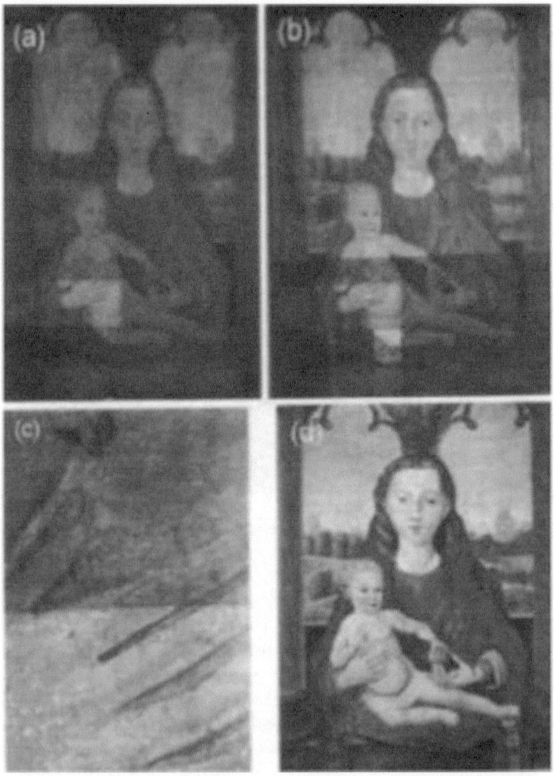

Fig. 9.11 A comparison of cleaning methods for Flemish egg tempera painting

Fig. 9.12 The comparison of oil painting before and after laser cleaning

damage to the substrate. Therefore, monitoring methods are of utmost importance. Currently, spectroscopic imaging [26], laser spectroscopy [27], holographic interference measurement [28], optical coherence tomography, and photoacoustic (PA) effects [29] have shown promising results for online monitoring during the cleaning of substrates such as stone, murals, and wood artifacts, allowing for real-time monitoring of the cleaning process.

Taking the photoacoustic effect monitoring as an example. The photoacoustic effect can be simply described as the generation of sound waves in a medium after absorbing intensity-modulated optical radiation. It is typically emitted by a pulsed laser source, and its magnitude is directly proportional to the applied energy flux and the effective optical absorption coefficient of the interacting medium. Tserevelakis et al. [29] conducted laser cleaning monitoring experiments using Naxos white marble slabs coated with uniform layers of acrylic black paint. They employed a Q-switched Nd:YAG laser with the following parameters: a wavelength of 1064 nm, a pulse duration of 6.5 ns, and a repetition rate of 1–10 Hz. The experiment used an amplifier to amplify the signal and a high-speed oscilloscope to record, without excessive impact on the cleaning substrate. Subsequently, diffuse reflectance measurements were conducted on the black graffiti and white marble samples using a spectrophotometer, yielding an absorbance value of 6.25. Photoacoustic effect monitoring yielded absorbance values of 6.33 and 6.45, which were very similar to the relative absorbance ratio calculated using independent characterization techniques, confirming the reliability of the monitoring method.

Internationally, laser technology for the cleaning and preservation of cultural relics has been widely applied in the cleaning and protection of museum artifacts and buildings, with mature technology. Many examples have demonstrated that laser cleaning is an effective means for cleaning cultural relics and artworks. For the cleaning of stone substrates, infrared Nd:YAG lasers are generally used, with the cleaning mechanism primarily based on the force generated by the interaction between light and matter, achieving the removal of surface contaminants. However, for painting works, the types and bonding modes of surface contaminants differ significantly from those of stone materials. Due to their longer wavelength and lower single photon energy, infrared lasers cannot effectively remove surface contaminants, and increasing laser power can easily cause damage to the substrate of paintings and calligraphy. Therefore, shorter-wavelength ultraviolet lasers are generally used for the cleaning and protection of paintings and calligraphy. The cleaning mechanism is based on the cold ablation effect of breaking the binding bonds, with minimal thermal impact that does not cause thermal damage to the artworks themselves.

In China, the work of laser cleaning of cultural relics has just started. China has a long history and a vast number of ancient artworks. For China, the development of laser cleaning technology for artworks is of special significance and will have broad prospects for development.

References

1. Zhang R. Unveiling the secrets of cultural relic restoration at the capital museum: national treasures restored with dental cleaning machines [EB/OL]. http://tech.sina.com.cn/d/2004-11-28/0934466098.shtml, 2004. (in Chinese).
2. Teule R, Scholten H, Vandenbrink OF, et al. Controlled UV laser cleaning of painted artworks: a systematic effect study on egg tempera paint samples [J]. Journal of Cultural Heritage, 2003, 4: 209–215.
3. Kane DM, Fernandes AJ, Hirschausen D, et al. Recent advances in UV and VUV cleaning of optical materials [C] // Australia, Sydney: Conference on Optoelectronic and Microelectronic Materials and Devices, University of New South Wales IEEE, 2003.
4. Martin C. Lasers in the preservation of cultural heritage: principles and applications [J]. Physics Today, 2007, 60(12): 58–59.
5. Rode AV, Freeman D, Baldwin KGH, et al. Scanning the laser beam for ultrafast pulse laser cleaning of paint [J]. Applied Physics A, 2008, 93(1): 135–139.
6. Li F, Ding H, Removing Cultural Relic Dirt with Lasers [J]. Laser and Optoelectronics Progress. 1993, 5: 29–30.
7. Conservation C, White C. Laser cleaning of buildings [DB/OL].
8. Pouli P. Laser cleaning studies on stonework and polychromed surfaces [D]. UK: Loughborough University, 2000.
9. Ortiz P, Antunez V, Ortiz R, et al. Comparative study of pulsed laser cleaning applied to weathered marble surfaces [J]. Applied Surface Science, 2013, 283: 193–201.
10. Marakis G, Pouli P, Zafiropulos V, et al. Comparative study on the application of the 1st and the 3rd harmonic of a Q-switched Nd:YAG laser system to clean black encrustation on marble [J]. Journal of Cultural Heritage, 2003, 4: 83–91.
11. Esbert RM, Grossi CM, Rojo A, et al. Application limits of Q-switched Nd:YAG laser irradiation for stone cleaning based on colour measurements [J]. Journal of Cultural Heritage, 2003, 4: 50–55.
12. Pozo-Antonio JS, Papanikolaou A, Philippidis A, et al. Cleaning of gypsum-rich black shells on granite using a dual wavelength Q-switched Nd:YAG laser [J]. Construction and Building Materials, 2019, 226: 721–733.
13. Qi Y, Zhou WQ, Chen J, et al. Laser cleaning of contaminants on the surface of Yungang grottoes [J]. Safety and Environmental Engineering, 2015, 22(2): 32–38. (in Chinese).
14. Ye YY, Qi Y, Yuan XD, et al. Laser cleaning of contamination on sandstone surfaces in YunGang Grottoes [J]. Optik, 2014, 125(13): 3093–3097.
15. Zhang XT, Zhang PY, Yang C, et al. Laser cleaning technology in the conservation of gilt bronze [J]. Sciences of Conservation and Archaeology, 2013, 25(3): 98–103. (in Chinese).
16. Donate-Carretero, Barrio MNJ, Medina-Sanchiiz M, et al. New advances in laser cleaning research on archaeological copper based alloys: methodology for evaluation of laser treatment [C]. Firenze: Nardini Editore, 2017: 279–296.
17. Froidevaux M, Platt P, Copper M, et al. Laser interactions with copper, copper alloys and their corrosion products used in outdoor sculpture in the United Kingdom [C]. Florida: CRC Press, 2008: 277–284.
18. Korenberg C, Baldwin A. Laser cleaning tests on archaeological copper alloys using an Nd:YAG laser [J]. Laser Chemistry, 2006 (5): 1–7.
19. Shen YJ, Zhou H, Shen JY. Applied research on the agar gel-mediated laser cleaning of bronze objects [J]. Sciences of Conservation and Archaeology, 2018, 30(3): 1–12. (in Chinese).
20. Lee H, Cho N, Lee J. Study on surface properties of gilt-bronze artifacts, after Nd:YAG laser cleaning [J]. Applied Surface Science, 2013, 284: 235–241.
21. Zhang LC, Zhou H. Restoration of a polychrome female pottery figurine from Han Dynasty using laser cleaning techniques [J]. Sciences of Conservation and Archaeology, 2017, 29(2): 67–75. (in Chinese).
22. Xu JW, Wu CW, Zhang X, et al. Influence of parameters of a laser cleaning soil rust layer on the surface of ceramic artifacts [J]. Applied Optics, 2019, 58(10): 2725–2730.

23. Zhao Y, Chen JM, Jiang MH. Laser cleaning of mildew on the Chinese painting and calligraphy [J]. Applied Laser, 2009, 29(2): 154–157. (in Chinese).
24. Chen JM. Application of laser cleaning in cultural relic protection [J]. OE Product & News, 2011, 2(9): 16–18. (in Chinese).
25. Song F, Su RY, Zou WF, et al. Laser cleaning of artworks [J]. Cleaning World, 2005, 21(10): 34–37. (in Chinese).
26. Pozo-Antonio JS, Fiorucci MP, Ramil A, et al. Evaluation of the effectiveness of laser crust removal on granites by means of hyper spectral imaging techniques [J]. Applied Surface Science, 2015, 347: 832–838.
27. Klein S, Hildenhagen J, Dickmann K, et al. LIBS-spectroscopy for monitoring and control of the laser cleaning process of stone and medieval glass [J]. Journal of Cultural Heritage, 2000, 1(1): 287–292.
28. Marton Z, Kisapati I, Torok A, et al. Holographic testing of possible mechanical effects of laser cleaning on the structure of model fresco samples [J]. NDT & E International, 2014, 63: 53–59.
29. Tserevelakis GJ, Pozo-Antonio JS, Siozos P, et al. On-line photoacoustic monitoring of laser cleaning on stone: evaluation of cleaning effectiveness and detection of potential damage to the substrate [J]. Journal of Cultural Heritage, 2019, 35: 108–115.

Chapter 10
Applications of Laser Cleaning in Other Areas

10.1 Laser Cleaning of Oil Stains

10.1.1 The Significance of Oil Stain Cleaning

The utilization of oil stain cleaning is extremely widespread, involving various aspects of production and daily life, such as oil extraction and transportation pipelines, machine components, and kitchen fumes in households and restaurants.

I. **Industrial Oil Stains**

Industrial equipment surfaces, and various internal components lack lubricating oil and engine oil. After a period, oil stain deposition occurs due to physical and chemical reactions, as well as the inclusion of dust and powder from the surrounding environment. This affects the use of equipment and components. It is crucial to perform industrial cleaning of equipment and components for the following reasons: (1) Extend the service life of equipment by cleaning the components; (2) Reduce wear and prevent explosions or ruptures caused by overheating, thereby reducing the occurrence of accidents; (3) Enhance production efficiency and restore the original precision; (4) Ensure production processes and guarantee product quality; (5) Create a healthy and clean production environment, effectively safeguarding the health of workers and reducing raw material pollution.

II. **Kitchen Oil Stains**

There is an old saying in China that “food is the paramount necessity of the people.” Cooking is a daily activity. In restaurants and homes, it is inevitable to have oil stains in the kitchen. The removal of kitchen oil stains holds significant importance for the following reasons: (1) Maintain a clean and tidy kitchen; (2) Eliminate the growth of bacteria and ensure bodily health; (3) Improve cooking efficiency [1].

F. Song and X. Lin, *Laser Cleaning*, https://doi.org/10.1007/978-981-97-6631-4_10

10.1.2 Theory Supplement for Laser Cleaning of Oil Stains

The mechanism of using laser to clean oil contaminants is based on the thermal effect of the laser, which causes the oil to evaporate and vaporize, thus separating it from the substrate surface. For example, the gold film on the metal substrate surface has a very high reflectivity to CO_2 lasers and does not absorb much, while oil contaminants start to evaporate and vaporize at temperatures as low as a few hundred degrees (e.g. dimethyl silicone oil at 573 K). This allows for the removal of oil contaminants without damaging the substrate. Therefore, from a theoretical perspective, the cleaning of oil contaminants on gold film surfaces using CO_2 lasers does not fall under the narrow definition of dry laser cleaning or wet laser cleaning, but rather represents a cleaning method that lies between the two.

There are two possible scenarios to consider. One is assuming that only oil pollutants absorb laser radiation, such as oil films being removed under CO_2 laser or partial molecular laser irradiation. The other is that the base absorbs the heat from laser radiation, causing the oil film to heat up, such as when nanosecond laser pulses heat the liquid above its boiling point through the substrate, leading to explosive evaporation at the interface [2]. The temperature distribution can be described by the heat conduction equation, as we have previously described. In the Cartesian coordinate system, the heat conduction equation is [3].

$$\rho c \frac{\partial T(x, y, z; t)}{\partial t} = k\left(\frac{\partial^2 T}{\partial x^2} + \frac{\partial^2 T}{\partial y^2} + \frac{\partial^2 T}{\partial z^2}\right) + W(x, y, z; t) \tag{10.1.1}$$

where ρ, c, and k are respectively the density, specific heat capacity, and thermal conductivity of the oil pollution; and $W(x, y, z; t)$ is the heat source function within the oil pollution. The heat conduction equation can be solved by using the laser heat source equation [3].

$$W = (1 - R)I_0 \exp(-\alpha z)F(t) \tag{10.1.2}$$

$$F(t) = \begin{bmatrix} -\frac{1}{t_p}\left|t - t_p\right| + 1, & 0 < t < 2t_p \\ 0, & 2t_p < t \end{bmatrix} \tag{10.1.3}$$

In the equation, t represents time, z represents the axial coordinate, c_p represents specific heat, W represents the heat source term from laser heating, α represents the absorption coefficient, R represents the reflectivity, I_0 represents the surface irradiance, $F(t)$ represents the dimensionless time pulse shape, and tp represents the pulse width. This equation assumes that all external boundary conditions are adiabatic. Ye et al. [3] used finite element analysis software to calculate the temperature distribution on the surface of a dimethyl silicone oil film after being irradiated by a high-power laser and found that the temperature at the center of irradiation increases with the increase of laser power and irradiation time. With the increase of laser power, the silicone oil starts to exhibit cleaning effects.

10.1.3 Research on Laser Cleaning of Oil Contamination

Cold rolling oil is a representative contaminant in laser cleaning of oil contamination. Kim et al. [4] used a Q-switched Nd:YAG laser with a wavelength of 1064 nm and a pulse width of 6 ns to clean the surface oil of low carbon steel. The beam diameter was 7 mm. According to EDX analysis, the carbon content on the original surface was 72%, which was reduced to an undetectable level after laser cleaning with a laser energy density of 120 mJ/cm^2. Laser pulses below 120 mJ/cm^2 reduced the carbon content but could not completely clean the surface, as the oil contained various components with different optical properties. For example, after laser cleaning with energy densities of 43 mJ/cm^2 and 53 mJ/cm^2, the carbon content was 20.1% and 9.1% respectively. The complete cleaning threshold under 10 pulses was (117 $\pm$ 10) mJ/cm^2, and the damage threshold was (140 $\pm$ 10) mJ/cm^2.

An Nd:YAG laser with a wavelength of 1064 nm and a pulse width of 6 ns, operated in Q-switching mode, was used by Ahn et al. [2] to clean cold-rolled oil from metal plates. The laser beam had a square shape with a side length of 1 mm. Optical microscopy and scanning electron microscopy (SEM) combined with energy-dispersive X-ray spectroscopy (EDS) were employed to detect the carbon content and surface damage after the cleaning process, thus calibrating the cleaning effectiveness. During the experiment, the carbon content in the carbon steel did not decrease to zero, as the carbon content in the carbon steel substrate was approximately 6%. The experimental results indicated that, for carbon steel, the removal of mineral oil began at a fluence of (160 $\pm$ 10) mJ/cm^2, as shown by the EDS results. As the laser energy density increased, the carbon content decreased and the oil film was completely removed at 210 mJ/cm^2, with a threshold for substrate damage at (410 $\pm$ 20) mJ/cm^2.

Different types of lasers show significant differences in their absorption rates for oil contamination. In addition, Ahn et al. [2] also utilized a KrF excimer laser with a wavelength of 248 nm and a pulse width of 25 ns to clean oil contamination on substrate surfaces. Through comparative experiments, it was found that for Nd:YAG lasers, most of the laser energy is absorbed by the solid matrix, leading to a significant increase in the temperature at the interface. The oil on the solid surface is removed through a fluid dynamic mechanism caused by explosive evaporation, with the thickness of the oil film having little effect on the experimental results. On the other hand, for excimer lasers, the laser is strongly absorbed by the oil film, resulting in an ablation mechanism. The thickness of the oil film has a greater impact on the experimental results, and this cleaning mechanism shows a lower removal rate compared to the explosive vaporization mechanism. For lasers with high absorption by the oil contamination, the thickness of the oil film has a significant influence on the laser cleaning process. It is known that the penetration depth of a KrF excimer laser is 11 μm. Therefore, if the oil contamination has a thickness of 10 μm, only a small portion of the laser energy is deposited onto the metal after burning off the oil. However, if the oil contamination has a thickness of 150 μm, the excimer laser will directly ablate the oil contamination from the surface.

There is a significant difference in the absorption rate of oil contamination for different types of lasers. Nd:YAG lasers, for example, have lower absorption rates, which can lead to different temperature increases due to varying laser reflectivity on different substrate materials, thus impacting the removal of oil contamination. Ahn et al. [2] compared the cleaning thresholds and damage thresholds for carbon steel, stainless steel, and copper when using a Q-switched Nd:YAG laser with a wavelength of 1064 nm and a pulse width of 6 ns to clean oil contamination on metal plates. The experimental results showed that for carbon steel, the removal of mineral oil, as indicated by EDS results, began at a fluence of (160 ± 10) mJ/cm^2. As the laser energy density increased, the carbon content decreased, and the oil film was completely removed at 210 mJ/cm^2, with a damage threshold for the substrate at (410 ± 20) mJ/cm^2. For stainless steel, the cleaning threshold and damage threshold were (130 ± 10) mJ/cm^2 and (420 ± 20) mJ/cm^2, respectively. However, for copper, both the cleaning threshold and damage threshold were much higher than those for carbon steel and stainless steel due to the higher reflectivity of the copper surface ($R = 0.85$) compared to the reflectivity of carbon steel ($R = 0.47$) and stainless steel ($R = 0.68$). The cleaning threshold and damage threshold for copper were (1500 ± 100) mJ/cm^2 and (3200 ± 200) mJ/cm^2, respectively.

Mateo et al. [5–7] suggest using laser cleaning as an alternative method for removing pollutants generated by oil spills. They utilized a Q-switched Nd:YAG laser to clean fuel contaminants (including saturated hydrocarbons, polycyclic aromatic hydrocarbons, resins, and naphthalene) on rocks and steel. The laser had a wavelength of the second harmonic generation at 355 nm, a pulse duration of 6 ns, a beam diameter of 9 mm, and a repetition rate of 10 Hz. The study showed that the damage threshold for the rocks was approximately 20.0 J/cm^2. A working energy density of 8.0 J/cm^2 was below the rock's destruction threshold and effectively removed the fuel. The damage threshold for the steel substrate was lower than the cleaning threshold for the fuel.

Ye et al. [3] investigated the influence of duty cycle on laser cleaning of silicone oil. A CO_2 laser with a frequency of 100 Hz was used to clean dimethyl silicone oil at a power of approximately 9 W and a duty cycle of 16%. The cleaning effect was observed when the irradiation time was around 0.5 s. Subsequently, a CO_2 laser with a wavelength of 10.6 μm, a beam diameter of 0.7 mm, and a power of 10.80 W was employed in the experiment. Under fixed frequency of 100 Hz and fixed irradiation time of 10 s, it was found that a duty cycle of 6% achieved better cleaning results. Figure 10.1 shows the cleaning effects under different duty cycles. At a duty cycle of 4%, evaporation of the silicone oil was observed, but there were still residual oil droplets in the irradiation area, indicating incomplete cleaning. A duty cycle of 6% resulted in better cleaning effects, with no residual silicone oil contaminants in the laser-irradiated region. The cleaning effects were even more pronounced at duty cycles of 7 and 9%. The study demonstrated that, under conditions that do not damage

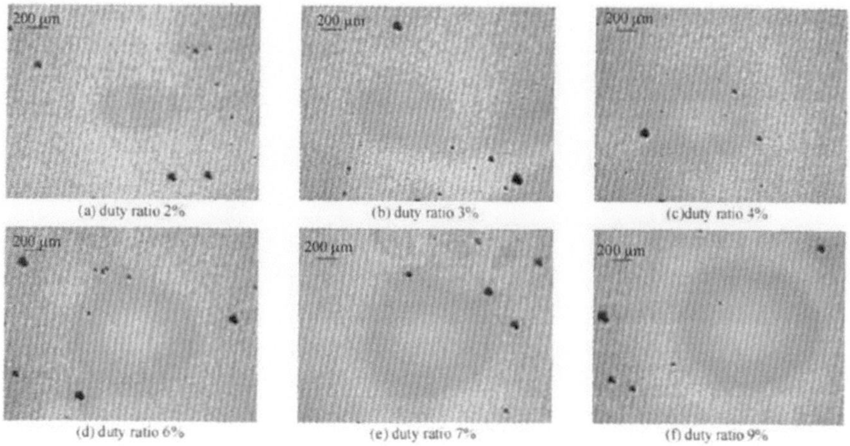

Fig. 10.1 The cleaning effects after 10 s of laser irradiation under different duty cycles: **a** duty cycle 2%; **b** duty cycle 3%; **c** duty cycle 4%; **d** duty cycle 6%; **e** duty cycle 7%; **f** duty cycle 9%

the substrate and gold film, the higher the laser power with the same irradiation time, the more significant the cleaning effect.

Under the fixed duty cycle of 16%, a laser irradiation time of 0.5 s showed the presence of laser cleaning effects, as demonstrated in Fig. 10.2. There was evidence of silicone oil evaporation in the irradiation area, but residual oil droplets remained. With an irradiation time of 1.0 s, better cleaning effects were observed, and there were no residual silicone oil droplets in the irradiation area. Therefore, under the condition of not damaging the substrate and gold film, a longer irradiation time results in a larger cleaning area.

The removal of oil stains from the surface of commonly used steel wire ropes in industry (Fig. 10.3) is a common operational procedure. Wu et al. [8] invented a laser cleaning equipment and method specifically designed for steel wire rope oil contamination. The equipment used for laser cleaning of the steel wire rope oil contamination includes an electrical cabinet, a speed sensor, and a laser cleaning system. The electrical cabinet consists of a computer control system and a laser. The specific method involves determining the laser cleaning process parameters and

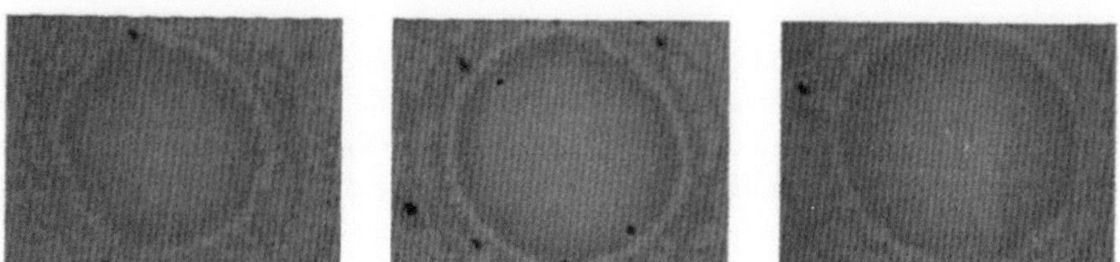

Fig. 10.2 The cleaning effects at different irradiation times under a duty cycle of 16%

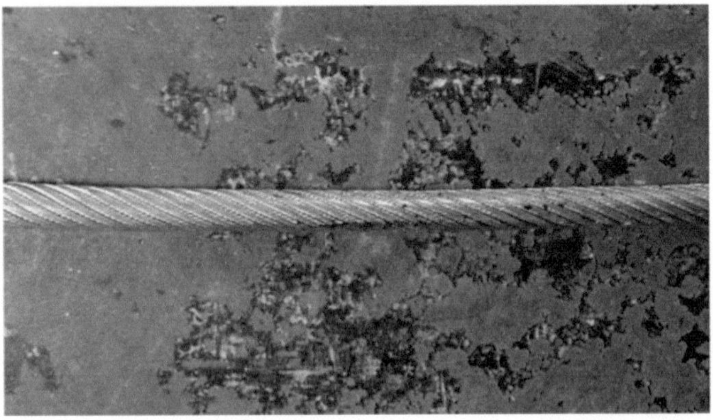

Fig. 10.3 A steel wire rope with oil contamination

aligning the steel wire rope. Then, the laser beam is shaped and focused into a linear spot to scan and clean the steel wire rope, ensuring that the laser spot used covers the 360° direction of a certain cross-section of the steel wire rope, thereby guaranteeing the removal of oil stains from the entire circumference of the steel wire rope cross-section.

The cleaning of the inner wall of oil pipelines is also crucial. Huang et al. [9] invented a laser-thermal water combined cleaning device for oil pipe inner walls. This device includes a laser generator, lens, connecting plate, oil pipe, reflector, water flow channel, laser channel, and base. The laser beam generated by the laser generator is collimated by the lens into a parallel beam. This beam enters the reflector through the laser channel at the center of the laser-thermal water combined cleaning device. Through reflection, the parallel laser beam is emitted through the bottom laser channel of the device and irradiates the inner wall of the oil pipe to heat the dirt. During the laser irradiation process, the oil pipe moves slowly along the axial direction and begins to rotate rapidly. The hot water flows out from the water flow channel on the outer side of the laser channel, flushing the position after laser irradiation. It also flushes the dirt that remains attached to the inner surface of the oil pipe after being melted by laser irradiation, together with the solid dirt that has been detached due to high temperature, achieving the purpose of cleaning the inner surface of the oil pipe. This method is more efficient, and less polluting compared to traditional cleaning methods.

The presence of oil and fuel stains on highways and airport runways greatly reduces the friction between tires and the road surface, leading to accidents, especially during turns and high-speed driving. The traditional method of burning these stains from top to bottom proves ineffective due to the forced downward direction of the flames. Mechanical cleaning methods are also inefficient due to surface roughness. Laser cleaning, on the other hand, provides an effective and fast solution [10–13]. By utilizing a 25 W continuous CO_2 laser as the light source, with an unfocused beam diameter of 1 cm and a power density of $I \approx 20$ W/cm^2, excellent cleaning results were achieved on samples of oil and fuel stains on the surface of asphalt and concrete roadways.

10.2 Laser Cleaning of Optical Substrate Surfaces

10.2.1 Significance of Laser Cleaning Optical Devices

With the development of laser technology and the demand for its applications, many countries have been researching and developing high-energy laser devices. In these high-power solid-state laser devices, there is a high requirement for the number and types of optical components, as well as their processing precision. During usage, environmental pollutants settle or adhere to the surfaces of optical components, affecting the performance of these components in various ways, ultimately impacting the output and efficiency of the laser device. The adsorption of contaminant particles on the surface of optical substrates is one of the critical factors affecting the performance of optical devices. As the performance requirements for optical devices increase, the impact of surface-adsorbed contaminant particles (such as diamond abrasive particles, SiO_2 soft particles, and certain dust particles) on the yield and performance of the devices becomes more prominent. For example, laser gyro ultra-smooth reflecting mirrors require a reflectance as high as 99.99%. Current manufacturing capabilities can achieve surface roughness levels of 0.2 nm or even better, but the reflectance only reaches 99.92%. The deposition of pollutant particles on the surface of optical components not only changes the quality of the laser beam, leading to variations in laser energy, but it can also result in the absorption of laser energy by subsequent device components, thereby reducing their damage threshold. Additionally, it can cause localized absorption of laser energy by the optical component or the chemical film coating on its surface, leading to a decrease in the damage threshold and rendering the optical component unusable. To ensure the normal and smooth operation of laser devices under high laser flux, it is essential to effectively clean the pollutants from the surfaces of optical components.

Common methods such as wiping with volatile liquids like ethanol or acetone yield average results. New cleaning methods such as ultrasonic cleaning and ion bombardment have been developed, but surface analysis and observation after cleaning reveal the presence of extremely small residual contaminant particles, which is the reason for the insufficient reflectance. Laser cleaning, on the other hand, can effectively remove the tiny particles from ultra-smooth optical substrates without causing damage to the substrate and without introducing secondary contamination. In recent years, laser cleaning has received attention and application in this field.

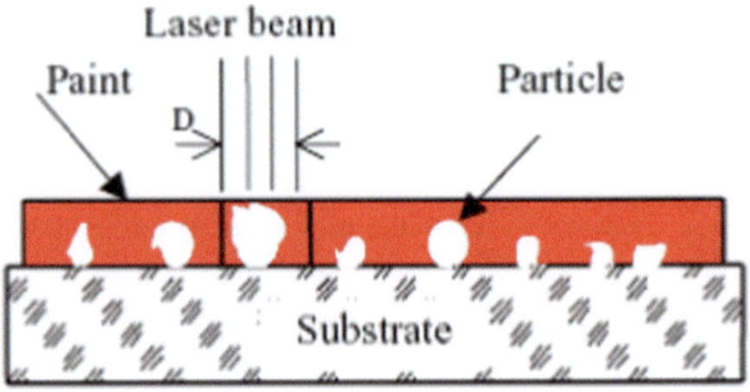

Fig. 10.4 Schematic diagram of laser cleaning of ultra-smooth optical substrates

10.2.2 *Experimental Research on Laser Cleaning of Optical Devices*

I. **Methods Used for Laser Cleaning of Optical Devices**

Contaminant particles on ultra-smooth optical substrates and the substrates themselves are often transparent and have low absorption rates for laser beams. It is difficult to uniformly apply a thin layer of liquid on such smooth optical substrates using dry or wet laser cleaning methods. Researchers from the College of Engineering and Engineering Management at City University of Hong Kong [13] proposed a novel method. This method involves covering the polished surface of the optical substrate with a few micrometers of volatile varnish, as shown in Fig. 10.4. Subsequently, a 1064 nm infrared Nd:YAG laser is used for cleaning, evaporating the adhered particles and residues on the substrate. After laser cleaning with an energy density of 16.4 J/cm^2, it was observed through a particle identification system that the maximum residual particle size on the mirror surface reduced from the original 2.57 to 0.17 μm.

For continuous CO_2 lasers, direct cleaning of transparent optical substrates can be challenging. Li et al. [14] first used a CO_2 laser to partially scan the area and then employed a vacuum plasma cleaning method for fine cleaning, effectively removing oil stains from transparent quartz substrates.

II. **Parameters to Be Controlled for Laser Cleaning of Optical Devices**

The impacts of different parameters on laser cleaning have been extensively discussed in previous studies on laser cleaning of electronic components, paint removal, and rust removal. Similar influences can also be observed in laser cleaning of oil stains.

Dry laser cleaning has wavelength selectivity. It can only achieve cleaning when there is a significant difference in the composition and absorption coefficient between the substrate and the particulate contaminants. Additionally, precise control of laser energy is required.

Xu et al. [15] used a 355 nm laser to clean larger particulate contaminants on K9 glass surfaces. During the cleaning process, the laser was directly irradiated onto the substrate surface, which increased the probability of substrate damage. Thus, it was necessary to control the laser energy carefully. As a comparison, laser-induced plasma shock wave cleaning was also used to remove dust particulate contaminants from K9 glass surfaces, as shown in Fig. 10.5. It can be clearly observed that as the working distance d decreases, the cleaning effect strengthens, resulting in a reduction in the number of particles. When d = 10 mm, there is almost no cleaning effect; when d = 4.0 mm, the number of larger particles is significantly reduced; and when d = 1.0 mm, no dust particles can be seen. Some extremely small particles are still present when observed under maximum magnification. Laser cleaning does not require wavelength selectivity and does not directly interact with the substrate, making it less likely to cause substrate damage.

Li et al. [14] studied the damage threshold of quartz substrates at different cleaning stages. The experimental CO_2 laser parameters used were a wavelength of 10.6 μm, a repetition rate of 5 kHz, a power of 10 W, a beam diameter of 4 mm, and a scanning interval of 2 mm. The contamination of silicone oil led to a decrease in the substrate's damage threshold. With an applied laser power of only 10 W, the substrate only experienced minimal heating, and no significant stress was observed using a self-made stress gauge. When the laser power was too high, the large temperature difference caused by the melting and cooling of the surface layer after laser irradiation could lead to excessive stress. Therefore, it is necessary to control the laser power carefully. The experiments showed that the damage threshold of the substrate decreased significantly from 4.73 to 3.77 J/cm^2 after being contaminated. After CO_2 laser cleaning, the damage threshold was restored to 4.75 J/cm^2. In other words, the

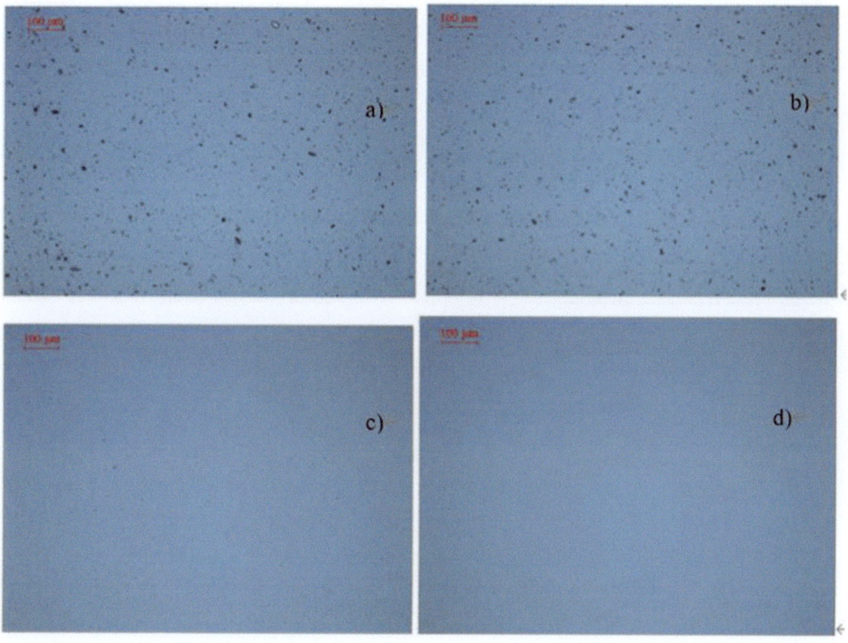

Fig. 10.5 **a** The optical images of the cleaned surface before cleaning, with different cleaning effects observed at various working distances (d): **b** d = 8.0 mm, **c** d = 4.0 mm, and **d** d = 1.0 mm

presence of contaminants on the mirror surface caused a significant decrease in the damage threshold from the original 4.73 to 3.77 J/cm^2. After CO_2 laser cleaning, the damage threshold was essentially restored. Combining vacuum plasma cleaning with laser cleaning resulted in excellent cleaning effects, with a damage threshold reaching 5.09 J/cm^2, exceeding the pre-contamination level.

Ye et al. [16] used a 1.06 μm CO_2 laser to clean silicone oil from gold-coated K9 glass surfaces. The laser operated at a frequency of 100 Hz for 10 s. The cleanliness and spectra of the cleaned surfaces were analyzed and characterized using an optical microscope and an FT-IR spectrometer, and the results are shown in Fig. 10.6.

The results showed that a laser power of 10.8 W and an irradiation time of 10 s had a significant cleaning effect. The relationship between the cleaning area and laser power (Fig. 10.7) indicated that when the laser power was below the damage threshold of the gold-coated K9 glass, the cleaning area increased with increasing laser power.

Li et al. [17] addressed the issue of surface contamination in a large-scale laser device during operation using a Nd:YAG triple frequency laser to clean the surfaces of coated K9 glass. The cleaning was conducted using a single-point multi-pulse method and a multi-pulse scanning method. Measurements of the damage threshold on the contaminated substrate surface were performed by gradually increasing the

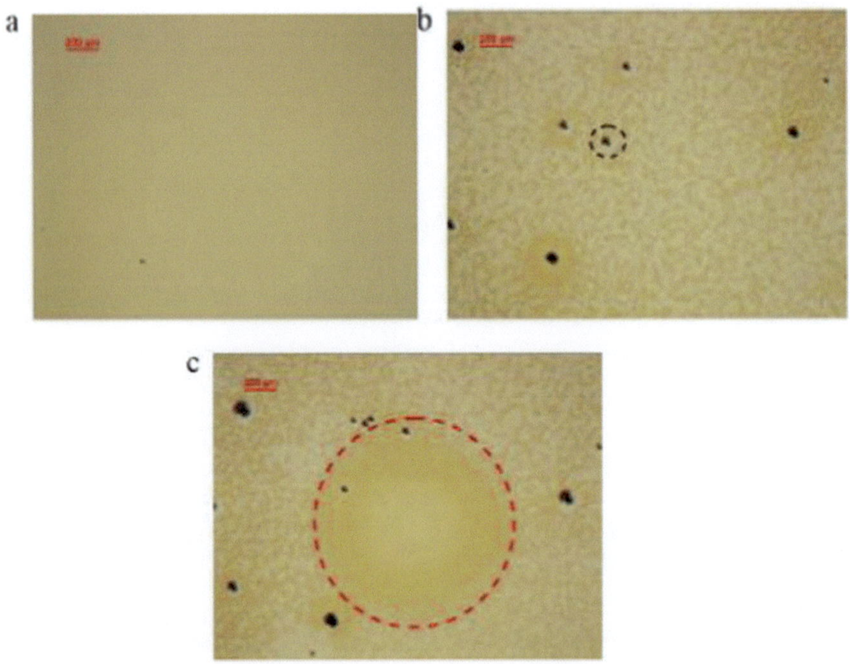

Fig. 10.6 CO_2 laser cleaning of gold-coated K9 glass surface: **a** surface morphology of gold-coated K9 glass before silicone oil contamination; **b** gold-coated K9 glass after silicone oil contamination; **c** gold-coated K9 glass after laser cleaning with a laser power of 10.8 W

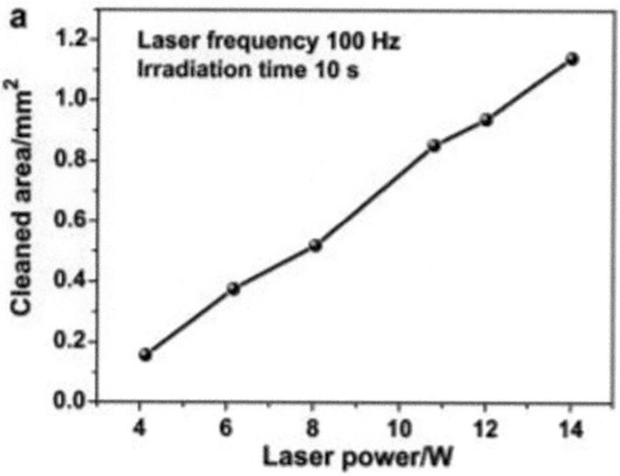

Fig. 10.7 Curve of cleaning area with respect to laser power

pulse energy until damage occurred at a specific point. Then, seven points above the damaged point were selected for cleaning experiments using different energy densities and numbers of pulses. The results of the damage threshold measurements indicated that no damage occurred on the substrate surface when the laser energy density was low, and damage only occurred when the incident laser energy density exceeded the damage threshold. Therefore, the laser energy density used for cleaning should not exceed this damage threshold. Before and after laser cleaning, measurements were conducted using a UV–visible spectrophotometer, an optical microscope, an FT-IR Raman accessory, and a surface tension measurement instrument. Prior to cleaning, the transmittance of the coated K9 glass decreased by 40% due to contamination. After cleaning, the transmittance was restored to its clean state.

Miao et al. [18] conducted tests on large-aperture fiber Bragg gratings (BSGs), which are prone to stimulated Brillouin scattering (SBS) in high-power laser devices. The damage threshold and damage morphology of BSG side surfaces irradiated by 355 nm lasers after chemical etching and UV laser cleaning were evaluated, and the degree of environmental pollution caused by quartz particle aerosols was analyzed. The results showed that UV laser treatment increased the damage threshold of contaminants on the side surfaces of optical components without affecting their performance. Microscopic morphology analysis and analysis of the impact on the aperture size demonstrated that UV laser cleaning had better safety and applicability than chemical etching.

III. Detection of Cleanliness in Laser Cleaned Optical Components

For optical components, there are various methods for measurement and characterization before and after cleaning. Common techniques include UV–visible spectrophotometry, optical microscopy, FT-IR Raman spectroscopy, surface tension measurement, and image analysis software, which can be used to measure transmittance, surface morphology, and stress, among other parameters. Here, we introduce a method that utilizes the wetting properties of solid–liquid interfaces for detection.

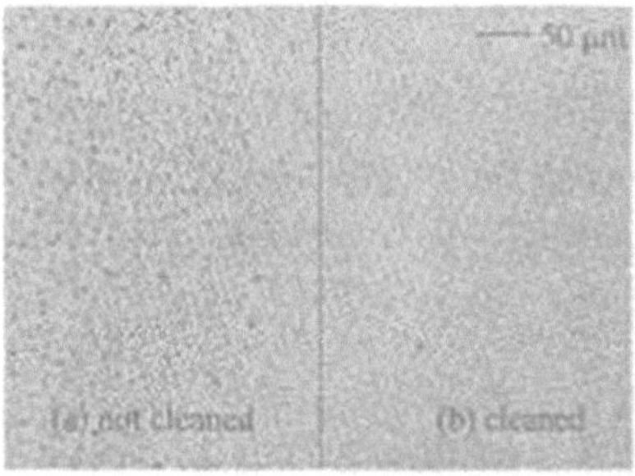

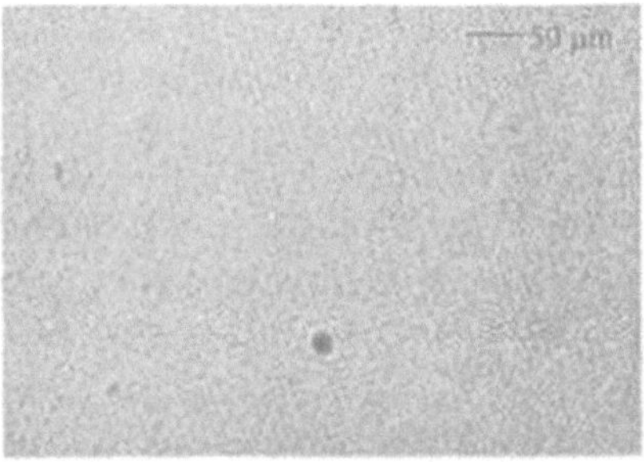

Fig. 10.8 Optical microscopy images before and after CO_2 laser cleaning, as well as the optical microscopy image after cleaning using vacuum plasma

Since optical components are primarily made of glass and quartz, they belong to hydrophilic substrate materials. If there is oil contamination on the surface, the hydrophilicity will decrease. By placing a small droplet of water on the surface of the optical component and measuring the contact angle of the droplet, the cleanliness of the substrate surface can be evaluated. A clean surface will result in a small contact angle, while the presence of oil contamination will increase the contact angle. Therefore, the size of the contact angle reflects the cleanliness of the substrate surface.

Li et al. [14] observed the surfaces of optical components cleaned using only CO_2 laser and those cleaned using laser combined with vacuum plasma, as shown in Fig. 10.8, using optical microscopy.

When characterizing using contact angles, it was found that the contact angle of the substrate was 41° before oil contamination. After oil contamination, the contact angle increased to 63°, indicating an increase in the degree of contamination. After CO_2 laser cleaning, the contact angle was 78°. However, the contact angle did not decrease after CO_2 laser cleaning, indicating that most of the oil droplets had melted and then solidified back on the substrate surface, indicating incomplete removal. After combined cleaning with CO_2 laser and vacuum plasma, the contact angle decreased to 4°, indicating almost complete wetting between the water droplet and the glass substrate surface, as shown in Fig. 10.9. The transmittance of the quartz substrate was measured afterward. It was observed that the transmittance of the substrate decreased after oil contamination. Prior to contamination, the transmittance was around 93% at 400 nm, while after contamination, it decreased to around 92.3%. After CO_2 laser treatment, the transmittance only recovered to about 92%, showing limited improvement. However, after plasma cleaning, the oil molecules were completely removed, and the transmittance at 400 nm increased to 93.3%. This indicates that the cleanliness of the substrate was further improved after vacuum plasma cleaning.

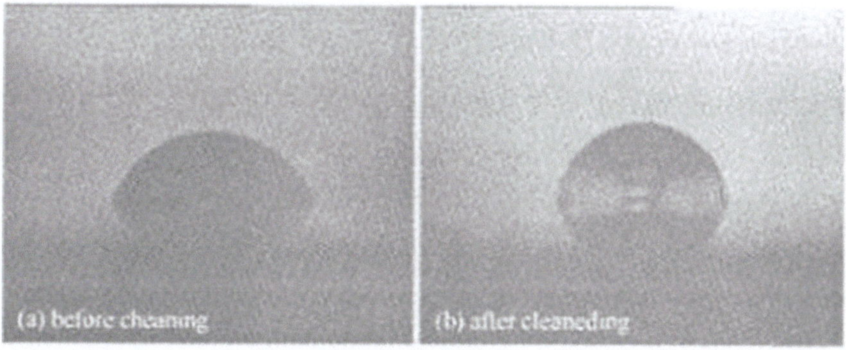

Fig. 10.9 Comparison of substrate wetting before and after CO_2 laser cleaning

10.3 Laser Cleaning of Buildings

Buildings, due to environmental influences, are inevitably exposed to various pollutants and can become unsightly due to graffiti and other forms of damage. With the rapid development of the economy, tall skyscrapers are being constructed all over the world, exacerbating the issue of cleaning the exterior walls. Currently, the main cleaning methods involve the use of chemical and mechanical techniques, commonly known as "spiderman" cleaning, as shown in Fig. 10.10. However, these methods pose safety hazards and can cause damage to the composition and structure of the walls. Laser cleaning offers a promising solution that is environmentally friendly, non-destructive to the walls, and capable of precisely removing paint and other contaminants from the wall surface [19–23].

Epoxy tile grout is a commonly used building material for walls and floors in buildings, as shown in Fig. 10.11. Due to the infiltration of various harmful substances through the porous surface and into the grout, the service life of epoxy tile grout is usually shorter than that of the tiles themselves. When it is necessary to renovate, the

Fig. 10.10 Illustrates the cleaning of buildings by "spiderman" using chemical and mechanical methods

Fig. 10.11 Depicts a schematic diagram of laser cleaning for tile grout

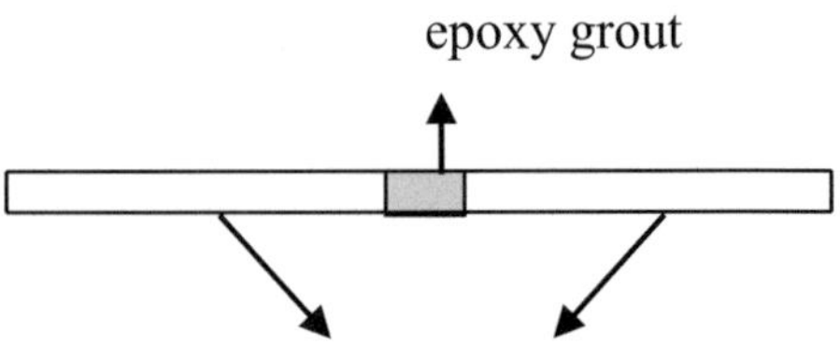

contaminated epoxy tile grout needs to be removed. Cleaning these contaminants can maintain the original appearance of the tiles and extend their service life. Currently, mechanical tools are commonly used to remove the contaminated grout, such as mechanical scraping and grinding. However, these methods can cause tile damage, generate noise, and produce toxic dust. Laser cleaning, on the other hand, offers advantages such as controllability, compact size, relatively low equipment energy consumption, and reduced health risks for operators through fiber optic transmission. Laser cleaning of tile grout does not require direct contact with the building materials, thus overcoming the limitations of traditional methods.

Manchester Metropolitan University researchers, led by Minami et al. [24], utilized a high-power diode laser with a wavelength of 810 nm and a maximum output power of 120 W to clean approximately 5 mm thick epoxy grout, including acrylic emulsion, limestone, and marble. The experiment demonstrated that the minimum power density required for effective cleaning was 3 W/mm^2. It was also observed that the laser energy applied to the epoxy grout was directly proportional to the duration of laser exposure. In summary, the minimum required exposure time for effective cleaning was determined to be 0.5 s.

Sandstone was commonly used as a construction material in ancient Europe, but over time, it can become brittle due to prolonged exposure to wind, sunlight, and rain. Siedel et al. [19] conducted laser cleaning experiments on a broken sandstone fragment of an unidentified vase or sculpture from the Saxony State Office for Monument Preservation using an Nd:YAG laser. The selected energy density was 0.47 J/cm^2. The energy densities for various test areas were 0.33 and 0.51 J/cm^2. Visible material loss on a macroscopic scale only occurred in severely damaged sandstone areas that were unable to consolidate under any circumstances. Laser cleaning effectively removed paint layers from weathered surfaces without mechanical impact on the sandstone. On a microscopic surface level, even in areas cleaned multiple times, no significant damage to quartz particles was detected. With the same energy density, increasing the number of cleaning cycles allowed for complete removal of the coating layer. Furthermore, in the field of stone carving, Beadman et al. [20] demonstrated the effectiveness of laser cleaning in restoring a discolored multicolored Romanesque carved band from Lincoln Cathedral.

Mazzinghi et al. [21] demonstrated that for marble buildings, a short pulse duration (20 μs) laser from a free running (FR) Nd:YAG laser is more effective in cleaning compared to traditional lasers with longer pulse durations (> 200 μs). They successfully cleaned components removed from the Florence Cathedral of Santa Maria del Fiore, marble on the exterior walls of the Cappella di Piazza del Campo in Siena, a layer on the exterior wall of Palazzo Rucellai in Florence, the S. Giovanni church in Viterbo, and the Baptistery in Pistoia. Rivas et al. [25] conducted laser cleaning on lichen (moss) crusts on granite. The effectiveness of the cleaning and its impact on the granite were assessed using optical and electron microscopes, infrared spectroscopy, and spectrophotometry. The research revealed that the outcome of laser cleaning depends on the extent and color of lichen coverage, with lighter lichen being more challenging to clean compared to darker lichen. However, combining laser cleaning with traditional mechanical cleaning methods yielded excellent results.

Zhu et al. [26] studied laser cleaning of urban wall graffiti using a TEA CO_2 laser with green bricks as the substrate. The laser parameters were as follows: wavelength of 10.6 μm, beam diameter of 26 mm, and pulse energy of 2.0 J. The experiment used a converging lens to reduce the beam area and varied the distance between the sample and the lens to change the laser power density. Unlike paint and rust removal, the main objective here was to remove paper adhesive contaminants. At the selected power density, multiple laser pulses on a specific spot of the sample could remove the contaminants. When the power density was too low, the contaminants could not be completely removed, requiring a significant amount of accumulated energy. Additionally, the difference in light absorption coefficients between the substrate and the contaminants could lead to thermal effects. Conversely, if the power density was too high, the cleaning area would be small, and atmospheric breakdown at the surface could consume a large amount of laser energy, resulting in a high energy requirement per unit area. The experiment determined that the minimum laser energy density required for cleaning per unit area was approximately 7 J/cm^2. The cleaning of contaminants by the laser was mainly based on the impact effect. When appropriate laser parameters were used, the TEA CO_2 laser achieved good results in graffiti removal. Fiorucci et al. [27] employed a high repetition rate nanosecond Nd:YVO_4 laser with third harmonic generation to remove graffiti from walls. Four different spray paints artificially aged with SO_2 were applied to the granite surface, and the ablation thresholds for the four paint colors were determined. The effects of pulse repetition frequency, beam diameter, and line scanning speed on the cleaning process were analyzed. The experiment also compared the results with mechanical cleaning using different micro abrasive methods.

10.4 Laser Cleaning in Nuclear Power Plants

Laser cleaning systems can also be applied for cleaning the pipelines inside nuclear power plant reactors. By utilizing fiber optic cables, high-power laser beams can be introduced into the reactor internals to directly remove radioactive dust. The cleaned material can be easily disposed of, and since the operation is conducted from a distance, it ensures the safety of the workers.

"Atomic Energy Perspectives" reported that the Japan Atomic Energy Agency (JAEA) has adopted laser cleaning technology to remove contaminants from surfaces of structural materials or machinery that have been contaminated with radioactive substances. This technology has been developed in combination with fiber optic and robotic technologies, resulting in a long-distance unmanned laser decontamination method. As the lifespan of reactors constructed in earlier years approaches its end, a large amount of radioactive waste will be generated. Therefore, the research and development of safe and efficient treatment methods become particularly important, making laser cleaning technology a promising approach.

Ivanova et al. [28] and Graf et al. [29] conducted laser cleaning on cobalt deposits in a tokamak device. They analyzed the cleaning effect using methods such as gas phase analysis, surface analysis, spectroscopy, ion beam analysis, surface profiling, and microscopy. The results showed that the cleaning was effective. However, laser ablation resulted in the generation of dust particles larger than 50 nm, which could potentially recondense on surrounding surfaces.

In publicly available literature, only a few laser cleaning devices have been used for nuclear decontamination, with some projects conducted in the United States and France. In the early 1980s, the U.S. Department of Energy (DOE) proposed using high-power lasers for nuclear facility decontamination. This was included in the Mems Laser Decontamination Project framework from 1992 to 1996 [30]. They used a 100 W KrF excimer laser (248 nm, 27 ns) and a Q-switched Nd:YAG laser (1064 nm, 100 ns) to study the removal of radioactive oxides from metal surfaces (aluminum, steel, copper, lead). They found that the efficiency of the ablation process was higher under higher ionization potentials of the ambient gas. Using a cylindrical lens to focus the beam resulted in good surface cleaning coverage and reduced particle redeposition. In the experiments, high-power infrared lasers such as CO_2 or Nd:YAG lasers were commonly used. The Argonne National Laboratory [31] also developed an Nd:YAG laser cleaning machine and may have tested it on decommissioned nuclear reactors. Since 1990, similar research and development approaches have been adopted in France, conducting experiments with Nd:YAG and excimer lasers. Framatome proposed a laser device for steam generator decontamination in nuclear facilities [32], and Cogema used an Nd:YAG laser for cleaning nuclear fuel and applied for a patent [33]. Lexdin used an XeCl laser with beam transmission through mirrors for purifying organic glass chambers [34]. In laser cleaning of nuclear contamination, a comparison between Nd:YAG and KrF excimer lasers found that the influence of laser wavelength was minimal in an air atmosphere. Compared to dry cleaning, water film decontamination factor was 30 times higher, and 0.5 mol/L nitric acid was 85 times higher, while 5 mol/L nitric acid was 650 times higher [35]. In a study conducted at the University of Liverpool in the UK, Nd:YAG lasers (6 ns) and xenon flash lamps (200 ms) were used to remove dust from tokamaks [27]. Some laser decontamination-related patents have been filed in the UK [36] and Japan [37]. For example, Delaporte and Lacour et al. [38, 39] collaborated with ONECTRA to conduct laser cleaning experiments on facilities at the Commissariat Energie Atomique (CEA), using two XeCl laser systems: Lambda Physics EMG203MSC with a pulse width of 25 ns and an average power of 80 W (400 mJ, 200 Hz); and CILAS UV635 with a pulse width of 70 ns and an average power of 1.2 kW (3 J, 400 Hz). The initial activity of the samples to be cleaned was approximately 30,000 Bq/cm^2, and the laser energy density ranged from 0.5 to 2.5 J/cm^2. At 0.5 J/cm^2, the laser energy was below the ablation threshold for oxides formed on steel surfaces. For laser energy density above 1 J/cm^2, the decontamination was highly effective in the first 100 irradiations but decreased thereafter. At 2.5 J/cm^2, the efficiency of removing remaining contaminants decreased after 500 shots, but after 1000 pulses, a cleaning efficiency of 80–97% was achieved.

Costes et al. [40] invented a technique that uses ultraviolet (UV) laser irradiation to purify metal or polymer surfaces contaminated with nuclear pollutants. With the use of a UV laser beam, along with mirrors and controllers, the controller and mirrors allow the beam to scan the inner surface of the object being cleaned. This technique can be used to remove nuclear contamination when nuclear power plants are decommissioned. The method does not cause melting of the metal or polymer surfaces, and the use of an inert gas atmosphere can prevent reoxidation and the formation of microcracks. The equipment is automated, avoiding operator exposure to nuclear contamination, and it also allows for the monitoring and filtration of the environmental gas.

10.5 Applications of Laser Cleaning in Other Fields

I. Laser Cleaning of Small-Sized Debris in Space

In the development of near-Earth space, the direct collision between orbital objects and space debris is a critical issue that must be addressed. The amount of space debris in near-Earth space is continuously increasing, with the collision probability doubling in the past decade. These small particles (less than 10 cm in size) of space debris are numerous and are not easily observable, posing a significant danger to space flights and satellites. Even particles as small as a fraction of a millimeter can cause damage to the optical properties of spacecraft windows and scientific instrument lenses. Collisions with larger debris can have catastrophic consequences for astronauts' sealed spacesuits or spacecraft [41].

Active removal of small-sized debris through long-distance laser action is a practical scientific solution. Complete removal of such objects requires a significant amount of energy. To vaporize an object with a thickness of 1 cm, the laser radiation needs an energy density of approximately 70 kJ/cm^2 on the target surface. Material evaporation would generate an ablation jet, resulting in significant reactive and recoil forces on the irradiated surface. The laser-induced reaction on small-sized space objects can alter their velocity. Using this method to decelerate small-sized space objects can transition them to lower orbits and ultimately burn them up in the upper atmosphere, thus clearing the small-sized debris in space.

II. Laser Cleaning of Optical Discs

Over time, dust and other particles can adhere to the surface of optical discs, affecting data reading. Laser cleaning can be used to achieve the purpose of cleaning. Leone et al. [42] used a 30 W Q-switched pulsed fiber laser to ablate contaminants on the metal substrate located between the polycarbonate layers. Through a series of experimental tests, process conditions were determined to ensure a 100% surface cleaning rate, and a corresponding evaluation of process time for polymer degradation was conducted. The experiments also observed and described the separation mechanism between the deposited layer and the polymer substrate.

III. **Laser Cleaning of Compressor Blades**

The presence of sulfide-based deposits on the surface of compressor rotor blades can significantly reduce their performance and shorten their lifespan. In preparation for subsequent processes such as plasma spraying, it is necessary to completely remove these deposits. During the corrosion process of FV(520)B stainless steel, the sulfide-based deposits are divided into two layers due to the different outward diffusion rates of chromium, nickel, and iron. Tang et al. [43] used a pulsed fiber laser to clean the sulfide-based deposits on the blades and studied the cleaning thresholds of the substrate surface and inner layer, as well as the damage threshold of the substrate. Scanning electron microscopy, energy-dispersive X-ray spectroscopy, and a 3D surface profilometer were used to analyze the two sulfide layers on the sample before, during, and after laser cleaning.

In addition to the applications mentioned above, laser cleaning can also be applied in many other areas. For example, it can be used to clean oil stains floating on the ocean and on road surfaces, remove adhesives, clean large mirrors (such as astronomical telescopes), eliminate mold from multi-core plugs on control cables in weapons and equipment, and clean screen rolls (with the capability to clean rolls with a mesh count of 472.44 lines per inch or higher without damaging the mesh walls) [44, 45]. Compared to traditional cleaning techniques, laser cleaning has its unique advantages. Although its application in industrial cleaning has only recently begun, its safety and reliability have been proven in industrial settings. Laser cleaning is a promising technology with great potential. With further scientific research and industrial practices, it is expected to be widely applied in various fields.

References

1. Wang MD, Liu JC, Pan Y, et al. Laser cleaning method and equipment for kitchen grease: CN108577786A [P]. 2018-09-28. (in Chinese).
2. Ahn D, Jang D, Park T, et al. Laser removal of lubricating oils from metal surfaces [J]. Surface & Coatings Technology, 2012, 206(18): 3751–3757.
3. Ye YY, Yuan XD, Xiang X, et al. Laser Cleaning of Silicon Oil Contamination on Gold Film Surfaces [J]. High Power Laser and Particle Beams, 2010, 22(5): 668–672. (in Chinese).
4. Kim D, Lim H. Laser Decontamination of Carbon Steel Surfaces [J]. ISIJ International, 2003, 43(8): 1289–1291.
5. Mateo MP, Nicolas G, Pinon V, et al. Laser cleaning: an alternative method for removing oil-spill fuel residues [J]. Applied Surface Science, 2005, 247(1/4): 333–339.
6. Mateo MP, Nicolas G, Piñon V, et al. Laser cleaning of Prestige tanker oil spill on coastal rocks controlled by spectrochemical analysis [J]. Analytica Chimica Acta, 2004, 524(1–2): 27–32.
7. Nicolas G, Mateo MP, Ctvrtnickova T, et al. Laser cleaning of varnishes and contaminants on brass [J]. Applied Surface Science, 2009, 255(10): 5579–5583.

8. Wu SQ, Fu YH, Wu GQ, et al. Laser cleaning equipment and method for wire rope oil contamination. CN106862177A [P]. 2017, 06, 20. (in Chinese).
9. Huang ZQ, Wang RH, Zhang WL, et al. Laser-warm water combined cleaning device for oil pipe inner wall. CN108816972A [P]. 2018, 11, 16. (in Chinese).
10. Meng SX. Laser Cleaning of oil stains on road surfaces [J]. Laser & Optoelectronics Progress, 1983, 10: 35–36. (in Chinese).
11. Chen JM. Laser cleaning technology [J]. Cleaning Technology: 2003, (9): 24–27. (in Chinese).
12. Zhang YJ. Laser cleaning technology and applications [J]. Cleaning World: 1999, 4: 20–22. (in Chinese).
13. Song F, Zou WF, Liu SJ, et al. Other applications of laser cleaning [J]. Cleaning World: 2006, 22(3): 38–41. (in Chinese).
14. Li XP, Zu XT, Yuan XD, et al. Enhancement of damage threshold of quartz substrate by CO_2 laser and plasma cleaning [J]. High Power Laser and Particle Beams, 2007, 19(10): 1739–1743. (in Chinese).
15. Xu SZ, Dou HQ, Han FM, et al. Laser Cleaning of particle contaminants on K9 glass surfaces [J]. Laboratory Research and Exploration, 2017, 36(6): 5–8. (in Chinese).
16. Ye Y, Yuan X, Xiang X, et al. Laser cleaning of particle and grease contaminations on the surface of optics [J]. Optik, 2012, 123(12): 1056–1060.
17. Li HP, Guo BL. Research on Nd:YAG pulsed laser cleaning technology [J]. Optoelectronic Technology and Application, 2018, 159(2): 67–71. (in Chinese).
18. Miao XX, Cheng XF, Wang HB, et al. Experimental study on side cleaning of large aperture optical components in high power laser devices [J]. High Power Laser and Particle Beams, 2013, 25(4): 890–894. (in Chinese).
19. Siedel H, Neumeister K, Sobott RJG. Laser cleaning as a part of the restoration process: removal of aged oil paints from a renaissance sandstone portal in Dresden, Germany [J]. Journal of Cultural Heritage, 2003, 4(4): 11–16.
20. Beadman K, Scarrow J. Laser cleaning Lincoln cathedral's Romanesque Frieze [J]. Journal of Architectural Conservation, 1998, 4(2): 39–53.
21. Mazzinghi P, Margheri F. A short pulse, free running, Nd:YAG laser for the cleaning of stone cultural heritage [J]. Optics & Lasers in Engineering, 2003, 39(2): 191–202.
22. Pini R, Siano S, Salimbeni R, et al. Application of a new laser cleaning procedure to the mausoleum of Theodoric [J]. Journal of Cultural Heritage, 2000, 1(2): S93–S97.
23. Chloros J, Salmon H, Talland V. Laser cleaning at the Isabella Stewart Gardner Museum, Boston, USA: Sixteen Roman sculptures, fourteen months, and three conservators [J]. Studies in Conservation, 2015, 6(1): S41–S48.
24. Minami K, Li L, Schmidt M, et al. Materials behaviour and process characteristics in the removal of industrial cement tile grout using a 1.5 kW diode laser [J]. Thin Solid Films, 2004, 453–454(1): 52–58.
25. Rivas T, Pozo-Antonio JS, Lopez DE, et al. Laser versus scalpel cleaning of crustose lichens on granite [J]. Applied Surface Science, 2018, 440: 467–476.
26. Zhu YF, Tan RQ. Laser cleaning applied to remove urban graffiti [J]. Laser and Infrared, 2011, 41(8): 840–844. (in Chinese).
27. Fiorucci MP, Lopez AJ, Ramil A, et al. Optimization of graffiti removal on natural stone by means of high repetition rate UV laser [J]. Applied Surface Science, 2013, 278: 268–272.
28. Ivanova D, Rubel M, Philipps V, et al. Laser-based and thermal methods for fuel Removal and cleaning of plasma-facing components [J]. Journal of Nuclear Materials, 2011, 415(1): S801–S804.
29. Graf J, Luk'yanchuk BS, Mosbacher M, et al. Matrix laser cleaning: A new technique for the removal of nanometer-sized particles from semiconductors [J]. Applied Physics A, 2007, 88(2): 227–230.
30. Pang H, Lipert R, Hamrick Y, et al. Laser decontamination: a new strategy for facility decommissioning [C]. Proceedings of the Spectrum'92 Conference of American Nuclear Society, Boise, ID, 1992: 1335–1341.

31. Pellin MJ, Leong K, Savina MR. Waste volume reduction using surface characterization and decontamination by laser ablation [R]. 1998 Annual Progress Report. United States: EMPS Project Summaries, 1998:4–5.
32. Clar G, Martin A, Cartry JP. Apparatus for working by laser, especially for the decontamination of a pipe of a nuclear reactor: US5256848 [P]. 1993-10-26.
33. Picco B, Marchand M. Nuclear fuel rod cladding decontamination using pulsed laser beam: FR277480(A1) [P], 1999.
34. Costes J R, Briand A, Remy B, et al. Decontamination by ultraviolet laser: the Lexdin prototype [C]. Proceedings of the Spectrum'96 Conference of American Nuclear Society, Seattle, 1996:1760–1764.
35. Dupont A. Aix-Marseille II University, Ph.D. Thesis No. 2079453, 1994.
36. Modern P, Steen W, Lin L. Laser decontamination method: W09513618 [P]. 1995.
37. Shimizu K, Fukui Y, Nemoto MSI. Laser decontamination method: US6056827 [P]. 2000.
38. Delaporte P, Gastaud M, Marine W, et al. Dry excimer laser cleaning applied to nuclear decontamination [J]. Applied Surface Science, 2003, 208: 298–305.
39. Lacour B, Brunet H, Besaucele H, et al. 500 watts industrial excimer laser high repetition rate [J]. Proceedings of SPIE the International Society for Optical Engineering, 1994, 2206: 41–45.
40. Costes J, Remy B, Briand A. Decontamination by ultraviolet laser: the LEXDIN Prototype [C]. Spectrum'96: international conference on nuclear and hazardous waste management, Seattle, WA, 1996.
41. Zhou DQ. Laser removal of space debris and asteroids [J]. Collection of Translations on Modern Science and Technology, 2006, 2: 25–25. (in Chinese).
42. Leone C, Genna S, Caggiano A. Compact disc laser cleaning for polycarbonate recovering [J]. Procedia CIRP, 2013, 9: 73–78.
43. Tang QH, Zhou D, Wang YL, et al. Laser cleaning of sulfide scale on compressor impeller blade [J]. Applied Surface Science, 2015, 355: 334–340.
44. Song GF, Li LC, Xia FJ, et al. Experimental research and exploration of the application of laser cleaning technology in ammunition repair [J]. Laser and Infrared, 2017, 47(1): 29–31. (in Chinese).
45. Lin WC. Application of laser cleaning technology in manufacturing radar T/R Components [J]. Electronic Process Technology, 2013, 34(6): 352–355. (in Chinese).

Postscript

It has been about twenty years since 2003 when we pay attention to laser cleaning. In 2003, the Journal of Clean World (in Chinese) invited us to write some articles to introduce laser cleaning and several short papers on laser cleaning were published, introducing the concept, principle, and application of laser cleaning. These short papers are relatively early in the field of laser cleaning in China and have also received attention from colleagues involved in cleaning and laser applications. Afterwards, we worked with graduate students to conduct research on laser cleaning for about twenty years, received some funding support, trained some master and Ph.D. students, published some papers, obtained some patents, and made some small achievements in the field of laser cleaning. Tiny progresses pile up gradually. In order to summarize the research results, provide valuable insights, and communicate with domestic and overseas colleagues better, I emerged the idea of writing a booklet on laser cleaning. Initial draft was completed in 2017 and financial support was obtained from the Tianjin Natural Science Press Fund and the National Science and Technology Academic Works Publishing Fund. However, in the process of improving the book draft, we encountered some difficulties, including content selection, logical progression, and chapter arrangement.

In 2018, Lu Yongfang, an editor at Tsinghua University Press, learned that I was writing a monograph on laser cleaning. She encouraged and invited me several times to put the book in the "Transforming Optical Science and Technology" series, which is funded by the National Publishing Fund of China.

The continuous pursuit for the ultimate goodness shall never stop. No matter how limited one's ability is, he will keep striving to achieve a best result. In the process of writing the monograph, extensive data collection was conducted, and the content became more and more complex, making it difficult for me to handle it with my own abilities. So, in order to clear up confusion and difficulties, I sought advice from the famous academicians and professors through various means such as phone calls and face-to-face consultations, and had discussions with my colleagues and postgraduate students, and received a lot of help. Especially at the turn of the spring and summer in 2019, a seminar was specially held to invite experts from Tianjin Association for Science and Technology, Science Press, Nankai University, Tianjin University,

F. Song and X. Lin, *Laser Cleaning*, https://doi.org/10.1007/978-981-97-6631-4

Northwestern Polytechnical University, et al., to earnestly seek advice. After discussion and under the guidance of academicians Yao Jianquan and Huang Wei, it was ultimately decided to divide the monograph into two volumes, one focusing on laser cleaning technology and applications, and the other on laser cleaning principles. The former, as one of the books in the "Transformational Optical Science and Technology" series; The latter is sponsored by the Tianjin Natural Science Publishing Fund and the National Science and Technology Academic Works Publishing Fund and published by Science Press.

After clarifying the ideas, the outline and table of contents were redefined, and Professor Lin Xuechun was invited to collaborate to speed up the writing process. Professor Lin is the Director of the All Solid State Light Source Laboratory of the Institute of Semiconductor, Chinese Academy of Sciences. He has been deeply involved in the field of solid state lasers for many years and has gained a lot of achievements. In the past decade, he has also focused on the application of solid state lasers in the field of cleaning. His joining has made the content of this book more comprehensive.

As the book is about to be published, Academician Yao Jianquan from Tianjin University, a renowned expert in the field of solid-state lasers in China, who was my Ph.D. supervisor and is forever my mentor, happily wrote the preface for this book. Thanks Prof. Yao for his continuous guidance and help. Graduate students Liu Shujing, Shi Shudong, Li Wei, Du Peng, Tian Bin, Zou Wanfang, Wan Wenshun, Wang Huan, Man Shijie, Zhang Zhiyan, and others (ranked in no particular order) have made contributions to the research and application of laser cleaning, and this book integrates some of their work. Li Zhenyao, Zhao Zejia, Yu Xi, Yang He, Lin Dehui, Pan Wenfeng, Liang Fei and others spent a lot of time and efforts on data verification and manuscript proofreading. Feng Ming, Liu Lisa, Li Weihua, and others provided a lot of help. The meticulous patience of the editor Lu Yongfang from Tsinghua University Press has given us a deep impression. Thanks to the project sponsored Ministry of Science and Technology (Project No. 2022YFB4601500 and 2017YBB0405100), Weihai Fengtai New Materials Technology Co., Ltd., and the National and Tianjin Natural Science Foundation for their financial support. Thanks to all those who have provided assistance on our twenty-year road of laser cleaning research.

My respected mentor wrote the preface of the book, then I make every effort to finish the postscript.

After twenty years of hardworking and five years writing, the book is finally finished.

Although I am far away from intelligence, the readers will be brilliant enough to point the drawbacks.

Laser cleaning is the morning sunshine, and will irradiate to the whole space.

Laser cleaning is like a rough jade wrapped in a stone. The authors, despite our lack of talent and shallow knowledge, dare to go ahead with the carving, hoping for the emergence of beautiful jade from stones. We hope that the knowledgeable experts and readers will not hesitate to give us comments and guidance. Let us together to pave the road of laser cleaning and let the light of laser cleaning shine in the world.

After a period of development, laser cleaning has received increasing attention. We believe that laser cleaning, like many applications such as laser welding, laser cutting and laser marking, has a brilliant future and can make greater contributions to the intelligent manufacturing and the industrial development.

Due to the limited knowledge of the authors, errors in the book are inevitable. Readers are kindly requested to criticize and correct them.

This book is dedicated to our parents and families. Thank you for your continuous support.

MIX
Papier aus verantwortungsvollen Quellen
Paper from responsible sources
FSC® C105338

If you have any concerns about our products,
you can contact us on
ProductSafety@springernature.com

In case Publisher is established outside the EU,
the EU authorized representative is:
Springer Nature Customer Service Center GmbH
Europaplatz 3, 69115 Heidelberg, Germany

Printed by Libri Plureos GmbH
in Hamburg, Germany